A MONOGRAPH ON

RICE BLAST DISEASE

Emerging conidiophores of the rice blast fungus (*Pyricularia oryzae = Magnaporthe grisea*) through stomata of the panicle of rice cv. Koganemochi as seen in a scanning electronmicroscope. Photograph : Courtesy from Prof. Dr. H. Koga, Tsukuba International Agriculture Training Centre (TIACT), Tsukuba-Shi, Ibarki-Ken, 35, Japan.

Rice Blast Disease

K. MANIBHUSHAN RAO
Centre of Advanced Study in Botany
University of Madras
Madras

1994

DAYA PUBLISHING HOUSE
Delhi-110035

Front cover : Typical leaf blast lesions on rice (L to R) resistant, moderately susceptible and susceptible reactions.
(Courtesy : HOKKU Chemical Industry Co., Ltd., Tokyo, Japan)

Back cover : Rice crop standing in the field.

ISBN 81-7035-130-8

Published by
Daya Publishing House
1123/74, Deva Ram Park,
Tri Nagar, Delhi-110035.
Phone: 7220593

Laser Composed by
Delite Printers
Tri Nagar, Delhi.

Printed by
D.K. Fine Art Press
Delhi.

Bound by
Book Warp, Delhi-110035

PRINTED IN INDIA

To My Mother......

ACKNOWLEDGEMENTS

I appreciate permission from authors, publishers and Institutes to reproduce plates, figures, tables and Copyright © material from the following sources : Mrs. S.Y. Padmanabhan for generous gift of her late husband's library collection especially on rice diseases and for use of his published materials (Plate 10 and Table 15) : Dr. K. Lampe, Director General, IRRI, Philippines to use the materials of his colleagues who are working on rice B1 disease (Plate labc; Figs, 2, 17, 19, and 28); Prof. N. Matsuyama (Table 10); Dr. J. Veeraraghavan (Table 5); Prof. H. Kato and his associates (Plate 4, Figs, 20 ab, 24ab and Table 1); Dr. S.Y. Cho (Table 2 and 4); Prof. T.S. Sadasivan (Fig. 14 and Table 7); Dr. R. Sridhar (Fig. 16); Dr. (Mrs.) Urmila Dhua (Fig. 22); Dept. Plant Pathology; Fac. Agric., Gifu Univ., Gifu, Japan (Plate 6ab); CAB International, UK (Figs, 1,2 and Table 14); Dr. H. Koga (Plates facing title page, 3ab, 8ab, 9ab, Fig. 21 and Table 13); Cambridge Univ. Press, UK and the author Prof. J. Webster (Fig. 4ab); Dr. T. Higashi (Table 12); Dr. R.J. Howard and his associates (Plates 11, 12 and Fig. 12) also from Copyright © 1988 by American Association for the Advancement of Sciences (AAAS) (Fig. 12) and Copyright © 1989 by Academic Press Inc. USA (Plate 11); Dr. Y. Takahashi (Fig. 18); Dr. Gouchang (Table 6); Dr. S. Kiyosawa and his associates (Table 11); Dr. M. Hori (Fig. 15ab) and finally my erstwhile colleagues and research students for reproducing from several of the joint publications which are listed appropriately.

Plate /Figure/Table numbers referred are from this Monograph.
Detailed citations are available by consulting bibliography.

FOREWORD

I deem it a privilege to be asked, by my erstwhile colleague Professor K. Manibhushan Rao, to write introducing this Monograph on Blast Disease of Rice to graduate students, researchers and teachers in the subject of Plant Pathology. I undertake this assignment with genuine pleasure.

Three decades ago when Professor Manibhushan Rao opted to join our group of plant pathologists at the University Botany Laboratory, Madras, we were already poised to an intimate understanding of some aspects of the host-parasite interactions in this complex blast (B1) disease of rice. Professor Rao's entry into this team was not so fortuitous as it turned out to be one of design. He has justified his selection to join this team of 'blast' pathologists.

It is a matter of surprise that this disease was first reported as early as 1637 from China and subsequently in 1704 from Japan. The causal organism was identified and named by Cavara in Italy as *Piricularia oryzae* only in 1891. Of course, a hundred years later we have seen many nomenclatural changes in the genus based on its many races and their virulence to rice infection. In Professor Rao's words: "Several areas of study such as genetic variability, karyology and physiology of the pathogen, host-parasite relationships, disease resistance and biocontrol are not yet well understood. Admittedly, for years to come this lacuna will remain". This is true of many of the widely distributed diseases of crop plants. For centuries scientists have worked on cereal rusts, their physiologic races, their etiology and syndrome but yet the last word has not been said about rust epidemiology.

The history, geographical distribution and symptomatology of the B1 disease have been well documented by Professor Rao and it will be redundant to say more about them here. There is, however, much new material, dealt in a lucid way by the author, is the Ultrastructure, Cell and molecular biology. DNA sequences have helped to indicate major pathotype groups and enabled to understand the structure of clonal linkages within and among pathotype groups. Further work in this area would seem to be rewarding as these techniques have been used and found promising in locating B1 resistance as well as mapping loci for quantitative B1 resistance.

The problem of host range on a global basis, presence of many physiologic races, host-parasite interaction, disease resistance and disease control are all ably presented. It is quite apparent that both basic and applied research would seem to be an exciting future possibility if we are to take a holistic view of the vexed problem of B1 in the many rice growing areas of the tropic and in the temperate regions.

I would like to commend for serious study of the chapter dealing with simulation models in forcasting epidemics of the B1 disease as it has a worldwide interest from the point of view of recommending control measures. Of particular interest in this regard is the chapter on 'Disease control' where the mode of action of fungicides and antibiotics, including biological control would seem to open up new resarch possibilities.

I believe this Monograph would be most useful to all those working on this model host-parasite system of *Pyricularia* and the rice plant. There is little doubt that in-depth studies from all angels would be rewarding and one can only look forward to more exchanges of ideas and newer techniques to be pressed into a global research effort to understand this disease better than we do today.

A word about the exhaustive bibliography of about 1200 references. It lends a weighty documentation to the Monograph.

MADRAS
January 20, 1994

T.S. SADASIVAN
Professor-Emeritus
Centre for Advanced Studies in Botany
University of Madras, Madras 600 025

FOREWORD

PREFACE

Considerable progress has been made in recent years in furthering the knowledge on rice blast (Bl), caused by the fungus pathogen *Pyricularia oryzae* syn. *P. grisea* (teleomorph = *Magnaporthe grisea*). Blast still remains the world's most important disease of this cereal crop. Particularly striking has been the information gained from studies on cell biology and molecular techniques to understand the mechanisms of pathogenesis and such efforts are timely from the stand-point of Bl control strategies that could improve crop productivity. Current research is being extended to recognition of field and quantitative resistance and use of system simulation approach in forecasting the epidemics.

Although 90 per cent of the world rice is grown mostly in the tropics, especially in Asia, to cater to the needs of about three billion people, relatively little work has been done on rice Bl and the problems that are encountered. For instance, much quantitative knowledge on the epidemiology is available for temperate rice Bl, but very little is known about the disease in the tropics.

In view of these efforts, there is a need for timely publications that draw together the most significant recent advances in Bl research being conducted by rice pathologists, breeders and molecular biologists. Such a role is aimed for this treatise. Besides, the knowledge I gained through close personal and professional working relationship with world's leading rice pathologists as well as my own contribution for which I was recently awarded a D. Sc. (University of Madras) is presented.

Since the first publication of Padwick's *Manual of Rice Diseases* in 1950, rice Bl was discussed with only 120 references while Ou in his 1972 edition book on *Rice Diseases* quoted 360 references and in 1985 revision of the same he increased this to 517. Further, there has been an enormous accumulation of Bl literature in the last few years and I felt it essential to survey and update the number till June 1993. As a result, almost 1200 references are given. This led to an increase of about 5.4 to 18.5% in important Bl citations from India by Ou (1985) to the present monograph (*Rice Blast Disease*) respectively. I intend to bring this monograph as an extended version of Ou's publication.

Particular attention has been given to include highly significant breakthroughs in both fundamental and applied research. The sixteen chapters in this book, deal among other things with history, yield loss, symptomatology, life cycle, disease resistance and disease control etc. with an emphasis on research highlights. The chapters dealt are not uniform as some are precisely brief and a few of them are handled more elaborately as per the available literature.

I am greatly indebted to my friends and various institutes/publishers for their permission to use or modify previously published/unpublished material and are acknowledged separately.

Thanks are due to the following, who read the draft of the various chapters : Dr. P. R. Day,

Director, Centre for Agricultural Molecular Biology, (CAMB), Rutgers Univ., New Jersey, USA on Physiological races; Dr. A. K. Sinha, Emeritus Professor, Dept. of Plant Pathology, Bidhan Chandra Krishi Viswavidyala, Kalyani, West Bengal on Nature of resistance and to some extent on Disease resistance; Dr. R. J. Howard, E. I. du Pont de Numerous and Co., Inc., Wilmington, DE 1988, USA on Cell and molecular biology (partly). I am also grateful to Dr. Roy Johnson, plant pathologist, AFRC Institute of Plant Science Research, Cambridge Laboratory, Norwich, U.K., for his critical reading of the entire manuscript and his valuable comments. I own complete responsibility for any errors, misinterpretations and deficiencies that may remain and welcome suggestions for improvement.

I take special pleasure in thanking my teacher Prof. T.S. Sadasivan, former Director, Center for Advanced Studies in Botany, University of Madras, who initiated me into this study on rice Bl over three decades ago and for writing the foreward to this book. I also take this opportunity to express my gratitude to late Prof. S. Suryanarayanan, who nurtured and helped me in those formative stages.

I wish to record my grateful appreciation to the University Grants Commission (UGC) and Indian Council of Agricultural Research (ICAR), New Delhi, for extending continued financial support to major research projects on rice Bl over these years. I also thank the University of Madras, Madras, for its support and to Prof. A. Mahadevan, Director, CAS in Botany for facilities. Further, I am grateful to Daya Publishing House, Delhi for their willingness to publish the manuscript at short notice. Lastly, I wish to emphasize that it would have been impossible to complete this study without the help and cooperation of my wife, Annapurna during this long endeavor.

June 10, 1994

K. MANIBHUSHAN RAO

Contents

Standard Abbreviations

°C	:	degree Celsius
cm	:	centimetre (s)
cv.; cvs.	:	cultivar; cultivars
d	:	day (s)
dia	:	diameter
g	:	gram (s)
h	:	hour (s)
ha	:	hectare (s)
kg	:	kilogram (s)
l	:	litre (s)
m	:	metre (s)
μ ; μm	:	micron (x 10^{-6}), micro meter
m^2	:	square metre (s)
mg	:	milligram (s)
min	:	minute (s)
ml	:	millilitre (s)
mm	:	millimetre (s)
mw	:	molecular weight
N	:	nitrogen
p.; pp.	:	page; pages
pH	:	hydrogen ion concentration
ppm	:	parts per million
RH	:	relative humidity
sec.	:	second (s)
sp.; spp.	:	species; species (plural)
t	:	metric tonne (s)
±	:	with or without
%		per cent (age)
/	:	per
σ	:	sigma
α	:	alpha
β	:	beta
γ	:	gamma
p	:	para
o	:	ortho
m	:	meta
r	:	correlation
y	:	observations

History

The rice blast (B1) disease was first reported in China by Soong Ying-Shin in 1637, and subsequently in Japan by Tsuchiya in 1704 (Goto, 1953), Miayanaga in 1788, Kojima in 1793 and Konishi in 1809. In Italy, it was reported as *brusone* by Astolfi (1828), Brugnatelli in 1833 and Gera (1846). The causal organism, *Pyricularia oryzae*,* was named by Cavara in Italy (Cavara, 1891) and subsequently in Japan (Shirai, 1896).

In the United States, Metcalf (1906, 1907) recorded the disease in South Carolina as early as in 1876 and was perhaps the first to call it 'blast', a name appropriate to the damage it causes. He could neither establish any relationship between the disease and weather conditions nor produce infection with pure culture. However, Fulton (1908) was successful in getting infection with the spores of the fungus *P. oryzae*. Miyake in 1910 referred to it as the *brusone*, although it was initially known in Japan as *imochi*.

In India, it was first recorded in Thanjavur (Tanjore) delta of South India in 1918 by McRae (1922) though it attracted attention only when a devastating epidemic occurred in 1919 (Padmanabhan, 1965 b).

Intensive studies on this disease have been made in Japan, India, the Philippines and the US. Extensive coverage on B1 appeared from time to time (Padwick, 1950; Ou, (1985). Further, Padmanabhan (1965d), Gangopadhyay (1983), Gangopadhyay and Padmanabhan (1987) and Manibhushanrao (1990) highlighted on rice B1 as the principal disease.

Although B1 literature is sprinkled with data on most aspects of the disease, our knowledge is still meagre. Several areas of study such as genetic variability, karyology disease resistance and biocontrol are not yet well understood. Admittedly, for years to come this lacuna will remain.

Rice B1 disease described over 350 years ago, is now the subject of research in molecular plant pathology laboratories all through the world ever since the fungus, (ascomycetales-teleomorph; and moniliales-anamorph) has been of considerable economic interest because of its global impact in inciting the disease. It is recognized as one of the most important cereal crop diseases. The recent significant findings of the above will suggest to redefine the early events of host-pathogen interaction.

---

* The correct name for the rice blast fungus is now given as *Pyricularia grisea* (Rosman et. al., 1990). Although *P. oryzae* and *P. grisea* are synonymized with the earlier name *P. oryzae* the voluminous literature appeared on B1 pathogen is retained by the name *P. oryzae* and in the recent literature, the teleomorph *Magnaporthe grisea* has been used and when such work is referred.

Geographical Distribution

Blast is the most widespread disease of rice and over 85 countries have reported its presence (CMI Distribution Map 51, Fig. 1). It is present in almost every rice growing tract of the world except California, in the US. The disease is adaptable to adverse environmental conditions of widely fluctuating temperatures and relative humidity. It appears in irrigated lowland or rain-fed upland rice as well as in rice grown submerged or in deep waters. In desert conditions, especially in the Middle-East, where high temperatures prevail with very low humidity, the B1 fungus infects the nodes just above the field water level. In Iraq, it is referred to as 'shara' and has assumed immense importance due to the heavy crop losses involved.

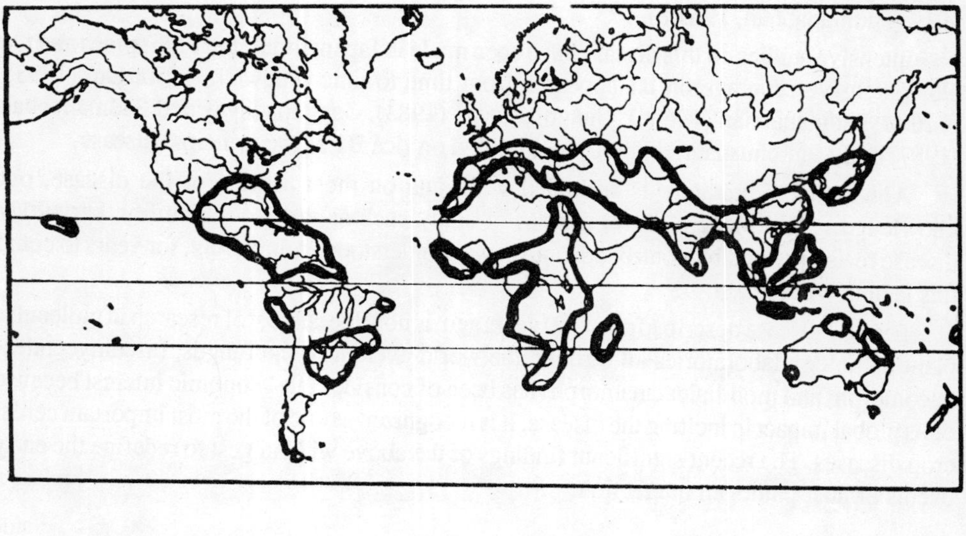

Fig. 1: Distribution of blast, *Pyricularia oryzae* on rice and other Gramineae. Map No. 51, CAB International , UK (formerly CMI, 1968).

Yield Losses

Blast is considered the principal disease of rice owing to its wide distribution and destructiveness and remains the world's most economically important malaise. It makes the cultivation of rice impossible in many areas despite intensive efforts to control it. Early seedling infection often results in death of the plants while neck infection leads to half-filled and totally unfilled panicles which reduce the yield.

Yield losses from B1 epidemics are either due to typhoons or due to individual or combined effects of pests, diseases, cold injury, floods etc. Reports on yield losses are available from Italy (Baldacci, 1947; Corbeta, 1961);India (McRae, 1922; Padmanabhan, 1967); Japan (Kuribayashi and Ichikawa, 1952; Goto, 1963); Korea (Crill et. al., 1982); Philippines (Ou, 1985) and Egypt (Reddy and Bonman, 1987).

Earlier Kuribayashi and Ichikawa (1952) estimated losses caused by B1 by applying the regression equation or non-linear curve as :

$$Y=0.69\ x\ +2.8$$

where, Y= yield loss ratio, and x = % panicle B1 incidence. They expressed the difficulty in studying leaf B1 separately from panicle B1. From the statistics, yield loss from 1953-1960 due to B1 varied from 1.4 to 7.3% (Goto, 1963). Kato (1974) reported losses of grain yield due to neck, node or panicle B1. Prabhu and De Faria (1982) studied the exponential relationship between time of neck infection and grain weight while, Shegimura et al. (1988) worked out a linear model for predicting yield loss. Kim and Park (1987) reported that grain yield loss was more significant for panicle B1 than for neck B1.

In India, Padmanabhan (1967) studied the relationship of yield with incidence of B1 using multiple regression and found that neck B1 significantly reduced the yield. It showed two fold (4%) loss due to B1 with 2% disease incidence. Despite the non-availability of organized surveys for B1 incidence for the entire country, he made an attempt to estimate the yield loss during 1960-61 as about 2,65,000 t. Subsequently, Reddy and Bonman (1987) estimated it as 140,000 t from just three states (Andhra Pradesh, Karnataka and Tamil Nadu). Also, between 1980 and 1987, seven epidemics of B1 occurred in four states, two each in Himachal Pradesh, Andhra Pradesh and Tamil Nadu and one in Haryana, causing heavy yield losses (Nagarajan, 1988).

Studies have been made to estimate the yield losses due to leaf and panicle B1 by many workers (Atkins, 1962; Ou and Nuque, 1963; Goto, 1965 a; Padmanabhan, 1965 a; Exconde and Raymundo, 1970; Awoderu and Esuruoso, 1974; Tien and Huang, 1975; Rodas and Rosero, 1975). Reports on yield loss due to neck B1 are also available (Katsube and Koshimizu, 1970; Kapoor and Singh, 1983; Sharma et al., 1988).

Supad (1982) observed that transplanting B1 infected rice seedlings resulted in a 30% yield loss in Malaysia while in Brazil, Prabhu and Moris (1986) reported losses up to 61% in upland rice. In Thailand, Loehken (1990) found yield losses ranging from 44 to 54% in irrigated plots. Teng et al. (1988) has suggested techniques for estimating yield losses caused by B1 and other rice diseases. They include quantification of the pathogen or disease, generation of epidemic of different duration and intensities, experiments to provide data for characterizing the disease-loss relationship, modelling and field surveys. The first three techniques are relatively well developed as compared to the other two techniques. The methodology used is being tested and the regression models are being investigated. Reddy et al. (1988) reported the linear relationship between severity rating and realized yield as :

$$Y = 4306.4 - 67.7\,x\ (r = 0.9427)$$

where Y = estimated yield and x = the % terminal B1 severity.

Surin et al., (1989, 1991) estimated yield losses at different crop stages (45, 60 and 75–d after seedling). An infected area of 1% corresponded to 3% yield loss (Y=0.606 + 3.08 x, r=0.96). Torres and Teng (1989, 1991) used modified single hill techniques to estimate yield loss due to leaf B1 (LB1) and Panicle B1 (PB1). The estimated yield loss (Y) computed as :

$$Y = 5.0725 + 0.3664\ LB1 + 0.3301\ PB1$$

$$r^2 = 0.72, \text{(significant at } p = 0.01)$$

The equation suggests that, in the absence of PB1 at maximum tillering causes approximately 0.4% yield loss, while the absence of LB1, 1% PB1 incidence at harvest results in approximately 0.3% yield loss.

Symptomatology

The B1 fungus attacks all aerial parts of the rice plant. The disease is most conspicuous when it attacks the leaf blades, nodes, panicle and neck, but it is also found on the leaf sheaths, rachis, the joints of the culm and even on the glume. On the leaves, the lesions/spots first appear as minute brown specks, then grow to become spindle-shaped, pointed at both ends, several cm long and about 0.5 – 1.0 cm wide (Plate 1a). The centre is greenish grey often showing a brownish margin. The size, colour and shape of the lesions, however, vary with different conditions and varietal response. On resistant cvs. or under unfavourable conditions the lesions become smaller, more brownish and narrow and sometimes may not develop beyond the size of specks. Under favourable conditions on a susceptible cv. several greyish spots may appear, become larger and broader, and coalesce, leading to withering of the whole leaf.

In severe attacks, seedlings and plants in the nursery may be completely blasted or killed. In older plants, the head emerges prematurely because of the attack and appears withered long before the normal time of ripening. The attack results in a chaffy ear head if it is in an early stage; if some grain has already filled, the panicle breaks the rotten neck owing to the weight of the panicle (Plate 1b). Affected lower nodes of stem turn black and if the culm is pulled out one can see the blackened area (Plate 1c). Brown to black spots are also seen on the rachis, particularly where it branches and on the glumes. Stunting of the plant too is known (Kuilman, 1940). Tokunago et al. (1953) studied the physiology of the stunting process while Kozaka and Teraoka (1977) suggested that ethylene phytotoxicity is responsible for stunting. Leaf B1 spots may sometimes be confused with herbicidal injury as well as other common leaf spot disease of rice Fig. 2.

Pathological histology

Three zones in the B1 infected region of the lesion can be distinguished; the venenate, necrotic and disintegrated (Yoshii, 1937). The venenate zone (Fig. 3, b, i) occupies the border of the diseased portion showing a light yellowish stripe that blends into the healthy tissues. The zone is infiltrated by toxic substance(s), now identified as α–picolinic acid, piricularin, pyriculol or tenuazoic acid. The necrotic zone (Fig. 3b, iii)is usually a brown, narrow streak along the inner side of the venenate zone. In the venenate zone, the pathological changes involved are the discoloration and decrease in the size of the chloroplast, degeneration of the cell membrane and vacuolar or granular disintegration of the protoplasm. With the spread of the hyphae into the venenate zone, the cell inclusions as well as the cell walls collapse thus, following the disintegrated zone (Fig. 3, b,ii). Typical leaf B1 symptoms (Fig. 3a) on rice and a transection of infected leaf showing conidia and conidiophores (Fig. 3c) are given (Manibhushanrao, 1992).

Further, Terui (1940) also found conidia to be formed internally in various tissues of the rice plant such as lysogenic cavities along the midrib of the leaf blades, leaf sheaths and in the hollow cavities of the culm about 5 mm below the neck. Subsequently, Koga and Yoshino (1988) used calcoflour to impart fluorescence in B1 fungal structures especially conidiophores and conidia formed on infected sites such as neck, rachis (Plate 2a, b), panicle branch, pedicel and husk of susceptible rice cv. Aschi-Asachi.

Recently, Koga (1989) studied under light microscope the infection process of *P. oryzae* isolate Ken 60-19 in leaves of susceptible cv. Koshihikari and highly resistant cv. Toride 1 for B1. The per cent successful penetration on leaves of resistant cv. was much less than the susceptible cv. The invaded hyphae are branched and filled the cell (Plate 3a) of susceptible host while it is not branched but granulation is observed on the epidermal cellular contents (Plate 3b) with the highly resistant cv.

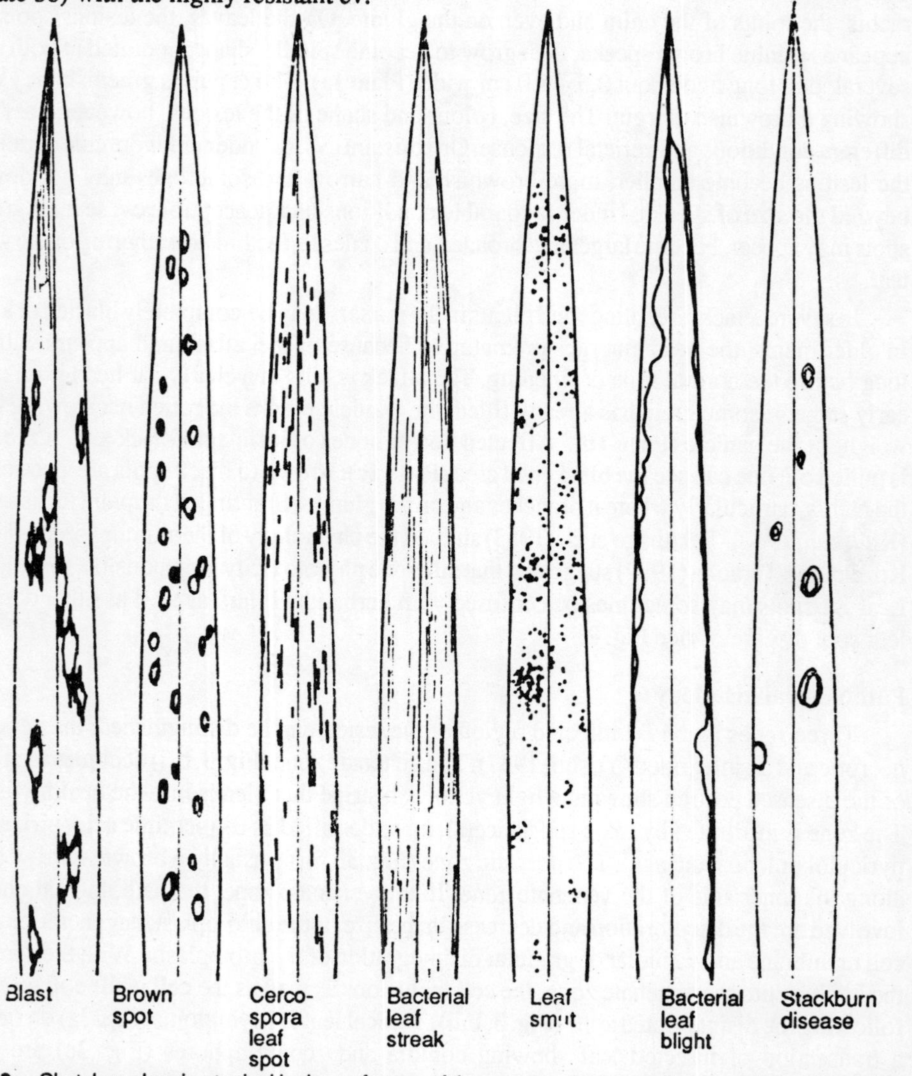

Fig. 2: Sketches showing typical lesions of some of the common leaf spot diseases of rice (Ou, 1985).

Plates 1abc : (a) Leaf blast; (b) Panicle blast; (c) Nodal blast (Ou, 1985).

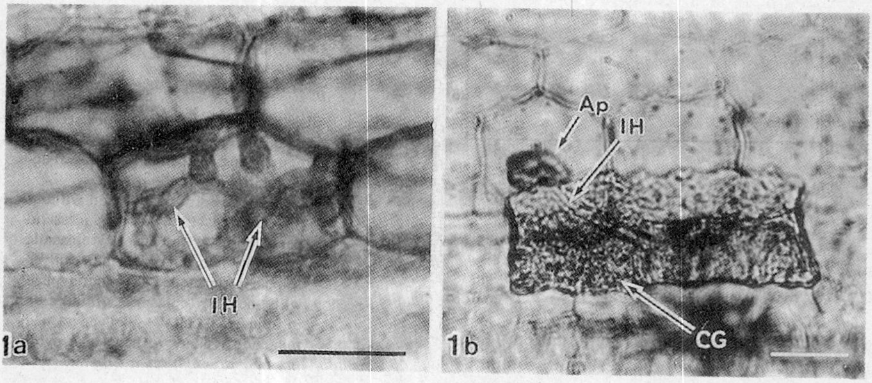

Plate 2ab: Photomicrographs of epidermal cells of rice leaves invaded by *P. oryzae* (Ken 60-19).

(a) Invaded hyphae is branched and rapidly filling the cell on susceptible cv. Koshinikari at 72 h after inoculation.

(b) Invaded hyphae is not branched but observed granulation on epidermal cellular contents. Ap : appressorium; IH : invaded hyphae and G: granulation. Bar : 20 µm (Koga and Yoshino, 1989).

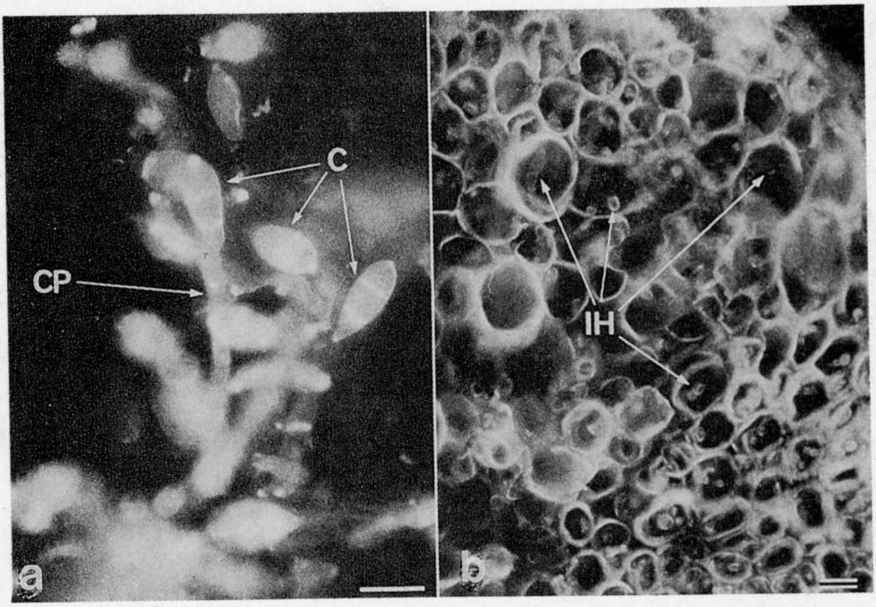

Plate 3ab : (a) Conidiophores (Cp) and conidia (C) formed on epidermis of rachis.
(b) Invading hyphae (IH) in cells of a rachis, Bar, 20 μm (Koga, 1989)

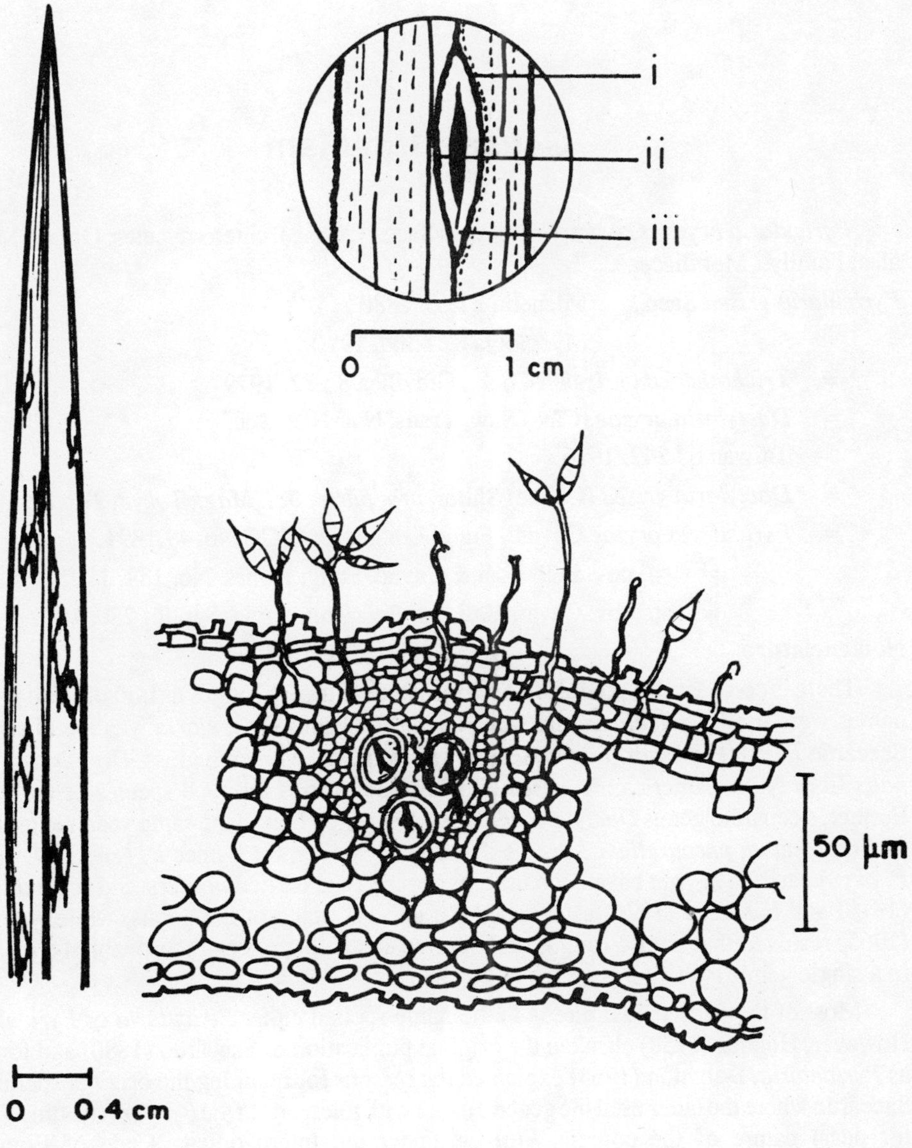

Fig. 3abc: (a) Leaf B1 symptoms on a susceptible rice cv. (b) lesion magnified showing different zones of necrosis (i) venenate zone, (ii) disintegrated zone and (iii) necrotic zone. (c) A transection of infected leaf showing conidia and conidiophores (Manibhushanrao, 1990).

Causal Organism

Pyricularia oryzae Cavara, (=*P. grisea* Sacc.) Class : Deuteromycetes; Order : Moniliales; Family : Moniliaceae.

Pyricularia grisea Sacc., Michelia 2 : 20. 1880

Mycologia 82 : 509. 1990

= *Trichothecium griseum* Cook., Grevillea 8 : 12. 1979

= *Dactylaria oryzae* (Cav.) Saw, Trans. Nat. Hist. Soc. Taiwan 6 : 242. 1916

= *Dactylaria grisea* (Cooke) Shirai, *pro. parte, Bot. Mag., Tokyo*, 29 : 19. 1905

= *Pyricularia oryzae* Cavara, Fungi Longb. Ex CICC No. 49.1891.

= sp. nov. in Brisosi & Cavara Fungi Parass. No. 188. 1892

= sp. nov. Cavara Atti. Ist. Bot. Unit. Pavia, Ser. 2 : 280. 1892

Nomenclature

There is considerable confusion over the name of the B1 fungus as initially two generic names were used, *Pyricularia* and *Dactylaria.* The genus *Pyricularia* was established by Saccardo in 1880 based on *Trichothecium griseum* Cooke on crab grass, *Digitaria sanguinalis* from North America. *P. grisea* (Cooke) Saccardo is the type species of the genus. Further, a separate genus *Dactylaria* was also erected by him in the same year and assigned as *Acrothecium purpurellum* Sacc. as the type. However, the names *P. grisea, D. grisea, P. oryzae* and *D. oryzae* have been used by some of the earlier workers. Subsequently, Ito (1918) and Nisikado (1926) established the general usage of the genus, while Asuyama (1965) resolved that the B1 fungus on both rice and crab grass can reasonably be included in a single genus *Pyricularia.*

Most of the earlier literature is in the name spelled either *Piricularia* or *Pyricularia.* However, Hughes (1958) checked the original publication of Saccardo (1880) and found it as *Pyricularia.* Deighton (1965) explained the reasons for retaining the original spelling of Saccardo where the latter used the generic name with reference to the *Pyriform* (Latin : *pyrum* for pear) nature of the conidia. Further, under the International Code of Botanical Nomenclature (ICBN) (1966) the original spelling *'Pyricularia'* was retained and is being used by most authors.

There is much confusion about the specific name of the fungus. Although Cavara (1891) first reported the disease on rice in Italy and named it as *P. oryzae,* an identical description of the fungus was published in 1892 as *P. oryzae* sp. nov. by Briosi & Cavara, as well as *P. oryzae* sp. nov. by Cavara. However, Padwick (1950) suggested that the first publication by Cavara alone should be retained and he alone should be cited as the author.

In Japan, Shirai (1896) described the rice B1 fungus which resembled *P. oryzae,* although it differed in the number of conidia borne on the fungus with American specimens referred to as *P. grisea* and maintained that *P. grisea* produced 3-5 conidiophores, each bearing, many conidia, while *P. oryzae* had one conidiophore with a single conidium at its apex. However, Kawakami (1901, 1902) considered that both *P. grisea* and *P. oryzae* were indistinguishable. Besides rice and crab grass, the fungus attacked Italian millet *(Setaria italica),* German millet *(S. italica* var. *germanica)* and yellow foxtail *(S. glauca).* Subsequently, Shirai (1905) and many other workers (Fulton, 1908; Duggar, 1909; Stevens, 1913 and Metcalf, 1913) confirmed the above observation.

Taxonomy

Nishikado (1917, 1927) compared the species of *Piricularia* from various hosts such as cereals and grass weeds from the morphological and physiological aspects. He concluded that, among the fungi studied, only four species (two of them identified as new) were described with distinct morphological characters as follows :

Piricularia oryzae Cav., on *Oryza sativa* L.

P. grisea (Cke.) Sacc., on *Digitaria sanguinalis* L.

P. setariae Nishikado, on *Setaria italica* Beauv. and S. *viridis* Beauv.

P. zingiberi Nishikado, on *Zingiber mioga* Rosc. and Z. *officinale* Rosc.

Sueda (1928), Ono and Nakezato (1958) and Yamanaka and Kobayashi (1962) reported that *Pyricularia* on rice is not always readily distinguished from other forms by the size and shape of the conidia alone. Similarly, Ramakrishnan (1948) failed to find any appreciable difference in morphological characters of *Pyricularia* from rice (*Oryza sativa*), ragi (*Eleusine coracana*), Italian millet (*S. italica*) and crab grass (*Digitaria marginata*) studied by him. Many suggestions were made to regroup the morphologically related forms into one species and subdivide it into specialized forms on the basis of pathogenicity. In agreement with that, Thirumalachar et al. (1956) described a fungus closely resembling *P. oryzae* on *Commelina benghalensis* as *P. oryzae* var. *commelineae* and on *Brachiaria mutica* as *P. oryzae* f. *brachiariae* (Govindarao and Koteswararao, 1958). Thus, to establish this system of classification, a systematic and comprehensive study is needed.

Further, Thirumalachar et al. (1956), Govindarao and Koteswararao (1958) and Veeraraghavan and Padmanabhan (1965) proposed *forma specialis* for those isolates which are morphologically alike but differ in pathogenicity. Latterell et al. (1960), also of the same opinion, suggested the old accepted binomial, accompanied by an appropriate *forma specialis* designation, i.e. *P. grisea* f. sp. *oryzae* for the rice B1 fungus.

Ranganathiah and Mathur (1976) isolated the B1 fungus from ragi seeds and identified it as *P. grisea* overlooking the importance of pathogenicity in confirming the identity. Against this background, Zuber and Manibhushanrao (1979) suggested that the isolate from ragi seeds should be named as *P. grisea* Sacc. f. sp. *eleusinae* (Linn) Derf. as they are not pathogenic on rice but morphologically indistinguishable from *P. oryzae.*

Padwick (1950) reviewed the earlier literature on rice B1 and its closely related forms. *Pyricularia* spp. reported from the Far East are :

On Gramineae

Pyricularia grisea (Cooke) Saccardo

P. oryzae Cavara

P. setariae Nishikado

P. panici-repentis (Sawada) Ito

P. panici-miliacei Hara

P. zizaniae Hara

On Zingiberaceae

P. zingiberi Nishikado

P. costi (Sawada) Ito

P. curcumae (Sawada) Ito

On Cannaceae

P. cannae Ronald

Dactylaria cannae (Sawada) Ito

Asuyama (1965) defined the following eleven *formae speciales* based on hosts such as cereals and grasses as follows :

(a) rice barley, barnyard grass (*Echinochola*) tall fescus (*Festuca*), sugarcane, *Leersia oryzoides*, giant reed (*Arundo donax*) and maize

(b) crab grass (*Digitaria*), St. Augustine grass (*Stenotaphrum*) and rice

(c) Italian millet, green foxtail (*Setaria*) and ragi (*Eleusine*)

(d) bajra (*Pennisetum*)

(e) proso millet (*Panicum*)

(f) sweet vernal grass (*Anthoxanthum*)

(g) *Brachiaria mutica*

(h) *Eriochioa villosa*

(i) *Leersia japonica*

(j) *Panicum repens*

(k) *Zizania latifolia*

A broad division of the B1 fungus on cereals and grasses is suggested, but the question arises as how to select a specific name, if the related forms of *Pyricularia* should be included in one species. *P. grisea* is the earliest name given to this group of fungi but most voluminous literature is found under *P. oryzae* on rice and other hosts including grasses. *P. oryzae* is a familiar and well-established name and hence can not be easily discarded. As a compromise, Sprague (1950) and Asuyama (1965) suggested that *P. oryzae* can be retained for the fungus occurring on rice, while *P. grisea* can be used for the species reported on other cereals and grasses.

Despite the availability of vast literature on the taxonomy and nomenclature of the B1 fungus, our present knowledge is almost rudimentary as the problem is still unresolved. The name *Pyricularia oryzae* Cavara is used in this book for the rice B1 fungus.

Morphology

Anamorph : Padwick (1950) gave the translated original description of the lesions and fungus *P. oryzae* by Briosi and Cavara (1892) as follows:

"Spots on the leaves, oblong, drying up, dark-margined; sporiferous hyphae usually epiphyllus, scattered, rounded to roundish-awl shaped, a little thickened at the base, septate at that part, not or not so clearly so above, 60-120 μm long, 4-5 μm broad, grey; conidial obclavate, tapering at the apex, truncate or extended into a short tooth at the base, 2-septate, slightly darkened, translucent, 20-22 × 10-12 μm."

The fungus on rice in Japan has been described by Shirai (1896), Hori (1898) and Sawada (1917). Nishikado (1917) gave full description of the fungus as follows:

"Mycelium in cultures aerial or submerged, hyaline or olivaceous, 1.5-6.0 μm in width, septate, branched. Conidiophores 1 to many, fasciculate, simple, or rarely branched, 2-4 septate, not or slightly constricted at septa; at first monosporic, then pleurogenous on sympodium; olivaceous to fuligenous; base swollen, dark coloured, becoming lighter colour towards apex."

"Conidia variable in size and shape, terminal, pyriform to obclavate, base rounded, apex narrowed: 2-septate, rarely 1-3 septate not or slightly constricted at septa, almost hyaline to pale olive, 14-40 × 6-13 μm in size, usually 19-23 × 7–9 μm, with small basal appendage. Basal appendage 1.6-2.4 μm (2 μm) in width; basal cell 4.8-10μm (6.3 μm), middle cell 4.8-12.8 μm (7.8 μm), apical cell 4.8-10.7 m (7.4 μm). Conidia germinate from apical or basal cell and less frequently from middle cell. Germinating hyphae hyaline, not or constricted at septa, branched, 3-5 μm in width."

A succession of obpyriform, two septate, pale brown conidia (Fig. 4a) are produced each attached to the conidiophore by protruberant hilum. The conidiophores (Fig. 4b) are slender and extend from the mycelium within the diseased leaf through stomata (Webster, 1980).

There is no report of sclerotial formation in *P. oryzae* in culture. However, sclerotia were formed by *P. higginsii* L. var. *poonensis* in culture (Thirumalachar et al., 1956). Ramakrishnan (1971) reported that *P. oryzae* also form typical sclerotia in culture in a medium containing leaf extracts of *Eleusine coracana.*

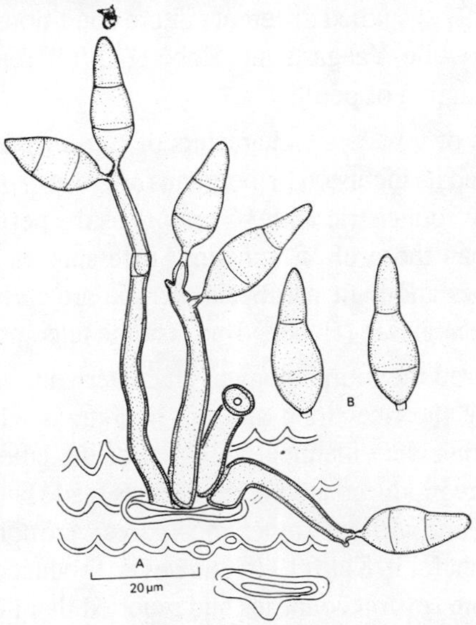

Fig. 4ab: (a) Conidiophores of *P. oryzae* emerging from a stoma on rice leaf; (b) detached conidia (Webster, 1980).

The swelling of the germ-tube (chlamydospores) was described by several workers (Hori, 1898; Suematsu, 1916; Sawada, 1917 and Nishikado, 1917) but Kawakami (1902) called it a resting spore. No chlamydospores are found in host lesions. Refatti (1955) reported chlamydospore-like structures in nature and this might help the fungus to survive through the hot dry summer months in plains, while Togashi (1961) studies the chlamydospore formation in *P. oryzae.*

Teleomorph: Sydow and Butler (1911) described *Sphaerella malinverniana* (Catt.) Sacc. on rice leaves collected in Bengal, and noted that *P. oryzae* was often seen associated with perithecia of *Sphaerella* spp. in India. These observations, however, do not establish the connection of *Pyricularia* with *Sphaerella* or *Mycosphaerella* (Hara, 1918). Asuyama (1965) also reported the lack of positive proof that they were the same fungus. Earlier, Webster (1965) found the perfect state of *P. aquatica* under natural conditions and classified it as *Ceratosphaeria grisea* Hebert. Subsequently Hebert (1971) described the teleomorph of *P. grisea* as *C. grisea.*

Ueyama and Tsuda (1975) reported formation of teleomorph resembling *C. grisea* from crosses of *P. oryzae* and a species of *Pyricularia* isolated from *Eleusine.* Kato et al. (1976) and Yaegashi and Nishihara (1976) produced the teleomorph by mating *P. oryzae* rice isolates with that of finger millet (*Setaria italica*), ragi and *E. indica* and with an isolate from *Phalaris arundinaceae*. Production of mature perithecia mated with *P. oryzae* and *Pyricularia* sp. from ragi was reported (Plate 4). Yaegashi (1977) could not succeed in mating rice isolates (78) from different regions of Japan as well as from isolates (272) obtained from 28 other hosts and concluded that species of *Pyricularia* are all heterothallic. He considered that absence of the teleomorph in nature was due to geographic isolation. Ueyama et al. (1977) produced the teleomorph by mating the *Pyricularia* isolates from ragi (finger millet) from Nepal and Japan and considered *P. oryzae* as heterothallic with type A and B but that the rice isolates belong to type A group and the failure to produce teleomorph should therefore be expected. Tsuda et al. (1977) studied different culture conditions affecting the formation of teleomorph in *P. oryzae* while Yaegashi and Hebert (1976b) reported that temperature and nutrients affect the production of perithecia.

Based on the views of taxonomic characters of *Ceratosphaeria grisea* Yaegashi and Udagawa (1978) proposed its inclusion in the genus *Magnaporthe*. Further, Barr (1977) felt *C. grisea* appeared to be congeneric with *M. salvinii* as the perithecia were similar though smaller in all aspects than those of *M. salvinii.* Since species of *Ceratosphaeria* Niessl, lignicolous hypersaprobes and their perithecial peridia are carbonaceous, he made a new combination *Magnaporthe grisea* (Hebert) Barr, for the teleomorph state of *P. grisea.*

Kato (1977) compared the morphological characteristics and the size of each sexual organ formed by mating the rice-strain and the ragi-strain with those of *C. grisea* and concluded that both strains were identical with that of the latter (Table 1). While a large number of perithecia were produced among the crosses made between rice and ragi isolates and rice isolates were crossed with each other, most cases were non-fertile and genetic factors were considered to account for it. Kato (1979), however, produced the teleomorph by mating rice isolates obtained from several countries and reported that the morphology was similar to that of *C. grisea* as reported by Hebert (1971):

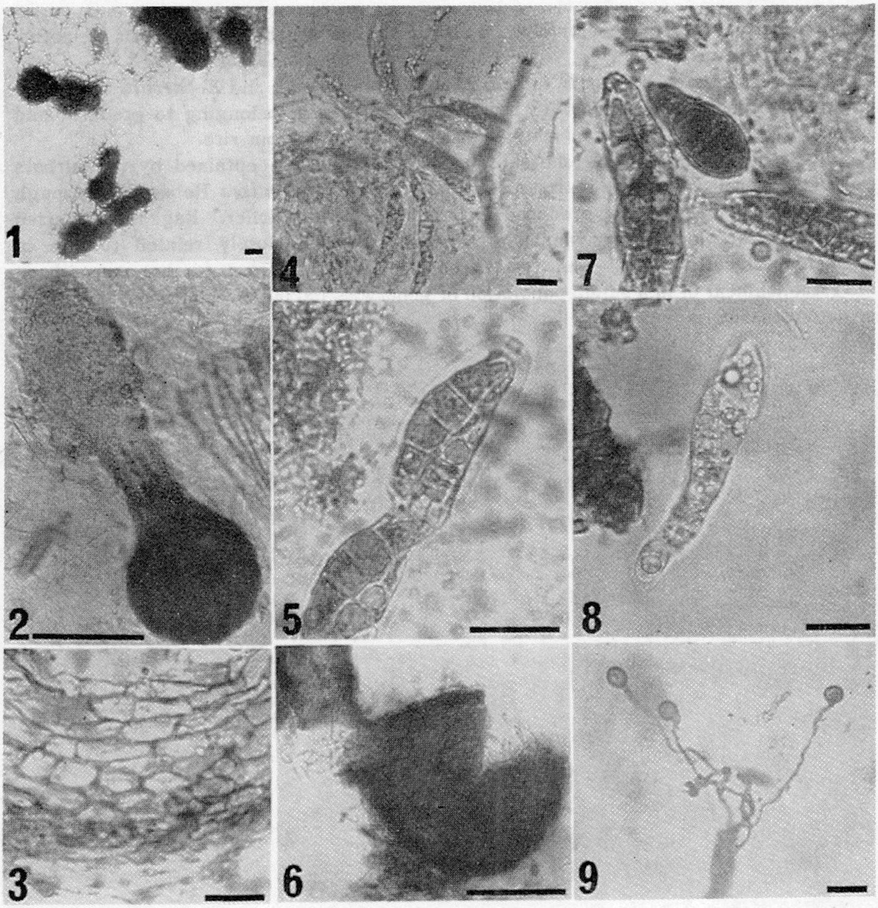

Plate 4 : Perithecium production in *Pyricularia oryzae* from rice mated with *Pyricularia* sp. from ragi, (1) Juvenile perithecia cultured for 7 d in modified Sach's nutrient agar with fragments of rice straw. (2) Mature perithecium with ostiolate beak. (3) Wall structure of peridium . (4) Asci in a fascicle cultured for 18 d. (5) Ascus and ascospores. (6) Forcible discharge of ascospores from perithecium cultured for 30 d. (7) A comparison between conidium and asci. (8) Juvenile ascus. (9) appressorium- like structures seen at the tip of germ tube. Bar : 10μm for 5, 7 and 8; 20 μm for 3,4, and 9; 100μm for 1,2,6 (Kato et al., 1976).

Table 1. Comparison of sexual organs produced in mating between rice-strain and ragi-strain of *Pyricularia* and in *Ceratosphaeria* [a] (Kato, 1977)

		Morphology	
		Rice-strain × Ragi strain[b])	*Ceratosphaeria grisea*[c]
	Peridium	110 (57–150) brown to dark brown	180 (60–300) dark brown to black
Perithecium		globose size of wall cells 9 × 16	thickness 15–20
	Beak	50 (33–95) × 160 (50–1000) hyaline at the tip brown at the base periphyses	90 (60–150) × 600 (100–1200) hyaline at early stage brown at later stage periphyses
Ascus		9 (8–12) × 60 (53–70) cylindrical to obclavate unitunicate[d]) refractive ring structure no stain with Melzer's iodine paraphyses disappeared at maturity short stipitate evanescent eight ascopores (8–0)	8.5 (7–10) × 70 (55–90) cylindrical to subclavate unitunicate refractive ring structure no stain with Melzer's iodine no paraphyses deliquescent eight ascospres (0–8)
Ascospore		5 (4–6) × 20 (14–24) 3 septa (1–3) hyaline spindle to fusiform	5 (4–7) × 21 (17–24) 3 septa (1–3) hyaline spindle of fusiform

(a) An unit of length is all µm. (b) Kato et al. (1976) (c) Hebert (1971), and (d) Fukutomi et al. (1976).

"The perithecia are single or in groups, particularly or wholly embedded with long beaks protruding from the surface, dark brown to black, about 180 (60-300) µm in diameter at the globose base; the beaks are hyaline , becoming brownish, 90 (60-150) × 600 (100–1200) µm, with long thread like deliquescent periphyses inside. Asci cylindrical to subclavate, uninucleate, thin-walled, with a refractive ring around a pore in the tip, 8.5 (7–10) × 70 (55-90) µm. Ascospores are hyaline, fusiform, 3-septate, with little or no constriction at the septa, usually slightly curved, 5 (4–7) × 21 (17-24) µm."

The perfect state of *P. grisea* is a heterothallic, uninucleate, Pyrenomycete, *M. grisea*. It is obvious that *P. oryzae* and *P. grisea* are not related to *P. aquatica* which has the perfect state the loculoascomycte *Massarina* (Webster, 1980), Further, Monod (1983) suggested that both *M. grisea* and *M. salvinii* (type species) belong in the genus *Phragmoporthe* Petrak, but synonym was not accepted as the anamorph is unknown.

Ryu et al. (1988) tested the pathogenicity of the ascospore progeny of 32 ascospores derived from 4 asci from perithecium-produced isolates between finger millet and Italian millet and were inoculated to their hosts. The segregation ratio was 1 : 1 and appeared to be conditioned by a single gene.

There is no report on the occurrence of perithecia *in vivo* and the low fertility of the field strains of *M. grisea* is possibly the limiting factor. However, Kato and Yamaguchi (1982) and Yaegashi and Yamada (1986) found that strains of both mating types exist in several countries. Further, coinoculation of compatible strains on rice cv. Maratelli resulted in the first perithecial production on rice plants (Silue and Notteghem, 1990). Nature might be existing with more fertile strains and if perithecia are found on the debris of rice plants it could be an important source of pathogenic variability which in turn perhaps explains why the breakdown resistance in some rice cvs. to B1 pathogen occurs.

Conidial variation

Enumeration of certain morphological differences from other species of Gramineae has been reported (Sawada, 1917; Nishikado, 1926; Goto et al., 1954). The size of conidia (average length × breadth in μm) of *P. oryzae* varied in different hosts, namely, rice (26.0 × 8.8); crab grass (23.6 × 8.1); *Leersia japonica* (27.0 × 8.6); *Panicum repens* (22.5 × 10.2); Italian millet (25.0 × 8.6); *Costus speciosus* (20.0 × 8.6); crab grass (22.5 × 7.2); Italian millet (21.0 × 7.5); green foxtail (23.9 × 8.2); *Zingiber mioga* (17.5 × 7.8) and *Zizania latifolia* (24.6 × 14.1).

The size and shape of conidia are affected by several factors such as type of host (Hemmi et al., 1949; Narita et al., 1956; Ono and Nakazato, 1958 and Asuyama, 1965), humidity (Nishikado, 1926 and Sueda, 1928), temperature (Sueda, 1928 and Yamanaka and Kobayashi, (1962), water surface (Ono and Suzuki, 1959 a), different nutritional media and cultural conditions (Sawada, 1917; Nishikado, 1926; Kulkarni and Patel, 1956 and Ono and Nakazato, 1958)

Asuyama (1965) gave a comprehensive account of conidial variation by comparing the measurements of the conidia on infected rice plants as provided by other workers.

Manibhushanrao (1971) described the variation of conidial shape and size of two pathogenic isolates of *P. oryzae* P1 (parent) and A1 (inositol-less mutant) and non-pathogenic M1 (pigmented from P1) and M2 (pigmented UV mutant of *P. oryzae*) from Baarn, Holland. Conidia of P1, A1 and M1 were similar in their shape and size which in general corresponded to that described by Padwick (1950) while M2 produces short conidia, variable in shape and size. Both P1 and A1 could be distinguished from M1 and M2 by the pigmentation of the colonies as the former were generally olive green in colour on most nutritional media tested in contrast to the brownish purple of the latter.

The appressoria produced vary in shape (globose, ovoid or oblong) and size (ranging from 5-15 μm in dia.) with certain strains much larger in dimensions, ranging from 13-20 μm (Suzuki, 1953). Appressorial formation by the four differentially virulent isolates (P1, A1, M1 and M2) *in vitro* showed that nearly 60% of the conidia of P1 formed by 24 has against 31 and 34% in the case of M1 and A1 while only 25% of the avirulent Baarn mutant M2 formed (Manibhushanrao, 1971 a). Appressorial formation by P1 on excised leaves was comparable while A1 tended to be higher than that formed *in vitro*. Further, avirulent M1 and M2 showed considerable reduction in growth on rice leaves as compared to virulent P1 and A1 (Fig. 5). These results clearly show host-parasite interaction at the host surface and the possible role of the wax coating on appressorial formation. Indeed Suzuki (1963) used the appressorial type as a genetic marker in the identification of physiological races of *P. oryzae*.

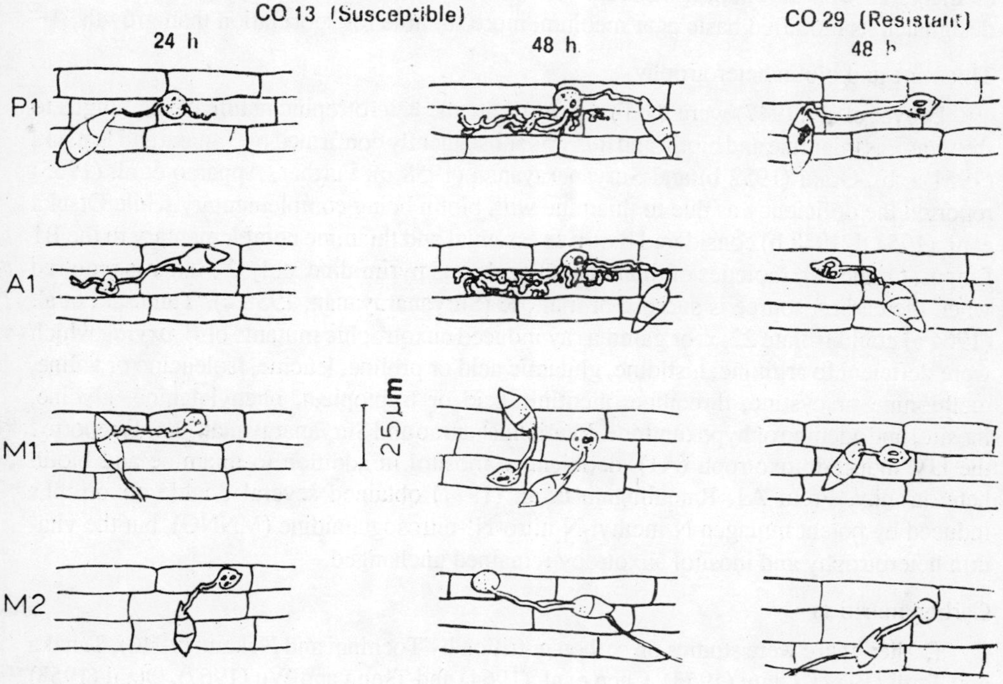

Fig. 5: Camera lucida diagrams of infection in leaf sheaths of inner epidermal cells of susceptible CO 13 and resistant CO 29 rice cvs. by parent (P1) and mutants (A1, M1 and M2) of *P. oryzae* at different h after inoculation. Note : The hyphal penetration of the cells and colonization of the fungus in the case of P1 and A1 on CO 13 only. (Manibhushanrao, 1971).

Ferault et al. (1971) detected the presence of virus like particles (25-32 µm dia.) in the mycelia of *P. oryzae* at 70,000 × using EM studies. No specific role has been assigned for its presence.

Physiology

Physiological studies on the B1 fungus are not of considerable interest because the isolates showed little difference in their mode of nutrition although they exhibited great genetic variability. Earlier, Yoshii (1936) reported that *P. oryzae* is amenable to anaerobic culture while Tochinai and Nakano, 1940 studied the nutritional aspects of the isolates. Although nitrate was readily utilized by the fungus, nitrite appeared to be harmful. Recent studies, however, focused on the breakdown of polysachharides, liberation of cellulolytic, pectolytic enzymes and other fungal metabolites by the B1 pathogen.

Nutritional studies

The fungus generally grows well in many semi-synthetic and natural media and media supplemented with plant extracts or decoctions. The difficulty to culture on pure synthetic media showed a heterotrophic nature. Earlier, Tomizawa (1953) studied the nutritional physiology of the rice B1 fungus using Tomizawa's medium while Tanaka (1965) developed three synthetic media A, B and C with the composition of C, N, Mg and macro and micro elements as well as vitamins. Suzuki (1976 b) recommended a semi-synthetic medium designated as modified basic agar medium, more suitable for sporulation than growth.

Thiamine and biotin heterotrophy

Leaver et al. (1947) were the first to attribute the heterotrophic nature of the fungus to the vitamins thiamine and biotin and this was subsequently confirmed by Tanaka and Katsuki (1951 a, b), Otani (1952 b) and Suryanarayanan (1958 c). Further, Apparao et al. (1955) reported the deficiency as due to thiamine with biotin being complementary while Otsuka et al. (1957 d, 1958 b) considered biotin as essential and thiamine complementary to the B1 fungus. Of the two moieties of thiamine, thiazole and pyrimidine, only the latter is required when the carbon source is sucrose or maltose (Suryanarayanan, 1958 c). Yamasaki et al. (1964 a) could isolate 22, x or gamma ray induced auxotrophic mutants of *P. oryzae* which were deficient to arginine, histidine, glutamic acid or proline, leucine, isoleucine or valine, methionine or cystine, threonine, nicotinic acid or tryptophan, phenylalanine, glycine, inositol and adenine or hypoxanthine. Manibhushanrao and Suryanarayanan (1971) reported the UV induced auxotroph (A1), deficient to inositol in addition to thiamine and biotin heterotrophy. From A1, Ramalingam et al. (1977) obtained several double auxotrophs induced by potent mutagen N-methyl-N-nitro-N'-nitrosoguanidine (MNNG), but the vitamin heterotrophy and inositol auxotropy remained unchanged.

Carbon nutrition

Earlier, there were studies on carbon nutrition by Tochinai and Nakano (1940), Tanaka and Tsuji (1952), Otani (1953), Chen et al. (1964) and Tseng and Wu (1967). Otani (1953) reported that maltose, sucrose, glucose, insulin and mannitol as well as organic acids such as succinic acid were the best carbon sources utilized while lactose and galactose were not suitable. Otsuka et al. (1957 a, b, c 1965) determined the relative efficacy of different carbon compounds on 47 isolates of *P. oryzae*. Among 32 different carbon sources tested, sucrose, glucose, maltose, fructose, lactose and xylose were found the most suitable while organic acids were not suitable. Only a few isolates* utilized sorbose, sorbitol, mannitol, insulin and glycerol. The degradation and utilization of cellulose was also reported (Inoue, 1939; Singh and Kunene, 1980). No correlation on pathogenicity and utilization of different carbon sources by the fungus isolates/mutants/races of *P. oryzae* was obtained (Otsuka et al., 1965; Tseng and Wu, 1967 and Manibhushanrao, 1971 b).

Nitrogen nutrition

Otani (1952 b) reported that nitrates (KNO_3 and $NaNO_3$), glycine, L-alanine, aspartic acid, DL-glutamic acid and asparagine were the best sources of N where as $NaNO_2$, cystine, taurine and creatine inhibited the growth of the B1 fungus. Besides, Tanaka et al. (1952; 1953) reported the utilization of ammoniacal N sources provided the pH of the medium was adjusted to 7.0. Apparao (1956 a) reported that asparagine, peptone, $NaNO_3$ and KNO_3 supported good growth. Nitrate could be utilized with ease and ammoniacal N very poorly owing to the attendant drop in pH. Further Otsuka et al. (1957 a, b, c) reported that among the several N compounds tested against 47 fungal isolates, KNO_3, $NaNO_3$, L-aspartic acid, L-asparagine, L-arginine, L-alanine, L-proline, L-serine, L-histidine, L-glutamic acid and L-

* For the purpose of uniformity the term isolate was used throughout irrespective of the terms culture/strain used in the publications.

glycine were the most suitable for growth while oxyproline, L-cystine and L-phenylalanine were not utilized. However, the utilization of L-phenylalanine, L-cystine, L-tryptophan and $NaNO_2$ varied markedly among the isolates. Chen et al. (1964) and Lee and Wu (1967) reported that glycine, glutamic acid, aspartic acid and KNO_3 were the best.

Based on the utilization of different C/N compounds as well as other physiological properties, Otsuka et al. (1957 d; 1958 a, b and 1965) classified *P. oryzae* isolates into 13 biochemical types. No appreciable differences were found among the isolates (IRRI, 1965) and mutants (Manibhushanrao, 1971 b).

It is evident that virulence/avirulence could not be convincingly correlated with the nutritional physiology of the B1 fungus although obvious differences were noticed. However, Goto and Yamanaka (1957), Suzuki (1953) and Otsuka et al. (1965) believed that pathogenic isolates are characterized by their inability to reduce NO_3 and assimilate NO_2.

Yamasaki et al. (1964) and Manibhushanrao (1971 b), while working with nutritional mutants with varied pathogenicity, pointed out that a nutrient requiring mutant behaves like the parent *in vitro* from the point of view of its virulence on the host plants especially in the early phase of infection.

In fact, Wood (1967), commenting on the nutritional theory of parasitism, rightly pointed out that studies with biochemical mutants of plant pathogens do not contribute to the understanding of virulence and resistance. Pathogenicity is not always restored when the particular growth requirements are met by an exogenous supply.

Requirements for trace metals

Heavy metals such as Fe, Zn, Mn and Cu are quite essential for the growth of B1 fungus (Apparao et al. 1955). Toxicity was more to Cu than Zn and its effect on growth was more pronounced in the absence of other essential elements and was counteracted by the presence of Fe and Mn (Apparao, 1959)

Toxic metabolites

Tamari and Kaji (1954, 1955) were the first to report α-picolinic acid ($C_6H_5NO_2$) and piricularin ($C_{18}H_{14}N_2O_3$) from the culture broth of the B1 fungus as well as from the severely B1 infected rice tissue. The former toxin interferes with the respiration by chelating the Fe^{+++} of respiratory enzymes while the latter totally inhibits respiration though it stimulates at extremely low concentration (0.6 ppm). Subsequently, Tamari and Kaji (1957) characterized piricularin as a colourless, crystalline substance (melting point 73.5°C and molecular weight 306), highly stable in water at acid pH. Piricularin on mechanically wounded rice tissue produced typical necrotic spots resembling those of B1 lesions (Ogasawara et al., 1961). Similar lesions can also be produced by the known metabolic inhibitors (Toyoda and Suzuki, 1958). Further, Tamari and Kaji (1959) reported that piricularin on rice plants stimulated the accumulation of coumarin in addition to its toxic effect. The fungus *in vitro* exuded piricularin—binding protein (PBP) mw. 69,000) into the culture filtrates which is again toxic to rice tissue although non-toxic to the fungus itself (Ogasawara et al., 1957 and 1961). Although much work has been done on the isolation and characterization of piricularin by the Japanese workers, surprisingly since 1954, no confirmed reports by other workers are available.

Iwasaki et al. (1969, 1972), Umetsu et al. (1972) and Narayanarao and Suryanarayanan

(1974 a) reported two other toxins pyriculol ($C_{14}H_{16}O_4$) and tenuazonic acid in cultures of *Pyricularia* sp. which are toxic and cause host cells death. Turner (1971) identified the latter as a heptaketide. Pyriculol also causes near natural B1 lesions and inhibits the growth of rice seedlings (Sato and Kozaka, 1970). Subsequently, Iwasaki et al. (1972) identified yet another toxic substance, namely 3, 4-dihydro-3,4 8-trihydroxy-1 (2H) naphthalenone from the culture filtrates of *P. oryzae* while Iwasaki et al. (1973) and Narayanarao and Suryanarayanan (1975) reported the isolation of many unidentified phytotoxic substances considered to be related to pyriculol either from B1 fungus/ races of *Pyricularia* spp. from different graminaceous hosts.

Both tenuazoic acid (Iwasaki et al., 1972) and pyriculol are considered *in vivo* toxins as they were detected *in vivo* in the infected rice and *Brachiaria mutica* leaves respectively. A norvalin-analog of tenuazoic acid was reported (Nukania and Sato 1992). Although a lot of literature is available on these toxins of *Pyricularia* spp., it is not yet possible to ascribe any concrete role for them in the B1 disease syndrome. More work is needed to confirm these results.

Other secondary metabolites

Otsuka et al. (1957 d; 1958 a, b and 1965) and Otsuka (1959 and 1961) reported the production of secondary metabolites such as amino acids and vitamins by *Pyricularia* spp. in culture media. Valine, serine, alanine, glutamic and aspartic acids are known to be produced in large quantities while heterocyclic, aromatic and thio amino acids have been produced in small quantities in synthetic media by *P. oryzae*. The amount again varied with the species isolate and the type of culture medium. Most isolates also produced the vitamins riboflavin, pantothenic acid, B-6 and folic acid in culture media. Further, indole acetic acid (IAA) was detected in the mycelial extracts of rice B1 fungus (Watanabe, 1957) and in culture filtrate ((Sridhar, 1967), while Wu and Shih (1967) studied on the fungal proteins.

In vitro production of enzymes

Tamari et al. (1965) and Otsuka et al. (1965) classified *P. oryzae* isolates on the basis of biochemical differences which showed no difference when assayed for amylase while the proteolytic activity varied markedly. Further, the succinic dehydrogenase, cytochrome and polyphenol oxidase (PPO) activities showed no correlation to the pathogenicity of the isolates. However, isolates of strong and weak pathogenicity showed less and more oxidase activity respectively.

Reports on the proteolytic enzymes (Otani 1955) peroxidase (PO) (Nuefeld et al., 1958); C_1 an C_x enzymes (Jothianandhan and Shanmugasundaram, 1969); proteinase (Jothianandhan and Shanmugasundaram, 1968); cellulase (Sudo et al., 1973) and a number of intermediate compounds of C_x enzymes (Hirayama et al., 1976); glucosidase (Hirayama et al., 1978) from the B1 fungus are available. No significant correlations were obtained with the geographic distribution of the two distinct isolate groups and their pathogenicity (Matsuyama and Kozaka, 1971). Further, Manibhushanrao (1971 c) could not get a correlation between the growth and polygalacturonase (PG) and cellulase activity among the virulent and avirulent mutants and virulent isolates of *P. oryzae* respectively. Similarly no correlation was obtained between virulence of the isolates and the enzyme activities (Lee et al., 1975). However, the virulent isolate P-2 could not produce the lytic enzymes.

Variability

Among all the rice diseases, probably B1 has the most extensive literature available over the years, but there are still some major gaps in understanding the variability of the B1 pathogen. The variability in cultural characteristics, nutritional requirements, and pathogenicity among B1 fungus isolates has increased the difficulties in breeding for B1 resistances. Further, the role of collateral hosts, perfect state, physiological races, differential resistance to leaf and neck B1 are some of the problems which have not been fully understood. Appearance of neck B1 in some crop seasons are some of the problems which have not been fully understood. Again, due to variability of the fungus the sudden outbreak of B1 on the high yielding resistant cvs. pose a serious threat to the breeding programme.

In India, Thomas (1941) was the first to report the breakdown of B1 resistance of the two selections 11348 and 10998. Now, reports on resistant cv. CO 20 (Krishnaswamy et al., 1969) in Tamil Nadu; IR 50 in Andhra Pradesh (Venkatarao et al., 1988) and Tamil Nadu (Loganathan and Ramaswamy, 1984), Intan in Karnataka and of the promising culture 30 in Kerala are available. Rice cv. Tongil in Korea had broken down resistance after being successfully grown for ten years while Lee and Cho (1990) reviewed the breakdown of B1 resistant rice cvs. in Korea (Table 2).

Table 2: Examples of breakdown of resistant varieties due to development of new race of *P. oryzae* (Lee and Cho, 1990)

Variety	Cross combination	Year of release	Year of breakdown	Duration (year)	Virulent races
Kwanok	Pungok × Kanto 55	1966	1969	4	KJ-105
Tongilchal	IR 888/IR 1317/IR 667	1972	1976	5	KI-413
Yushin	IR 667/IR 1317	1976	1976	1	KI-413
Nopung	IR 93/Tongil	1978	1978	1	KI-315, KI-307
Raegyeong	Suweon 231/IR 24	1978	1978	1	KI-315, KI-307
Jinju	HR 769-16/Palgeum/BLI	1980	1980	1	KJ-301
Youngpoong	H 23/YRI 501 FI	1983	1983	1	KI-315a, KI-315b
Pungsan	K 23/IRI 545-339-2	1982	1983	2	KI-315a, KI-315b
Seonam	Fuji 280/BLI	1983	1985	3	KI-413, KJ-301
Cheonma	BL 7/Nongbaeg	1985	1985	1	KI-201, KJ-301

Recent advances in microbial genetics show that variability can also be due to the 'transportation' or 'cassette model' and 'interconversion' of genes (Buck, 1987). Transposable elements may be involved in the activation/inactivation of genes involved in race specificity/pathogenicity in *P. oryzae*.

Induced mutations

Mutations were reported in B1 fungus for nutritional requirements by UV irradiation (Tomizawa, 1956; Loprino, 1964; Kuo et al. 1967; Manibhushanrao and Suryanarayanan, 1971; Wu and Tsai, 1974; Genovesi and Magill, 1976; Crawford et al., 1986); chemicals viz. $CuSO_4$ and H_3BO_4 with reference to aromatic H_2S, sucrose and lactose metabolism as markers (Yamasaki and Niizeki, (1965); MNNG (Ramalingam et al., 1977; Manibhushanrao and Ou, 1979). Recently, Hamer et al. (1990) induced mutations at the SMO genetic locus which affect the shape of the diverse cell types.

UV dose survival

Manibhushanrao and Suryanarayanan (1971) obtained an exponential dose-survival curve for UV irradiated *P. oryzae* conidia similar to the response seen in other fungi. However, a sigmoidal curve is more common in several fungi. Although the significance of these two types of curves is debated, studies on *Neurospora crassa* indicated that the exponential type is characteristic of uninucleate spores. The reports regarding the number of nuclei in the conidium of *P. oryzae* are contradictory. Chu and Li (1965) and Suzuki (1967 a) found that the conidia were multinucleate while others (Mizusawa, 1959; Horino and Akai, 1965; Wu and Tsao, 1967) reported that the cells of the conidium were mononucleate. The exponential nature of the dose-survival curve of *P. oryzae* with less than 1% of the conidia surviving in 150 sec of UV irradiation suggests that most survivors are likely to have only one viable nucleus.

Genetical studies

Suzuki (1965a, 1967 a, b) reported that heterokaryosis is the main mechanism involved as cells of conidia, appressoria and mycelia are multinucleate, and anastomosis is common. Yamasaki and Niizeki (1965 a, b) synthesized four different heterokaryons from paired auxotrophic mutants induced in several strains of *P. oryzae*, some of which carried other markers. Resistant to $CuSO_4$, these strains behaved as a dominant and production of H_2S as recessive. All these heterokaryons generated recombinant phenotypes. Heterokaryon-I gave rise to markers not seen in either of the parents. The explanation was that the variability resulted from a parasexual recombination involving diploid nuclei that are very unstable and segregate by haploidization (through nondisjunction) or mitotic crossing -over or both mechanisms. Later, Hashioka and Inagaki (1968) reported a heterokaryon resulted from hyphal anastomosis of *P. oryzae* and *P. zizaniae* isolates while Ginovesi and Magill (1976) and Manibhushanrao and Ou (1979) studied in *P. oryzae*. Reports on intra isolate heterokaryosis (Fatemi and Nelson, 1977), the heterokaryotic and vegetative diploid phases (Crawford et al., 1986) are available.

Parasexuality has been established through forced hyphal fusion (Caten, 1981) as well as by protoplast fusion (Biossonnet-Mennes and Lecoq, 1976). The isolation of protoplasts in *P. oryzae* was made by snail gut enzyme treatment (Giatogong and Frederiksen, 1969) as well as helicase (Biossonnet-Mennes and Lecoq, 1976). The rate of protoplasm fusion was greater with polyethylene glycol.

The level of fertility through genetic analysis has been studied with field isolates that infect rice as female-sterile (Itoi et al. 1983; Valent et al. 1986) hermaphroditic strains (Yaegashi and Asaga, 1981; Kato and Yamaguchi, 1982; Valent et al. 1986) and melanin deficient non-pathogenic mutants (Chumely and Valent, 1990). In general, it was too low, but with an exceptional isolate which produced viable ascospores with other rice pathogen field isolates (Leung et al. 1988). Fertile hermaphroditic rice pathogens have also been produced in the *in vitro* conditions through back crossing involving an isolate from rice and highly fertile pathogen of *Eragrostis curvula* (weeping lovergrass) (Valent and Chumley, 1989; Valent, 1990; Valent et al., 1991). These studies revealed the identification of genes that control specificity towards particular rice cvs. and genes that control specificity towards weeping lovegrass.

Parsen et al. (1986) developed a system for the genetic transformation of rice B1 fungus (*M. grisea*) to colonize its host on rice. Arginine-less auxotroph was shown to contain a mutation (*arg 3 - 12*) that abolished ornithine carbomolytransferase activity. *M. grisea* isolates that contain *arg 3 -12* were used as recipients in the transformation of plasmid pMA2, which carries the *argB*⁺ gene from *Aspergillus nidulans*. Stable protrotopic transfornants arose at a frequency of about 35 μg of plasmid DNA. Integration of single or multiple plasmid copies occurred at a single site in the genome of each transformant; rearrangements were often created during integration. When *M. grisea* genomic segments were incorporated into pMA2, the presence of any one of five different *M. grisea* segments did not greatly affect the efficiency of transformation. Integration via homologous recombination occurred when the donor plasmid was linearized by cleaving at a unique restriction site within the fungus segment. These results also suggested that gene cloning systems based on transformation with *argB*± should serve in the isolation on many genes that are responsible for the general pathogenicity and host-specificity phenotypes in the ubiquitous B1 pathogen.

Reports on crosses between isolates of *P. oryzae* pathogenic to Poaceae other than rice are fertile are available (Hebert, 1971, Yaegashi and Hebert 1976, Kato et al., 1976, Valent et al., 1986). However, Silue and Notteghem 1992 successfully obtained crosses between two isolates of *M. grisea* Guy 11 and CD 128 both pathogenic to rice. Further, all fertile crosses between isolates pathogenic to rice involving Guy 11 were reported (Leung et al., 1988, Ellingbob et al., 1990). This might be due to accumulation of antagonistic fertility genes in rice isolates as Guy 11 is considered an exceptional isolate. Isolate CD 128 or avirulent ascospores from cross Guy 11 × CD 128 was used for analyzing resistance in cv. Kusabue since one gene for avirulence and another (*pi–k*) for resistance (Kiyosawa, 1968b) are involved. This has helped to demonstrate Flor's 1956 gene-for-gene hypothesis as the relationship exists in the pathoystem rice-*M. grisea* (Silue et al., 1992).

Karyological studies

Suzuki (1953; 1965a; 1967b) and Chu and Li (1965) reported that the cells of mycelia, conidiophore initials, conidia and appressoria of *P. oryzae* are multinucleate and that hyphal anastomosis and nuclear migration are common. On the other hand, Chen (1964; 1965), Yamasaki and Niizeki (1965), Wu and Tsao (1967), Giatgong and Frederiksen (1967; 1968; 1969), Manibhushanrao and Ou (1980) are of the opinion that most of the cells are uninucleate and only a small percentage contains 2-6 nuclei.

Subramanian et al. (1978) made an attempt to study karyology of conidiogenesis of race ID-1 of *P. oryzae*. The nuclei in the cells of the conidia, conidiophores and mycelia showed marked variation in size, shape and density. An interphase nucleus migrates (Fig. 6, 1 and 2) from a hyphae into a conidiophore which is then delimited by a septum from the parent hyphae. During initiation of the first conidium the nucleus of the conidiophore divides mitotically and one of the daughter nuclei migrates into the elongated tip of the conidiophore (Fig. 6, 3). As the conidium initial elongates further, the nucleus divides and a septum appears between the nuclei immediately after the division (Fig. 6, 5 and 6) thus making it 2-celled. Subsequently, the nucleus of either the distal or the proximal cell of the bi-cellular young conidium divides after further elongation of the cells followed by the formation of a second septum (Fig. 6, 7 and 9) and dividing the conidium into three uninucleate cells. Before or after maturation of the first conidium, the daughter nucleus which remained in the conidiophore divides; one of the resulting daughter nuclei moves into the initial of the second conidium formed as a lateral proliferation just below the first conidium (Fig. 6, 10 to 12). A repetition of this process takes place during the development of successive conidia on the same conidiophore. Conidial cells are generally uninucleate, but cells with two nuclei do occur (Fig. 6, 14 to 18). The number of cells per conidium is mostly three and occasionally 2, 4 or 6 (Fig. 6, 16 and 17).

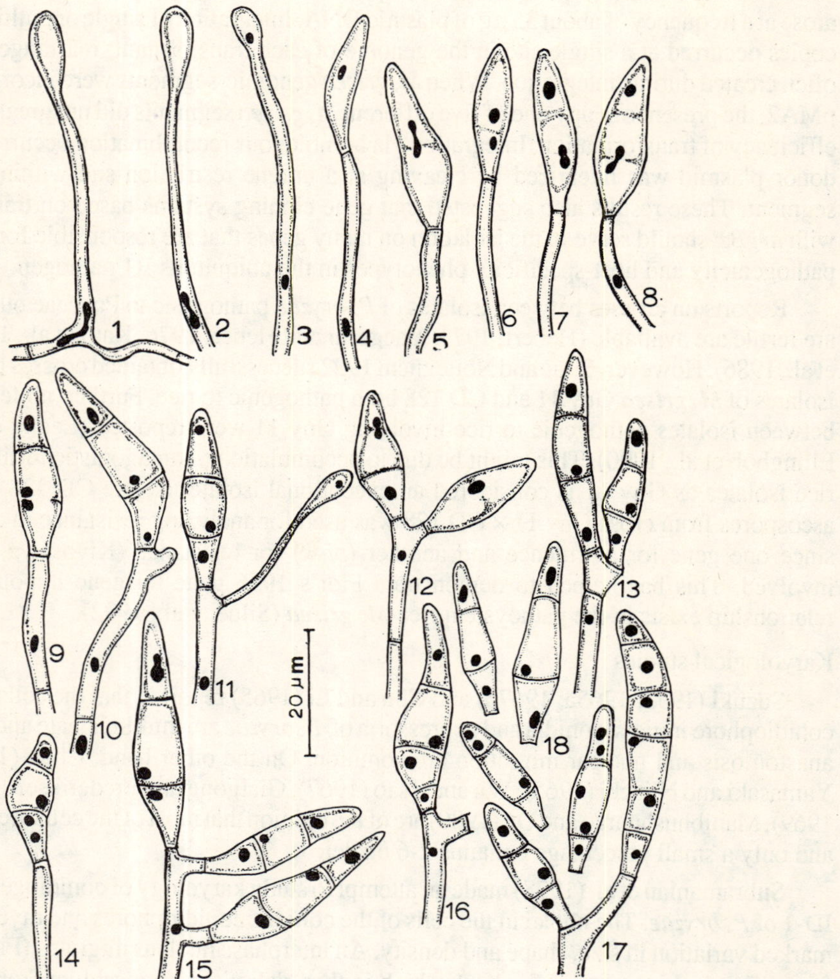

Fig. 6: Karyology of conidiogenesis of rice ID-1 of *P oryzae* (1-17). Bar : 20 µm (Subramanian et al., 1978).

Manibhushanrao and Ou (1980) also studied nuclear behaviour in 12 Philippine races/isolates of *P. oryzae* (Plate 5, A-H). Within 24 h of germination of the conidium, short club-shaped mycelial cells were observed, with a majority of them being multinucleate (2-6 nuclei). Shortly after 24 h these multinucleate cells produced mycelia ranging from 1.1 - 1.9 µm in width made of uninucleate cells where an appressorium formed the nucleus, divided mitotically and the number increased from one to sixty four. A few mycelial and conidial cells were seen without nuclei and this condition may be due to the migration of nuclei without any mitotic division. The number of nuclei in each type of mycelial cell, conidium and appressorium thus varied considerably. This confirms the results of earlier workers (Yamasaki and Niizeki, 1965; Gaitgong and Frederiksen, 1969). Yaegashi and Hebert (1976) considered the multinucleate state to be common in older mycelial cells.

Suzuki (1965; 1967) reported the conidial cells to be multinucleate with varying number

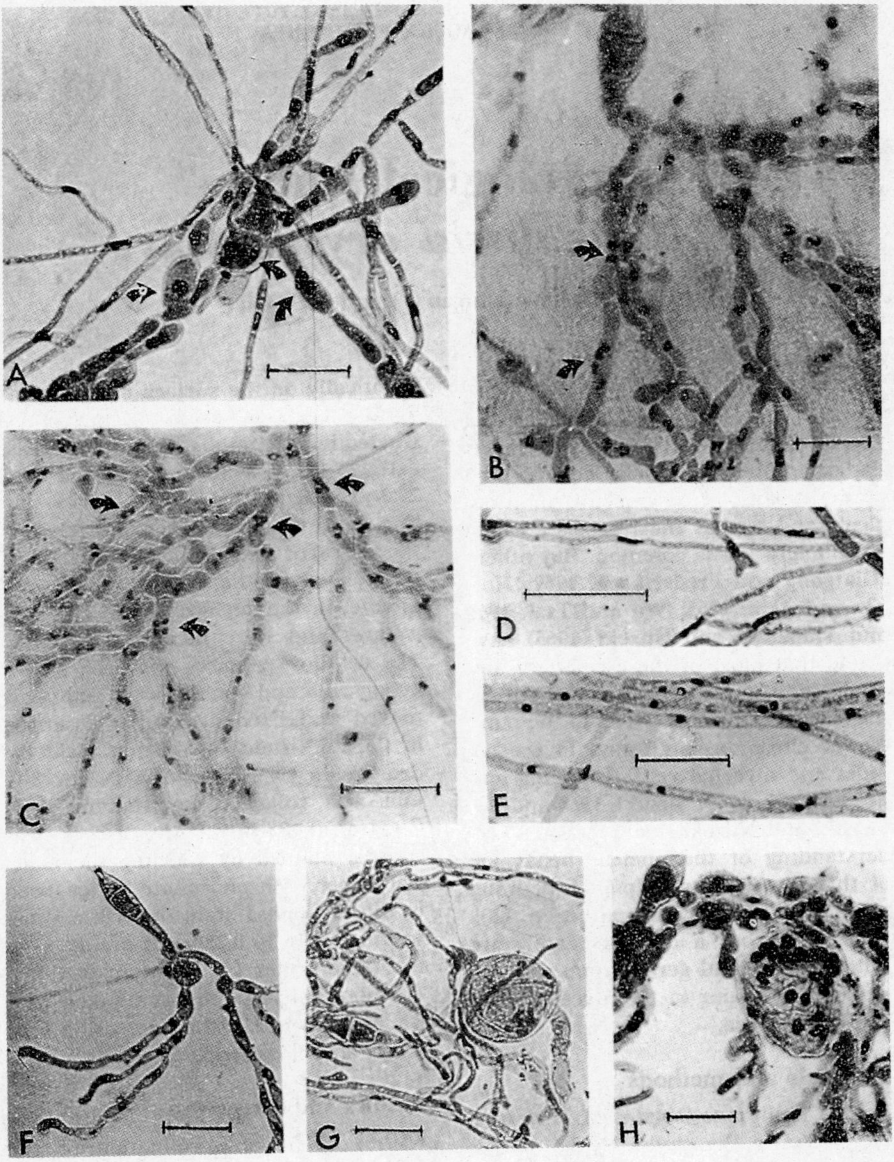

Plate 5a-h: Nuclear behaviour in *P. oryzae.* (A) Club - shaped, multinucleate mycelial cells development within 24 h after conidium germination. Note : Some of the nuclei (arrows) are larger than others. (B-C) Multinucleate club shaped mycelial cells with 2,3 and 4 nuclei (arrows). (D) Uninucleate slender mycelial cells with elongated nuclei. (E) Round nuclei on the border mycelial cells. (F-H) Multinucleate appressoria containing up to 32 nulcei. Bar : 20 µm (Manibhushanrao and Ou, 1980).

of chromosomes 2-5 in heterokaryotic and 3-4 in homokaryotic state. Reports on the chromosomal number in a nucleus are contradictory. Four (Chu and Li, 1965); 6, but occasioally 2, 3, 4, or 5 (Giatgong and Frederiksen, 1968, 1969); 2-12, with 3 and 6 having higher frequency (Lee et al., 1968; Tanaka and Yamasaki, 1974, 1975) and 6 only (Yaegashi and Hebert, 1976; Tanaka et al., 1979). Row and Crill (1979) correlated the size of the nucleus with the number of chromosomes. The chromosome number in the hyphal cells (2-12 but mostly 3 or 6) and the three celled conidia from top, middle and bottom cells (2-10, but mostly 3 and 6) varied in their chromosomal number. Considering the basic chromosome number to be n=3, the varying chromosome number could be due to asynchronous division, non-disjunction and lagging chromosomes during nuclear division. Tanaka et al. (1979) confirmed lagging chromosomes in both meiotic and mitotic nuclear divisions during the development of the perithecium. Singh (1983) has studied somatic nuclear division throughout typical mitosis and observed n=5 which is not a variable while Row et al. (1985) reported mitosis in the B1 fungus and suggested possible implications for pathogenic variability.

A correct knowledge of the nuclear behaviour is quite essential to explain the variability in the fungus but controversies exist on the very mode of nuclear division itself. How reliable is light microscopy for determining chromosome numbers in fungal mitotic behaviour is again questionable. More critical studies are needed to confirm the cytology of the B1 fungus.

Observations made by Manibhushanrao and Ou (1980) seem to show that the fungus always passes through a transitory or temporary multinucleate stage as club-shaped mycelial cells or as appressoria (Fig.7). The multinucleate condition of mycelial cells and appressoria is probably the critical stage at which genetic material of the nuclei may exchange, leading to heterokaryosis. If the conidia, are homokaryotic to begin with, that is all nuclei are identical, how does heterokaryosis occur in appressoria? A proper explanation is lacking.

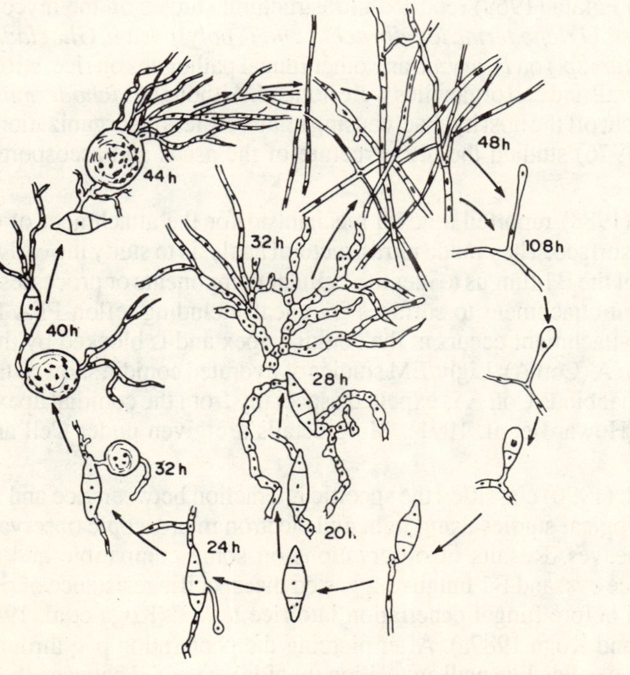

Fig. 7: Transitory multinucleate state following the germination of conidium of *P. oryzae*. Note : The diagram is not to scale. (Manibhushanrao and Ou, 1980).

It appears that both karyological and genetical changes play a major role in the variation of the B1 pathogen while the constant change in chromosome numbers in the cells of conidia, mycelia and appressoria due to non-disjunction asynchronous division and lagging due to nuclear division may be the primary factor. Suzuki's (1965; 1967) observation of multinucleate cells is perhaps due to a faulty staining procedure. Further, critical studies on nuclear fusion or quantitative estimation of DNA are necessary to confirm the chromosome number. Tanaka and Yamasaki (1973) correlated between DNA content and the area of the nucleus in conidial and mycelial cells. However, the reconstruction of serial sections of electron micrographs of the nuclei, developed by Moens and Perkins (1969), can be used to confirm the chromosomal number, while combined electron and optical microscopy (Girbardt, 1965) on karyological features may be a better way of resolving the question of chromosome number in *P. oryzae*.

Ultrastructure

Horino and Akai (1965), Hashioka et al. (1967 a, b) and Wu and Tsao (1967) confirmed the earlier observations of Mizusawa, 1959, Horino and Akai, 1965 and Yamasaki and Niizeki, 1965 by EM studies on the single nucleus per cell, that can be long, elliptical, globular or more or less irregular in shape. Earlier, Mizusawa (1959) reported at least three layers of cell membranes and added that the nucleus has a nuclear membrane. Hashioka et al. (1967 a) made precise observations of the fine structure of longitudinal and transverse section of the conidium showing the cell having a single nucleus (centre), epispore liquid particles, mitochondria etc. (Plate 6 a, b). Further, Hashioka et al. (1968) observed the mode of invasion of *P. oryzae* into rice epidermal cells while the conidium has lacinated or serrated epispores which may be characteristics of the genus *Pyricularia* (Hashioka and Ikegami, 1969).

Hashioka and Fukita (1969) reported ultrastructural studies on the mycoparasitism of potential antagonists (*Trichoderma longibrachiatum, T. polysporam, Gliocladium deliquescens* and *Acremonium sp.*) on *P. oryzae* and other fungal pathogens on rice, strong disorganization of host cell wall and cell organelles were noticed. Although *Trichoderma* spp. invaded, coiled around and cut off the host hyphae, causing relatively less disorganization of host cells. Fukutomi et al. (1976) studied the ultrastructure of the ascus and ascospore walls in the perfect state.

Hamer et al. (1988) reported a novel mechanism for the attachment of conidia of *P. oryzae* to rice leaf surface. They made ultrastructural analysis to study the early stages of the infection process of the B1 fungus to identify cellular components or processes as targets for rapid and persistent attachment to surfaces (rice leaf including teflon-PFA film) prior to germination. The attachment occurs at the conidial apex and is blocked by the addition of lectin concanavalin A (Con A). Light/EM studies of hydrated conidia showed that a spore tip mucilage (STM) that binds Con A is expelled especially from the conidial apex before germ tube emergence. (Howard et al. 1991). More details are given under Cell and molecular biology.

Yaegashi et al. (1990) classified the specific interaction between rice and B1 fungus on the basis of karyological studies using light and electron microscopic observations of cells reactions in rice leaves. Results of observations on some compatible and incompatible combinations of rice cvs. and B1 fungus suggested that specific resistance of rice cvs. to B1 was not expressed before fungal penetration into rice leaves (Koga et al. 1986; Peng and Shishiyami 1988 and Koga 1989a). After piercing the penetration peg through the cuticle layer and cell wall, a collar-like wall apposition (papilla) appeared between the cell wall and plasma membrane of the susceptible host, but did not appear in the resistant combination (Koga, 1988). The wall apposition may be different from the papilla-like structure observed

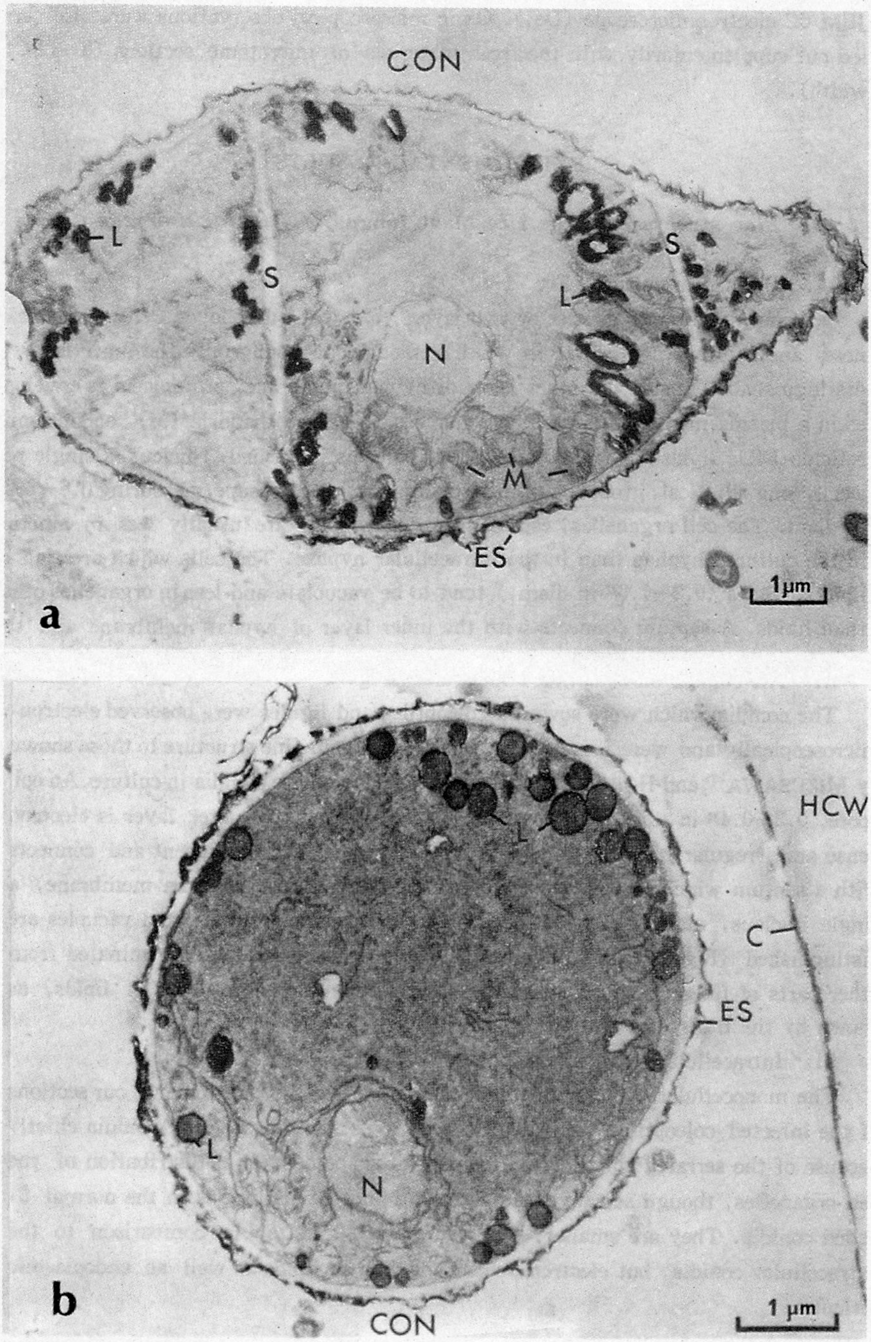

Plate 6ab : (a) Fine structure of a longitudinal section of the conidium of *Pyricularia oryzae* on rice.

(b) Transection of a conidium cell. ES : epispore; N : nucleus L : lipid; M : mitochondria; S : septum; CON: conidium cell C:cuticle and HCW: epidermal cell wall of a rice coleoptile. Note : The characteristic serrate epispore, a single nucleus and ample lipid particles (Hashioka, et al., 1967 a).

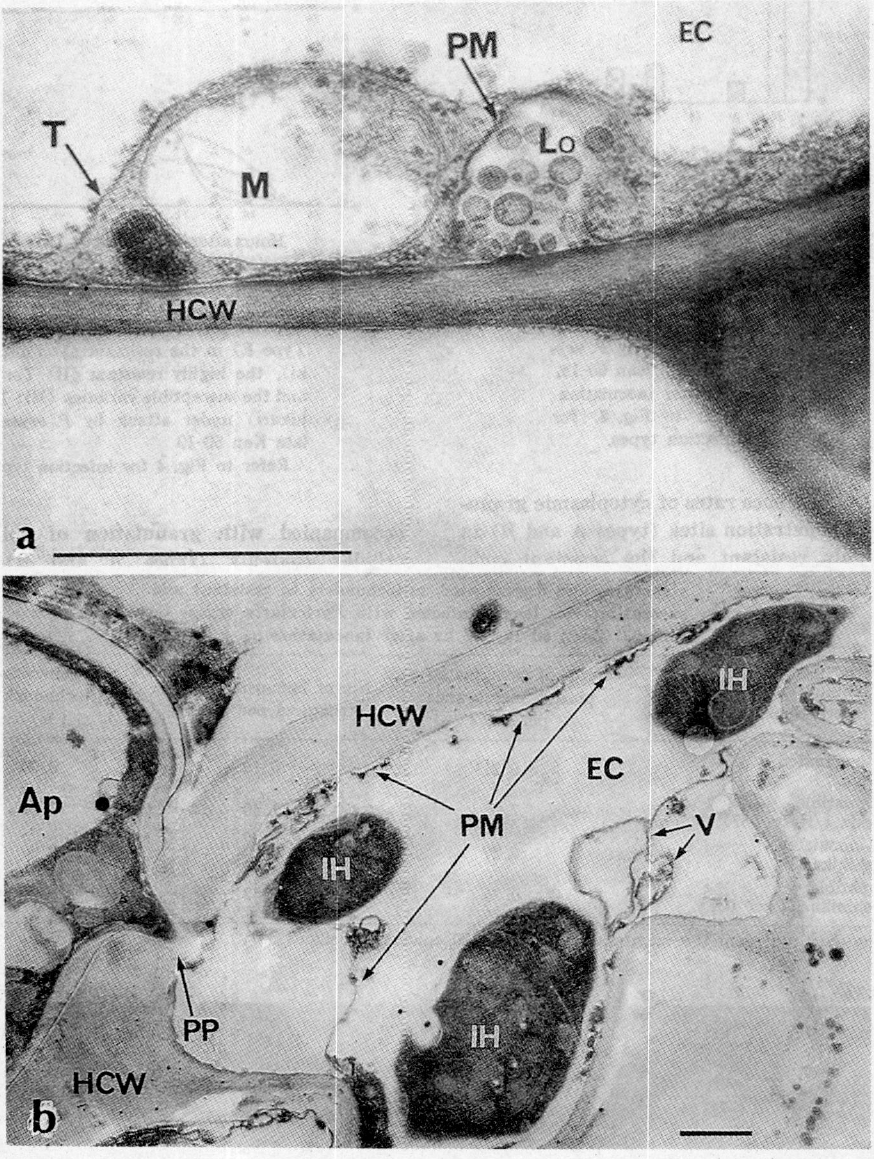

Plate 7ab : (a) Ultrastructure of a rice leaf of Toride 1 (highly resistant cv.) inoculated with *P. oryzae* , isolate Ken 60-19, 48 h after inoculation. Note : A swollen mitochondrion (M) and lomasome-like structure (Lo) observed in an epidermal cell. HCW : Host cell wall; PM: Plasma membrane; T : Tonoplast.

(b) Fragmented, highly electron dense plasma membrane observed in a cell on an inner leaf sheath. Note : To determine if the penetrated epidermal cell was alive or not, the microscopic observation of the material, was made after 27 h inoculation. PP : Penetration peg; IH : Invaded hyphae; V: Vesicular structure; Ap: Appressorium; EC: Epidermal cell. Bar : 1µm . (Koga, 1989a).

with a light microscope (Peng and Shishiyama 1989). Since the diameter of the wall apposition was smaller (<4 μm) than that of the appressoria (about 10 μm), the appressoria usually prevents the wall apposition from being observed under a light microscope.

In the resistant combinations, the destruction of the unit membrane structures and the fragmentation of the plasma membranes were observed at the attacked epidermal cells that did not plasmolyse (Koga, 1989a). Appearence of lomasome-like structures were prominent even after 24h of inoculation (Plate 7 a, b). The cytoplasmic granulation observed under a light microscope (Plate 3b) may be the fragmented and vesicled plasma membranes and other debris originating from cell organelles in the hypersensitively dead epidermal cells. In these cells of the leaf sheath, the invaded hyphae have undergone a compartmentalization process that lasted 27 h after inoculation and have appeared to be alive. However, most of the invaded hyphae lost fragmentation of plasma membranes and destruction of unit membrane cells observed in the infected epidermal cells suggesting that cell death occurred earlier than that of the invaded hyphae.

Recently , indepth studies on rice panicle and spikelets infected with *P. oryzae* using scanning (SEM) and transmission (TEM) electronmicroscopy was made (Koga and Yoshino 1990,1991 a, b). Conidiophores were abundant between empty glumes and rachilla and rarely on rachilla and on rudimentary glumes. They also formed among the row of tubercles, trichomes and directly from cuticular layer of epidermis and outthrushed from stomata (Plate 8a, 1) and from minute protuberance on the panicle axis as well as between primary axis and primary branch (Plate 8a, 2) as shown in SEM studies. Further, TEM studies of the blast infected panicle showed clear invasion of the pathogen hypae in the selerenchyma cells (Plate 8b, 3, 4, 5) degenerated host cell wall of the mesophyll.

In the case of spikelets invasion by the B1 pathogen, the invaded hyphae in the epidermal cells penetrate the walls, between epidermal cells and large trichomes. SEM studies indicate that conidiophores are abundant in the space between the former and latter while traces on large trichomes (Plate 9a, 1). On the other hand they are readily formed at the inner side of empty glume (Plate 9a, 2). Cross section of the invaded hyphae in sclerenchyma cells of lemma by TEM studies showed no change with the host cell wall (Plate 9b, 3 & 4) while denatured host cytoplasm is seen clearly in mesophyll cells (Plate 9b, 5).

In the susceptible combination, the ultrastructural changes observed in resistant combinations were rarely observed in plasma membranes of attacked epidermal cells that maintained plasmolysis. Moreover, the invaded hyphae were surrounded with the cytoplasm of the epidermal cell 40 h after inoculation.

Although much work has been in progress in recent years on the ultrastructural studies of rice B1 pathogen interactions, precise information about the mode of invasion and colonization is still lacking. Such critical studies are essential in interpreting the pathogen reaction to both resistant and susceptible hosts.

Antigenic comparison

Manibhushanrao et al. (1971) studied the serological affinities between the virulent P1 and A1 and the avirulent M1 and M2 mutants of *P. oryzae* by double diffusion and immunoelectrophoresis. The prototrophic P1 and auxotrophic A1 were identical serologically in possessing six common antigens (Fig. 8 a, b). The affinity of M1 (Pink avirulent mutant) to P1 and A1 is indicated by the five antigens which it shares with the latter two. M1, however, differs from P1 and A1 by the absence of the fifth antigen and also by the presence of a seventh one not present in the latter (Fig. 8 c). The Baarn mutant M2 (avirulent) appears to be further removed serologically from the rest, since this showed only three antigens in common with P1 and A1 and a fourth with M1 and was distinct from others in possessing the eighth antigen

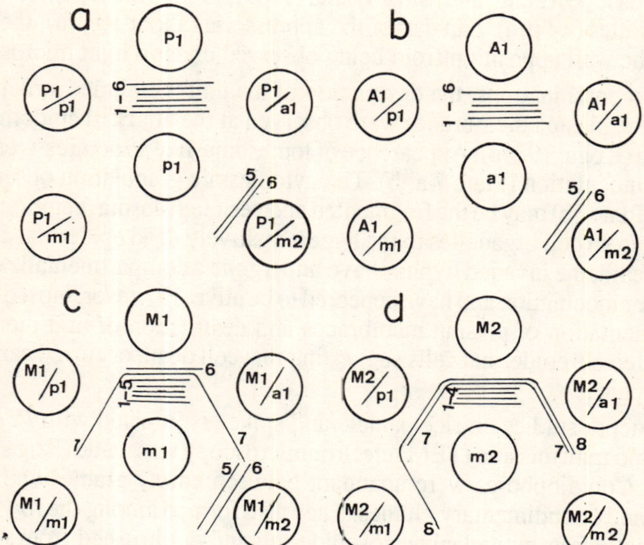

Fig. 8 abcd: Antigen-antibody reaction of four cultures of *P. oryzae* after adsorption of the antigen with homologus or hetetogenous antiserum. Capital letters indicate antigens and small letters antisera. Antigen/antisera indicates the type of adsorption (Manibhushanrao et al., 1971).

not present in the other three isolates (Fig. 8 d). The occurrence of a seventh antigen in the avirulent M1 and M2 but not in the virulent P1 and A1 may be significant. Avirulence may be reflected serologically by the absence of a particular antigen present in the virulent ones. The difference in antigenic composition between virulent and avirulent isolates may afford a valuable clue in understanding host-parasite compatibility.

Serological studies suggest that common antigens may play a role in pathogen specificity. They are based on the affinity that exists between the pathogen and appropriate antibody. Xia et al. (1992) developed monoclonal antibodies in the host, specific for the rice Bl pathogen *P. grisea*. However, these studies may not provide any base for the support or rejection of any model of gene action in pathogenesis.

Conidium ontogeny

Subramanian et al. (1978) studied conidiogenesis in the B1 fungus using race ID-1 of *P. oryzae*. Incidentally, race ID-1 corresponds to the US race 8 and had its origin in Lonavala, Maharashtra, India (Padmanabhan, 1965c). Both in the type species of *P. grisea* and *P. oryzae*, the conidia arise by a process of budding (blastic) and are, therefore, blastospores. Typically each conidium is also perched on a distinct denticle which separates the conidium from the conidiophore. The conidiophore proliferates sympodially and the successive conidia are all produced in the same way. The presence of a denticle is not only characteristic but so obvious that it has been interpreted even as a separating cell with a probable function in the ejection of the conidia (Ingold, 1965). On the other hand, a few of them exhibit different conidiogenesis as they produce successive conidia, which are gangliar in their ontogeny and,

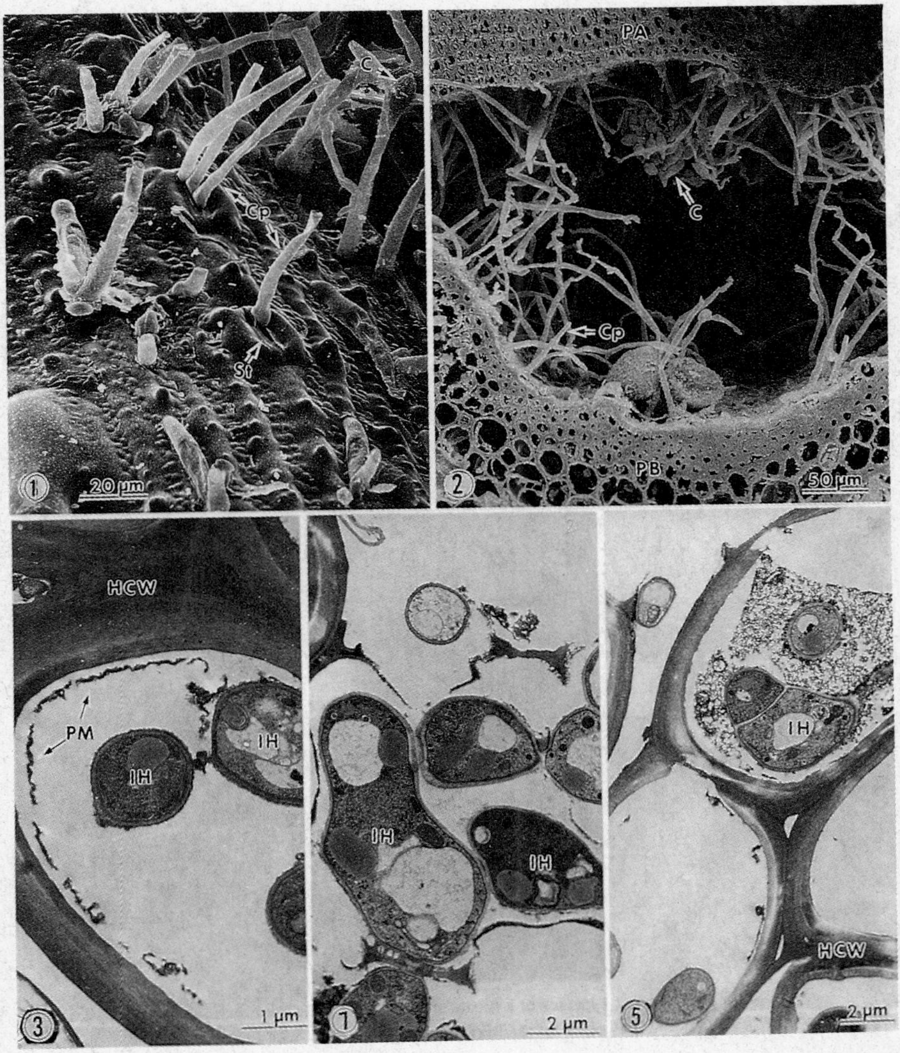

Plate 8ab : (a) SEM studies on rice panicle infected with *P. oryzae*. (1) Conidiophores (Cp) outthrushed from stomata (St) and also from cuticle of epidermis of the secondary branch of the rice panicle. (2) Conidiophores and conidia (C) formed between a primary axis (PA) and a primary branch (PB). (b) TEM studies on panicle infected with *P. oryzae*. (3) Cross section of invaded hyphae in sclerenchyma cell of a primary axis. (4) Sclerenchyma cells of panicle neck and (5) degenerated host cell wall in mesophyll of primary axis. HCW : host cell wall; IH: invaded hyphae; PM: plasma membrane (Koga and Yoshino, 1991a).

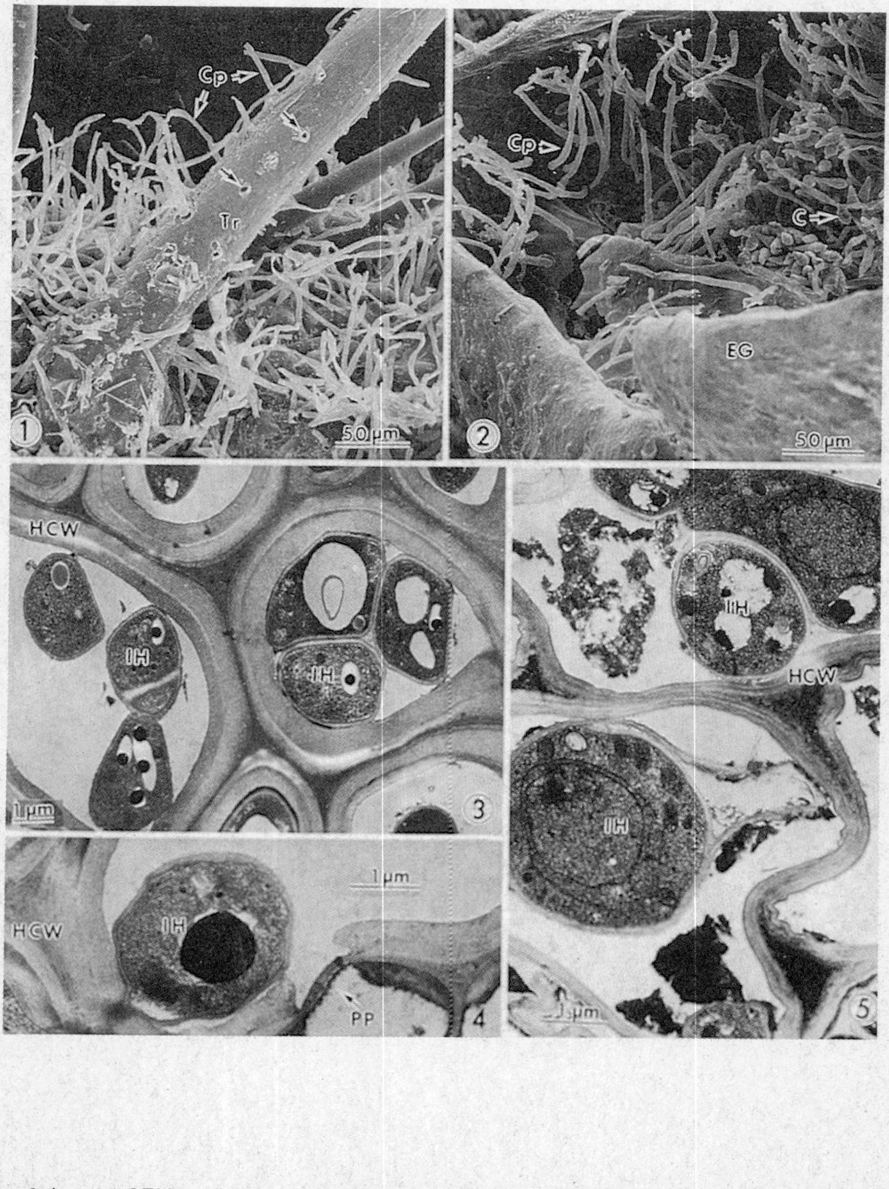

Plate 9ab : (a) SEM studies on rice spikelets of panicle infected with *P. oryzae*. (1) conidiophores around large trichome. (2) conidiophores and conidia formed at inner side of empty glume. Cp: conidiohphore; Tri : trichome; C: conidia and EG: inner side of empty glume.

(b) TEM studies on rice spikelets infected with *P. oryzae*. (3 & 4) Cross section of invaded hyphae in sclerenchyma cells of lemma (Not much changes observed in the host cells). (5) invaded hyphae and denatured host cytoplasm in mesophyll. HCW : host cell wall; IH: invaded hyphae (Koga and Yoshino, 1991b).

having a truncate base, are not perched on denticles. Consideration of the conidium morphology and ontogeny of the fungus studied by Subramanian et al. (1978) is out of place in the genus *Pyricularia*. However, it is quite possible that this culture has changed genetically, shifting from a blastic to a gangliar morphogenetic pattern by mutation. The conidium ontogeny of race ID-1 of *P. oryzae* showing the gangliar development is illustrated in Fig. 6 ,1-17). (See Karyological studies).

Manibhushanrao and Jesudhas (1984) studied in detail the conidiogenesis of 23 isolates of *Pyricularia* which included international races, some nutritional, coloured mutants and isolates obtained from cultivated and wild species of Gramineae. Of these, 16 isolates showed typical blastic ontogeny, characteristic of the genus while three coloured mutants and two international races (ID 14 and IA 111 Buff) showed the gangliar type of conidiogenesis. However, two races and one nutritional mutant showed both types (Table 3). Most isolates tested by them were found to be morphologically indistinguishable and potentially pathogenic with respective hosts on different cvs. Conidium ontogeny in *P. oryzae* races [blasto (Plate 10, 7-11; Fig. 9a), gangliar (Plate10, 7-11; Fig. 9b), and blasto-gangliar types (Fig 9c)] is given. The morphogenetic pattern of conidiogenesis can be altered by mutation and temperature (Bhat, 1970).

Although different isolates behaved differently in conidiogenesis it is possible that some races and mutants changed genetically and showed a different conidium ontogeny. As far as these wild types, races and mutants of *Pyricularia* are concerned, too much emphasis need not be given to the mode of conidium development in the identification since most isolates are morphologically alike and behave differently in pathogenicity tests on rice.

Sporulation

Kuribayashi (1928) and Sueda (1928) reported that sporulation of *P. oryzae* in culture occurs at 10 to 35°C with an optimum at 28°C (Tochinei and Shimamura, 1932) but Leaver et al. (1947) failed to sporulate in liquid culture. Although Henry and Andersen (1948) reported optimum sporulation at 28° C, it tends to decrease after nine days while at 16, 20 and 24° C, it continues to increase even after 15-d. There was heavy sporulation on oat-sorghum mixture 1 : 1 with forced aeration. Suryanarayanan (1958 c) could not obtain any beneficial results by aerating the culture grown on sucrose-nitrate medium. However, Apparao (1962) showed that sporulation in culture is not only governed by the initial concentration of carbon source but is profoundly influenced by the quantity of vitamins and heavy metals in the medium. He studied the effect of heavy metals such as Fe, Zn, Cu and Mn of which requirements of Cu and Mn were greater for sporulation than for growth.

Sporulation is aided not only in complex media containing straw extract or rice polishing or a mixture of barley grains and chopped rice leaves but also in natural media like banana shoots or whole barley grains (Atkins et al., 1967). The effect of various media (Suzuki, 1967b) and N source as aspartic acid and glutamine (Honma et al., 1968) and glycine by an avirulent (M2) mutant (Manibhushanrao, 1971 a) were studied. Glycine (0.001% N level) promoted maximum sporulation by 12-d while, higher concentration (1% N level) inhibited the sporulation (Fig. 10). Further Bhattacharya and Bose (1981) standardized the condition of sporulation in different media under different physiological conditions.

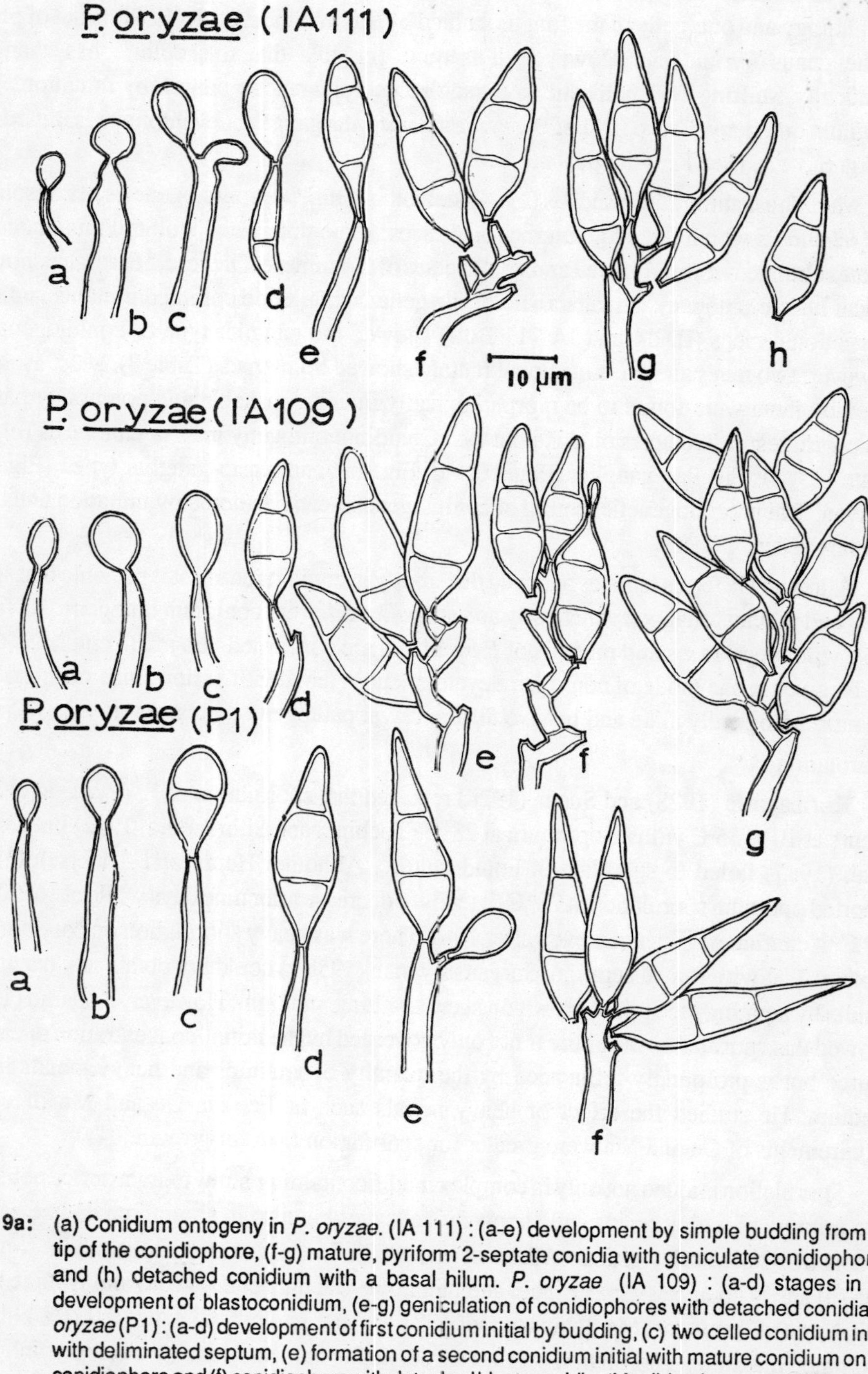

Fig. 9a: (a) Conidium ontogeny in *P. oryzae*. (IA 111) : (a-e) development by simple budding from the tip of the conidiophore, (f-g) mature, pyriform 2-septate conidia with geniculate conidiophores, and (h) detached conidium with a basal hilum. *P. oryzae* (IA 109) : (a-d) stages in the development of blastoconidium, (e-g) geniculation of conidiophores with detached conidia. *P. oryzae* (P1) : (a-d) development of first conidium initial by budding, (c) two celled conidium initial with deliminated septum, (e) formation of a second conidium initial with mature conidium on the conidiophore and (f) conidiophore with detached blastoconidia. (Manibhushanrao and Jesudhas, 1984)

Plate 10 : Conidium ontogeny in *Pyricularia*. (1-4.) Development of blastoconidium by simple budding from the tip of the conidiophore. (5) Formation of the second conidium initial. (6) Typical geniculate conidiophore. The mature conidia are pyriform and two-septate with a basal hilum. (7-9) Development of gangliar conidium. (10) Conidiophore bearing conidia in clusters. (11) Mature gangliar conidium with a truncate base. No constriction at the base of the conidal initial. Note: Gangliar conidia with a truncate base are attached to the conidiophore. Bar : 20 µm (Manibhushanrao and Jesudhas, 1984).

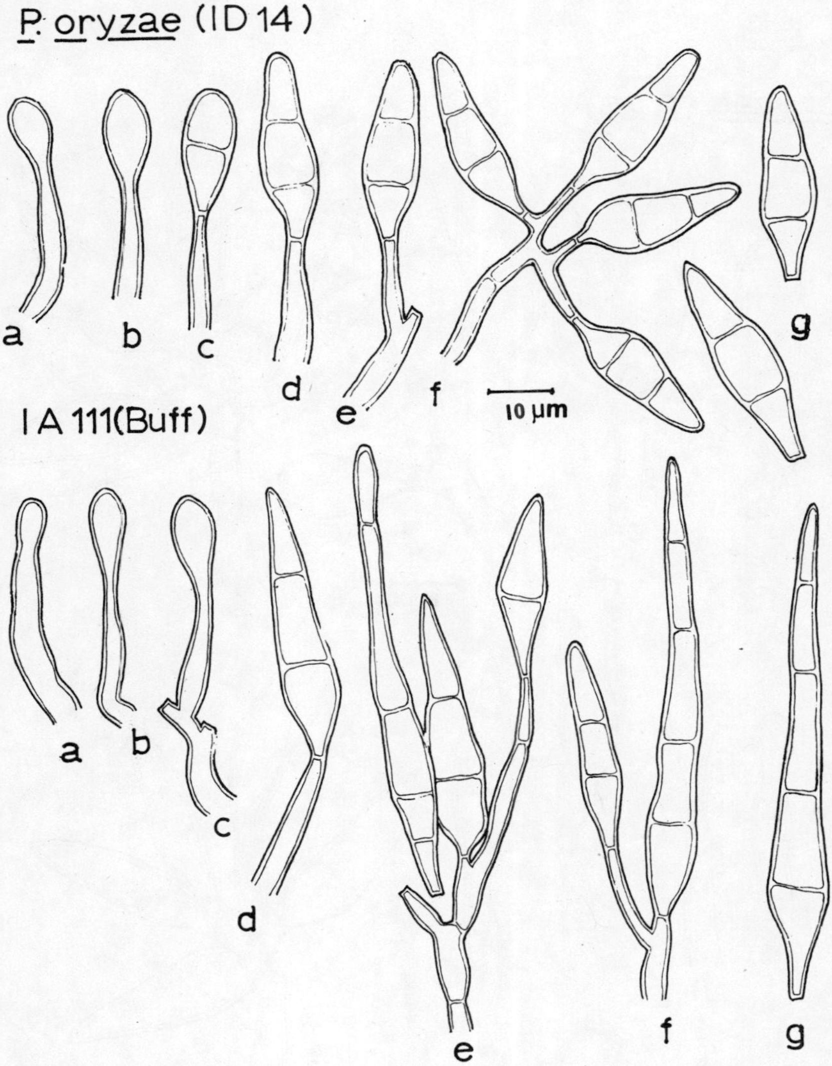

Fig. 9b: Development of gangliar conidia in two isolates. *P. oryzae* (ID 14) : (a-e) various stages in gangliar conidia development without any constriction at the base of the conidium, (f) conidiophores bearing conidia in clusters (conidia mostly attached to conidiophore), (g) detached conidia with truncate base. *P. oryzae* (IA 111, buff) : (a-d) developmental stages in gangliar conidia, (e-f) conidiophores with gangliar conidia are narrow, elongated, slightly bulged just below the middle part, 2-4 septate with a truncate base. (Manibhushanrao and Jesudhas, 1984).

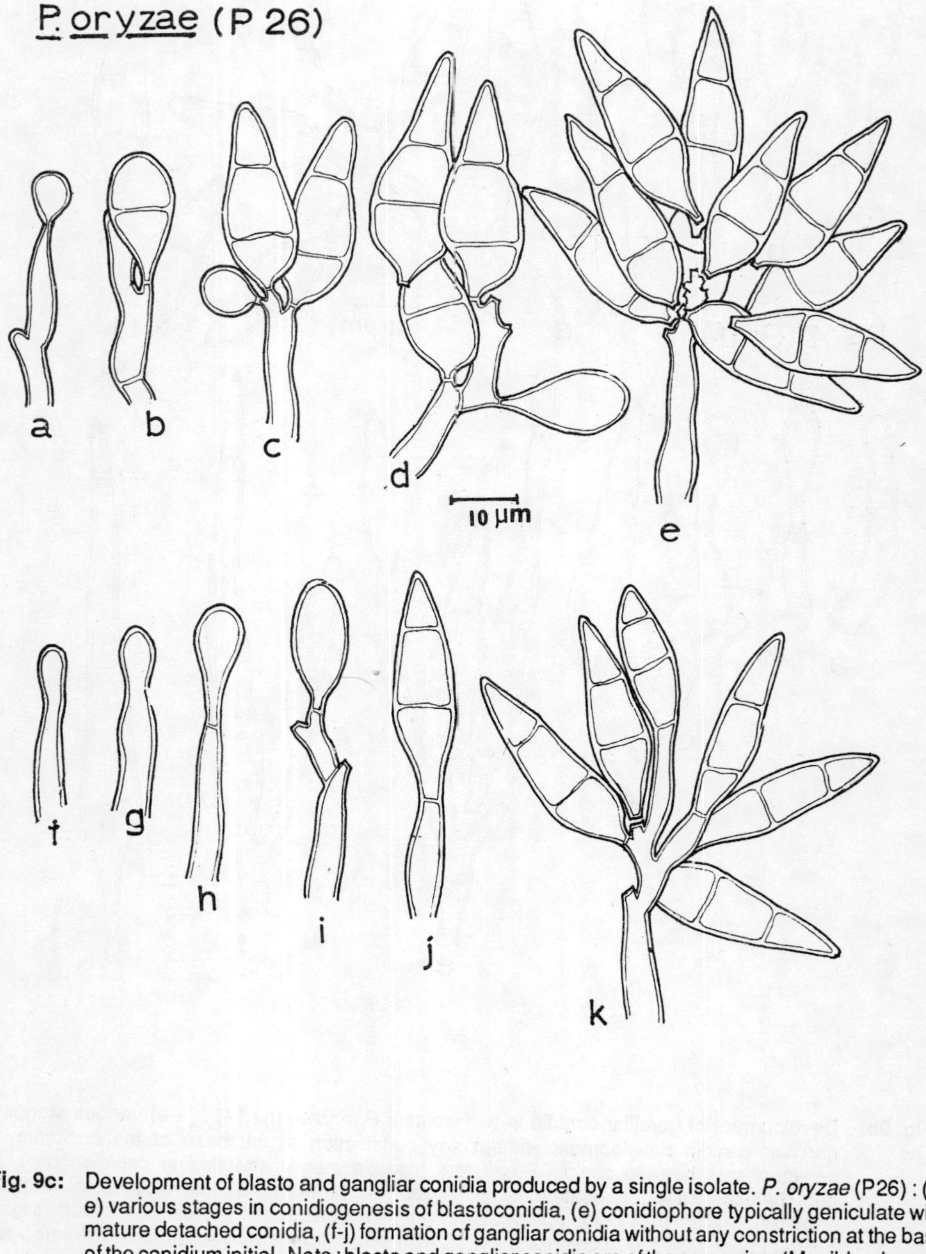

Fig. 9c: Development of blasto and gangliar conidia produced by a single isolate. *P. oryzae* (P26) : (a-e) various stages in conidiogenesis of blastoconidia, (e) conidiophore typically geniculate with mature detached conidia, (f-j) formation cf gangliar conidia without any constriction at the base of the conidium initial . Note : blasto and gangliar conidia are of the same size. (Manibhushanrao and Jesudhas, 1984).

Table 3 : Types and size of *Pyricularia* conidia (Manibhushanrao and Jesudhas, 1984)

Culture	Isolate	Conidia formation			size of condia
		Blastic	Gan-gliar	both	length × breadth (μ)
P1	*P. oryae* (wild type)	+			28—36× 9—11 (10)
M 1	Coloured mutant of P1		+		36—40 (34) × 5—7 (6)
55	*P. oryzae*	+			24—29 (26) × 8—11 (9)
186	*P. grisea* f. sp. *eleusinae*	+			24—34 (30) × 8—12 (10)
436	*P. oryzae*	+			28—36 (32) × 9—11 (10)
K 1	*P. oryzae* (P 150)	+			28—34 (30) × 8—13 (11)
K 10	Methionine-less mutant from P 150			+	[a]28—36 (32) × 8—10 (9) [b]26—32 (30) × 9—12 (11)
K 3	Pink mutant from P 150		+		34—40 (38) × 9—10 (10)
K 17	*P. oryzae* (P 26)			+	[a]30—34 (32) × 8—10 (9) [b]28—32 (30) × 8—10 (9)
K 18	*P. oryzae* (IB 33)			+	[a]28—38 (33) × 9—12 (10) [b]26—36 (30) × 9—12 (10)
K 19	*P. oryzae* (IE 1)	+			24—30 (26) 7—10 (8)
K 20	*P. oryzae* (ID 14)		+		28—56 (48) × 6—8 (7)
K 21	*P. oryzae* (IA 111)	+			20—28 (24) × 6—10 (8)
K 22	*P. oryzae* (IA 111, buff)		+		48—64 (52) × 6—10 (8)
K 24	*P. oryzae* (IA 109)	+			24—32 (29) × 8—11 (10)
K 25	*P. oryzae* (IC 1)	+			24—28 (26) × 8—10 (9)
K 26	*P. oryzae* (IF 1)	+			25—29 (27) × 9—11 (10)
K 27	*P. grisea* f. sp. *eleusinae*	+			24—30 (26) × 7—9 (8)
K 28	*P. grisea* f. sp. *setariae*	+			18—24 (20) × 10—13 (12)
K 29	*P. grisea* f. sp. *setariae*	+			30—36 (32) × 8—10 (9)
K 30	*P. grisea* f. sp. *panicae*	+			26—32 (30) × 8—10 (9)
K 31	*P. grisea* f. sp. *panicae*	+			24—30 (28) × 8—10 (9)
K 32	*P. grisea* f. sp. *brachiariae*	+			26—32 (28) × 8—11 (10)

[a] gangliar conidia; [b] blastic conidia.

Light seems to have a major effect on sporulation. In nature most conidia disseminate at night (Kuribayashi and Ichikawa, 1941). Barksdale and Asai (1961) reported that lesions on detached leaves kept at 100% relative humidity (RH) released conidia at night but fewer conidia were released when the lesions were subjected to continuous light (L) or dark (D) and an alternating L/D cycle seems to be necessary. Suzuki and Yoshimura, (1963), Kato and Diamond (1966) and Chakrabarthi and Wilcoxson (1970) studied the effect of L on sporulation in culture. Sporulation in *P. oryzae* could be initiated by UV light (Leach, 1962) and near-UV light (Ohmari and Nakajima, 1970 and Ahn and Chung, 1974) could enhance the sporulation as well as conidiophore initiation (Namai et al., 1977).

Narayanarao et al. (1972 a, b) studied the effect of a few aromatic/phenolic compounds on sporulation of six *Pyricularia* isolates from cereals and wild grasses. Chlorogenic acid promoted the best sporulation of the isolates from *Oryza sativa, Setaria italica* while the isolate from *Eleusine coracana* responded equally well to phenol or shikimic acid. None of the 13 phenolic compounds tested stimulated sporulation of the isolates from *S. italica* and

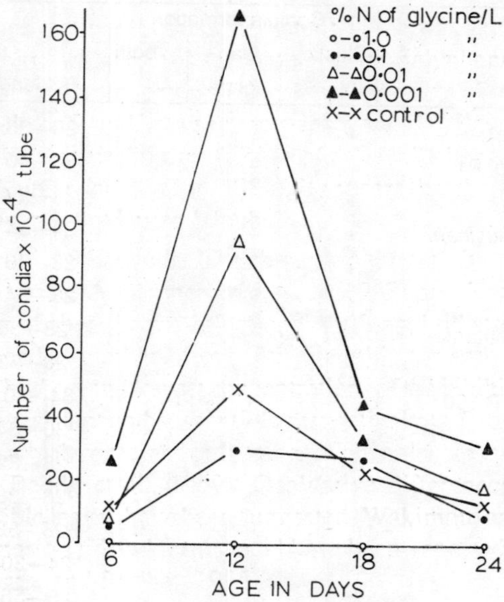

Fig. 10: Effect of glycine concentration on sporulation of *P. oryzae* mutant (M2) in minimal medium (Manibhushanrao, 1971).

Leersia hexandra. However, caffeic acid and guaicol stimulated sporulation of *P. oryzae* at 0.5 mm concentration. An isolate from *Brachiaria mutica* failed to sporulate consistently on various agar media as well as with other stimuli. However, it sporulated abundantly on detached injured leaves of the host, on *S. italica* and profusely on *Triticum vulgare* leaves. *A. priori* it would appear that these substrates contain specific compounds that initiate sporulation (Narayanarao et al, 1972 c). Subsequently, Rao and Rao (1981) reported sporulation of *Pyricularia* spp. stimulated by detached leaves and leaf extracts of *Commelina benghalensis*.

Failure to form conidia in culture has been assumed to be associated with vegetative genes. Mutation for non-sporulation in a single nucleus could establish a heterokaryotic state and selection pressure of culture cónditions and may lead to homokaryotic non-sporulating culture. Heterokaryosis is well established in *Pyricularia* (Suzuki, 1965, 1967 a and b) and this would seem to account for the varied sporulation responses of *Pyricularia* on detached leaves as well.

Spore germination

Conidia germinate freely in water drops (rain, dew and guttation) from the apical cell or basal cell and less frequently from the middle cell. They germinate in 4-6 h and require RH of 92-96% for maximum germination (Hemmi and Abe, 1932). Sueda (1928) and Suzuki (1969 b) reported the optimum temperature for germination of conidia as 25-28°C and Liang (1979) as 16-32° C while no germination occurred at 10-15° C (Nishikado, 1927). Direct sunlight inhibited the germination of conidia (Sueda, 1928; Suzuki, 1969 b) while in diffused light, it was reduced to 50% than that occurring in darkness. Besides, light also inhibited germ tube elongation (Abe, 1931).

Rain drops loaded with conidia adhered closely to rice plants and perhaps this was one of the reasons for the high rate of conidial germination and appressorial formation on rainy days (40%) as compared to fair days (2%) (Suzuki, 1968). Occurrence of differentially matured conidia on lesions also affected germinability. Dispersed conidia germinated better than those in lesions (Suzuki, 1973)

Other factors affecting conidial germination were density (Suzuki, 1969b) and compression of the conidia (Ikegami and Takagi, 1972) as the germination rate declined from 97 to 16% at 0 to 1800 kg/cm for 10 min.

Cell and molecular biology

An understanding of the molecular basis of inheritance requires knowledge on the nature of the gene, its replication to complement genetic materials to the parent cell, its expression to determine the chemical, metabolic and morphological characteristics of the individual cell. Much work has been done in recent years on the molecular genetics and cell biology of the B1 fungus, *M. grisea*. Recently, Mendegen and Deising (1993) reviewed the events occurring during differentiation of fungal infection structures with special emphasis on *M. grisea*. Specific strategies adopted for specialized modes of penetration to host cells were discussed.

Spore and germ tube

Hamer et al. (1988) reported that the conidia of the B1 pathogen adhere to the rice plant surface by means of mucilage released from the spore tip which probably helped in attaching the conidium to the hydrophobic surface of the host. The spore tip mucilage (STM) is gel-like and forms a thin layer which may enable a firm attachment to the substrate. Fluorescein isothiocynate (FITC) - conjugated Con A revealed a faint image of substance at the conidial apex (Plate 11a, b). TEM studies of the freeze substituted conidia revealed that unhydrated dormant conidia contained a prominent deposit of material at the conidial apex (Plate 11c) while those hydrated showed the presence of STM released through a rupture of the outer spore wall (Plate 11d). This group (Bourett et al. 1993) successfully identified specific Con A binding sites. Here, they could report new biochemical information concerning fungal cell physiology, not readilly available by other analytical methods.

Recently, Bourett et al. (1993) successfully identified specific intracellular Con A binding sites in the cells of *M. grisea*. The post embedement labelling of intracellular Con A binding sites revealed the unexpected cytoplasmic binding sites in the apex and cortical cytoplasm of the hyphae, and in conidia before and during germination (Plate 12, 1-5). Further, a short germ tube initiates appressorium development followed by host cell penetration (Fig. 11). The penentration peg (infection hypha) emerges from the appressorium and penetrates the host cell resulting in colonization of the plant tissue (Heath et al. 1990, 1992). This attachment of the conidia and germ tube to the host surface is one of the prerequisites for a successful infection process by *P. grisea*. This appears to be the first report inset here on the novel mechanism for attachment of phytopathogentic fungal spores to a plant surface.

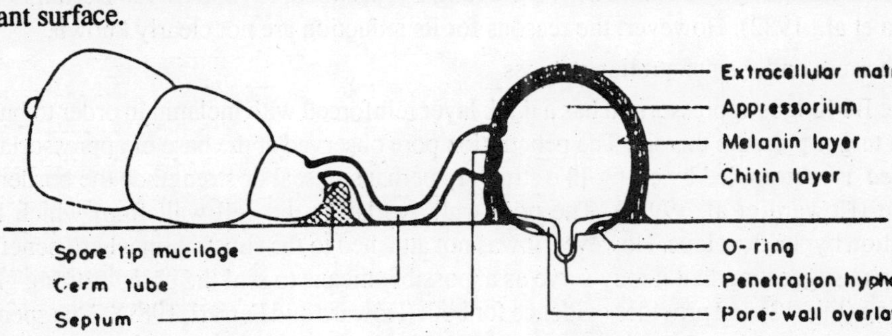

Fig. 11: Infection structure of *Magnaporthe grisea*. Courtesy : R.J. Howard (cited by Mendgen and Deising, 1993). Note : The drawing is not to scale.

Melanin involvement in hyphal penetration

Howard and Ferrari (1989) studied the function of appressorial melanin using wild type and melanin deficient strains of the rice B1 pathogen and the melanin biosynthesis inhibitor tricyclazole. Melanin seems to play an important role in the primary mechanism by which the fungus penetrates the host surface. They proposed that the function of melanin in appressoria is to provide for an osmotically generated hydrostatic pressure.

Bourett and Howard (1990) studied the morphogenetic stages associated with appressoria of *M. grisea* using light, fluorescent and electron microscopy of freeze- substituted specimens. According to them numerous membrane cisternae and cytoplasmic vesicles were present in developing appressoria and the vesicles were concentrated near the substratum where the cell wall was very thin or absent. During melanization of appressoria, vesicles and cisternae were absent. Before the onset of penetration peg formation a bilayered pore wall overlay was deposited over the appressorium pore. High chitin content of outer wall layer was indicated by labelling with wheat germ agglutinin (WGA). The mature penetration peg, formed within the cellophane substratum, was uniform in width, about 7μm in length and contained predominantly putative cytoskeletal elements. Infection hyphae seemed to emerge as lateral branching from the distal end of the peg. The peg rigidity may be increased by the cytoskeletal elements as the rigid peg can facilitate the mechanical breaking of most cell wall by *M. grisea*.

Further, vesicles play a role in the deposition of the outer appressorium wall layers in the early stages of infection structure formation, but are not involved in the final stages of melanization. The synthesis of melanin in the absence of vesicles has revealed that *M. grisea* has a non-vesicular mechanism for the delivery of some precursors and (or) enzymes to the sites of synthesis. This observation was confirmed as melanin-deficient mutants of *M. grisea* can only form a melanin layer when exogenous scytalone was added (Chida and Sisler, 1987). If the melanin biosynthesis was blocked by treating with the antipenetrant tricyclazole, appressoria lacked the melanin layer and failed to infect the host plant (Woloshuk et al. 1993). The authors surmised that this was due to the lack of rigidity of the unmelanized appressorial wall. Melanin was not involved in appressorium adhesion if any substrate modifications occur prior to melanization. This is in agreement with the findings of Howard and Ferrari (1989) but contrary to that of Inoue et al. (1984, 1987). Appressorium induction by artificial substrata is well known in the B1 fungus (Hashioka, 1972; Uchiyama et al., 1979; Hirooka et al., 1982). However, the reasons for its induction are not clearly known.

Appressorium and the penetration process

The B1 fungus appressorium has a thick layer reinforced with melanin in order to build up high turgor pressure therein. The penetration pore observed at the base of appressoria of *M. grisea* is surrounded by a ring-like structure perhaps to seal or strengthen the border of the pore (Howard et al. 1991a). The pore wall-overlay is the cell wall from which the penetration hypha develops. However, it was not attached to the substratum where penetration occurs suggesting that it may serve as a possible means to seal the pore, a strong glue and 0-ring (Fig. 12) and provides evidence for both (Howard and Ferrari, 1989). This special hypha, the penetration or infection peg, is very rigid and smaller in diameter compared to the pore (Howard et al. 1991a). This suggested that the forces other than turgor pressure alone are responsible for the extension of the penetration hyphal tip.

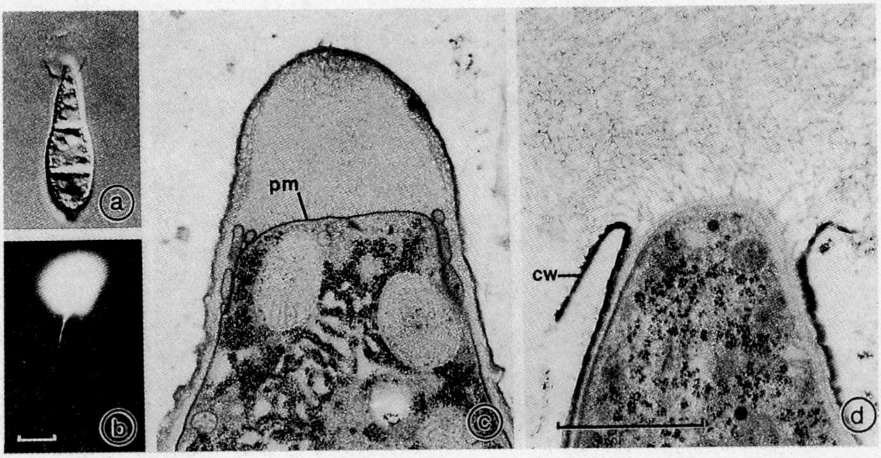

Plate 11: Light (a and b) and electron (c and d) micrographs of *M. grisea* conidia before (c) and after (a, b, and d) hydration. (a and b): The same FITC- Con A-labelled conidium imaged with differential interference contrast optics and epi-fluorescence, respectively. Bar : 5μm . (c) : A dry conidium apex contains a deposit outside the plasma membrane (pm). (d) : When hydrated, the cell wall (cw) is broken at the apex and an extracellular fibrous matrix is released. Bar. : 1μm (Hamer et al., 1988).

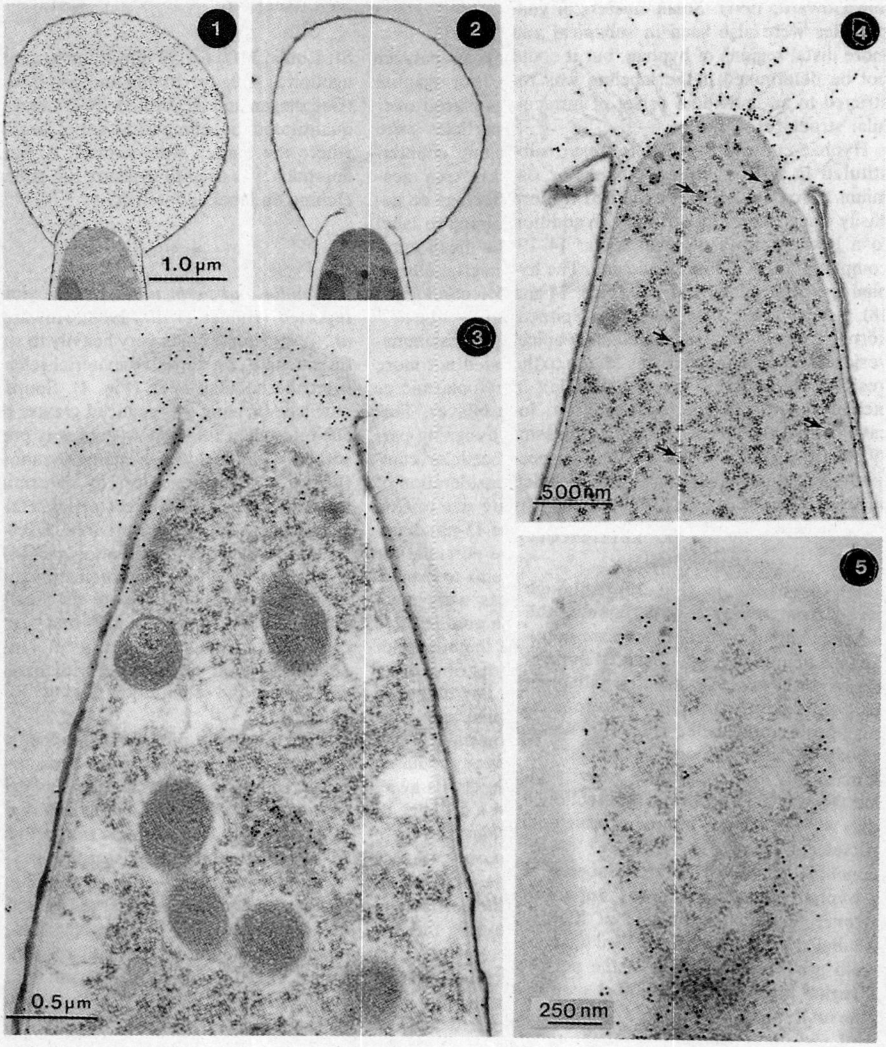

Plate 12 : A postembedement labelling of intracelluar Co A binding sites in freeze-substituted *M. grisea* cells. 1-2 : Con A binding sites in conidia pr ior to germination . Labelling of spore tip mucilage association with the conidial apex. Heavy labelling by Con A alone (1) was reduced dramatically when an adjacent section was labelled using Con A preincubated with the specific hapten methy 1-D-mannopyranoside (2); section through the apical cell of a conidium illlustrating the consistency of labelling associated with smooth membrane cisternae (3); Con A binding sites in germinating conidia. The plasma lemma is clearly labelled with Con A in transverse (4) and especially tangential section (5). Apical vesicles (arrows) are labelled in nascent. Note the many other Con A, binding cites in the apical region. (Bourett et al., 1993).

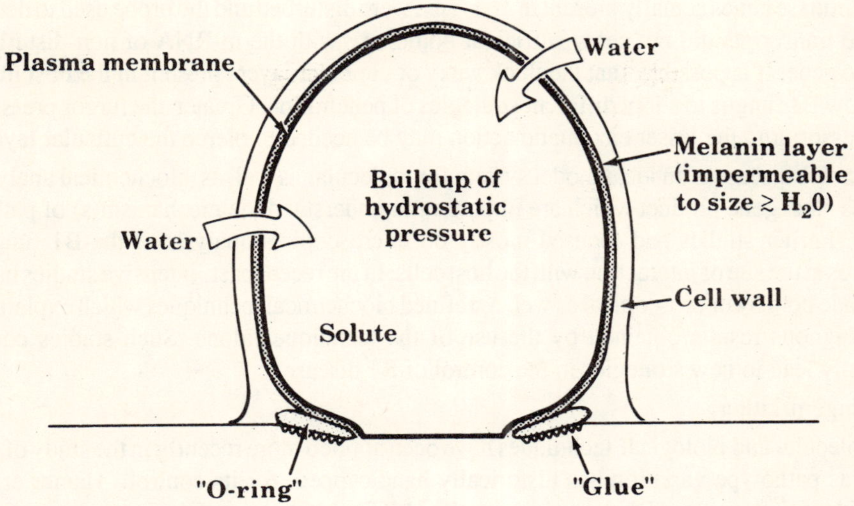

Fig. 12: Appressorium function. Melanin traps solute inside the appressorium but is permeable to water. As the concentration of solute is increased, more water is absorbed. The result is high internal hydrostatic pressure that subsequantly serves as an essential force during penetration peg emergence. Since melanin is absent across the appressorium pore there must exist a tight seal, against the substitution perhaps involving the action of a glue and an O-ring (i.e., the pore ring) (Howard and Ferrari, 1989).

Melanin is responsible for changes in the permeability of walls and acts to increase the concentration of glycogen degradation products which result in development of pressures up to 80 bars in the interior of appressoria of *M. grisea* (Howard et al. 1991b). Such pressure seems to contribute significantly to the indentation of the cuticle or cellulose/ artificial membranes by appressoria.

M. grisea uses its penetration peg to pierce the surface of the rice leaves and an examination of penetration pegs by electron microscopy and immunofluorescence microscopy and freeze-substitution preparation of appressoria at the early stages of penetration showed that peg cytoplasm consisted primarily of a zone of exclusion, excluding even ribosomes and was continuous with a similar region in the appressorium (Bourett and Howard, 1992). Further, they concluded that the penetration peg contains actin.

Role of enzymes in penetration

M. grisea produces cutinase when grown on media containing cutin as sole carbon source (Chumley and Valent 1992a). Recently, the cutinase gene of the rice B1 fungus was cloned (Sweigard et al. 1992 a) and the gene transcripts were not detected in two strains of *M. grisea* (Sweigard et al. 1992b) which are pathogenic on weeping lovegrass (*Eragrostis curvula*) and barley or rice. However, enzyme activity was detected with tritiated cutin *p*-nitro-phenyl butyrate. Mutants and wild-type strains were used to infect the three host plants. The tissue produced were indistinguishable from those of the parental strains. However, this work clearly implies that cutinase gene of *M. grisea* is not essential for penetration.

Since the mutants showed little cutinase and esterase activities, it is presumed that not all the cutinase genes actually present in *M. grisea* were disturbed and the probe used to detect cutinase transcripts did not show sufficient homology with the m-RNA of non-disturbed cutinase gene. It is possible that the thick waxy or cuticular layers present in the host plant will allow B1 fungus to adopt different strategies of penetration. Greater the turgor pressure of appressorium, the lesser enzymatic action may be needed to pierce the cuticular layer.

Rice B1 disease is an ideal model system for molecular as well as biochemical analysis of genes and gene product which are important to understand the mechanism(s) of pathogenesis. Earlier studies had focused mostly on microscopical analysis of the B1 fungus structures at the site of interaction with the host cells. In the recent past, intensive studies have been made both by microscopical as well as refined biochemical techniques which explanied the ambiguous results obtained by the use of the techniques alone. Such studies could eventually lead to new strategies in the control of B1 disease.

DNA fingerprinting

Molecular and biological techniques have been applied more recently in the study of B1 fungus as pathotype variation has historically handicapped for its control. Hamer et al. (1989) identified repeated DNA sequences called MGR (Methionine, Glycine and Arginine) in the pathogen and Valent (1990) used DNA sequences to differentiate strains of B1 fungus found on wheat and rice. Further, the MGR - DNA fingerprints enabled differentiation of the major pathotypes in the US, and defined the structure of clonal lineages within and among pathotype groups. (Levy et al. 1991) The histories of these isolates indicated that stable linkage pathotype associations have been maintained from 1967 to 1989. Such techniques offer a highly precise means of examining variation in the B1 pathotype stability in the rice B1 fungus.

Further, use of restriction fragment length polymorphisms (RFLPs) can detect and monitor genetic variation at the mitochondrial deoxyribonucleic acid (mt-DNA) level (Leung et al. 1989) and both mt-DNA and nuclear DNA levels (Ko et al. 1993) used as genetic markers. However, later studies indicated that the relationship between RFLPs in nuclear DNA and virulence of *M. grisea* races from Korea was inconclusive. Hunst et al. (1986) found some polymorphism for the lengths of double stranded (ds) RNAs among rice pathogen isolates, but this variation was not correlated with pathotype. Valent (1990) made such an attempt to characterize populations of B1 fungus using RFLPs. This technique was also used directly to the rice plant, and a library of mapped clones is now available (McCouch et al., 1988) which can be used to map the location of B1 resistance genes as well as mapping loci for quantitative B1 resistance.

Pathogen transformation

Leung et al. (1990) reported a transformation system for *M. grisea* based on selection for resistance to the antibiotic hygromycin-B. Successfully transformation though at low frequency was achieved using the plasmid vector pAN7-1 developed for *Aspergillus nidulans*. Using chromosome walking techniques, clones can be isolated that encode putative pathogenicity and virulence genes. Verification of cloning can now be easily accomplished simply by transforming these clones into the appropriate isolates of *M. grisea* and assessing the phenotype of the incitant.

Variability in pathogenicity among B1 pathogen isolates is well established. Althougn various causes of variation namely storage and subculture (Goto et al., 1964; Veeraraghavan,

1975a; Matsuyama and Yamaguchi, 1981; Kumar, 1988), heterokaryosis (Suzuki, 1965), parasexualism (Yamasaki and Niizeki, 1965; Wu and Tsai, 1974; Genovesi and Magill, 1976), nutritional requirements (Kuo et al., 1968), mutation (Katsuya and Kiyosawa, 1969), intra-isolate heterokaryosis (Fatemi and Nelson, 1977), 'interconversion' of genes (Buck, 1987) and chromosome number (Row et al., 1985) and variation in dsRNA (Hunst et al., 1986), have long been proposed. It is quite essential to understand the molecular basis of variation and to delineate the host-pathogen interactions at the molecular level to get an insight into the mechanism(s) of resistance. Such studies have been attempted recently (Hamer et al., 1989; Valent, 1990) on quantitative resistance. Molecular methods for enhancing the biocontol of rice B1 using *Pseudomonas* and *Bacillus* strains is also gaining attention (Gnanamanickam et al. 1993). Despite the availability of information in recent years on MGR-DNA fingerprinting which reliably indexes genetic and pathotype diversity in B1 fungus, our knowledge is still incomplete.

Host Range

Hori (1898) reported the successful infection of *P. oryzae* from rice to crab grass (*Digitaria sanguinalis*) and Kawakami (1901, 1902) from vice versa and considered that both the fungi were identical. Shirai (1905) was of the same opinion after the fungus was tested on various hosts although he considered the fungus to be a species of *Dactylaria*. Other workers (Fulton, 1908; Duggar, 1909; Stevens, 1913; Metcalf, 1913) also shared this view. Sawada (1917) found that *D. oryzae* (Cav.) Sawada infected rice, Italian millet, barley and wheat and *D. grisea* only crab grass and Italian millet.

Cross inoculation studies by several workers (McRae, 1922; Nishikado, 1927; Thomas, 1940; Giatgong, 1961 and Veeraraghavan and Padmanabhan, 1965; Venkatarao and Apparao, 1972) indicated that the pathogenicity of isolates from different hosts was largely restricted to their own hosts as they seldom attack other hosts. However, Sawada (1917) and Andersen et al. (1947) reported that rice isolates also attack other cereal crops. McRae (1922) and Thomas (1941) found that an isolate from *Panicum repens* was non-pathogenic to rice but Krishnaswamy et. al. (1959) reported this isolate could infect rice. Similarly Krishnaswamy et al. (1959) could get infection on rice with the isolates from *B. mutica* and *L. hexandra* but Giatgong (1961) obtained negative results with the same isolates. Asuyama (1965) also pointed out such discrepancies in the studies of Japanese workers.

Earlier, B1 incidence on wild grasses had minor importance as it had no bearing on disease cycle. However, Hemmi et al. (1949) suggested that *Pyricularia* on crab grass might infect rice while Suzuki and Hashimoto (1953) had convincing evidence that rice fungus could occasionally infect barnyard grass, crab grass as well as barley, wheat and Italian millet. There are reports of isolates from several grasses infecting certain rice cvs. (Kuribayshi et al. 1953) and vice versa (Iwata et al. 1955), that is by exposing general grass species in a rice B1 nursery.

Thyagarajan (1971) made an attempt to reinvestigate the problem of collateral hosts under controlled conditions of temperature and light. He found that the three grasses *P. repens*, *B. mutica* and *L. hexandra* might not be the real carriers since isolates from these hosts do not readily pass through rice. However, it appears that more bridging hosts are involved in the selection of suitable biotypes of the fungus that are virulent on rice.

Two possible explanations may be offered for the conflicting results regarding the host range of the rice B1 fungus *P. oryzae* as well as *Pyricularia* spp. from different hosts, that is, variation in the genetic background of the fungus as well as the environmental conditions such as relative humidity (RH) and low temperature along with soil fertility under which the tests were made. These factors apparently differ from one investigator to another, although they may handle the same isolates.

Much global work has been done on the collateral hosts but the results are inconclusive. The common weeds in the rice fields, namely, crab grass in the US and Brazil, St. Augustine grass in Brazil and *P. repens* in India are the most suspected carrier hosts of the fungus. The eradication of such weeds will reduce the B1 disease incidence to a great extent.

Several workers (Padwick, 1950; Asuyama, 1965; Veeraraghavan and Padmanabhan, 1965; Kato and Yamaguchi, 1980; Mackill and Bonman 1986) have studied the host range in *Pyricularia* spp. on rice and other collateral hosts from time to time and the available information is summarized below:

Cereals

Wheat - *Triticum vulgare* (Sundararaman, 1921; Thomas, 1940; Anderson et al., 1947; Suzuki and Hashimoto, 1953), *T. aestivum* - (Mcrae, 1922; Anstead, 1924; Malik and Khan, 1943) pearl millet (bajra) - *Pennisetum typhoides* (Buckley and Allen, 1951; Mehta et al., 1952; Wells et al. 1969), *P. purpureum* (Krishnaswamy et al., 1972), Italian millet - *Setaria italica* and green foxtail - *S. viridis (*Ferraris, 1913; McRae, 1922; Nozu and Yokogi, 1924; Nisikado, 1926; Thomas, 1940; Kuribayashi et al., 1953; Suzuki and Hashimoto, 1953; Narita et al., 1956; Goto and Yamanaka, 1960a), German millet (*S. italica* var. *germainica*) and yellow foxtail (*S. glauca*), Kawakami (1901, 1902), *S. intermedia* (Pawer and Kulkarni, 1954), finger millet (ragi) - *Eleusine coracana* (Sundararaman, 1921; McRae, 1922; Thomas, 1930, 1940, 1941; Park, 1937; Hansford, 1943; Ramakrishnan, 1948; Wallace, 1950; Thirumalachar and Misra, 1953; Kato et al., 1977), barley - *Hordeum vulgare* (Hara, 1916; Thomas, 1940; Anderson et al., 1947; Kuribayashi and Terazawa, 1956; Suzuki et al., 1953; Narita et al., 1956; Ono et al., 1961), oats - *Avena sativa* (Sundararaman, 1921; Thomas, 1940; Narita et al., 1956), Japanese barnyard millet - *Echinochloa crus-galli* (Togashi et al., 1946; Kuribayashi and Terazawa, 1953), proso millet - *Panicum miliaceum* (Nisikado, 1926; Kuribayashi et al., 1953; Narita et al., 1956; Simmonds, 1958), corn - *Zea mays* (Thomas, 1940; Anderson et al., 1947; Hemmi et al., 1949; Bailey and Van Eijnatten, 1961), sorghum - *Sorghum vulgare* (Goto and Yamanaka, 1960 a).

Grasses and sedges

Sweet vernal grass - *Anthoxanthum odoratum* (Goto and Yamanaka, 1960 a), crab grass - *Digitaria sanguinalis* (Kawakami, 1901, 1902; Ferraris, 1913; Sawada, 1919, McRae, 1922; Nisikado, 1926; Faweett, 1935; Tullis, 1951; Hemmi, 1949; Revilla, 1953; Narita et al., 1956; Veeraraghavan and Padmanabhan, 1965), *D. marginata* (Ramakrishnan, 1948; Krishnaswamy et al., 1959 and Padmanabhan, 1965 c). *D. setiera* (Veeraraghavan and Padmanabhan, 1965), *Digitaria* spp. (Hansford, 1943), *Leersia hexandra* (Krishnaswamy et al., 1959), *L. japonica* (Sawada, 1917; Narita et al., 1956; *L. oryzoides* (Narita et al., 1956); Veeraraghavan and Padmanabhan, 1965), *Panicum repens* (Sawada, 1917; Thomas, 1941; ICAR, 1954; Krishnaswamy et al., 1959; Veeraraghavan and Padmanabhan, 1965) *P. barlinodes* (Revilla, 1953), *P. ramosum* (McRae, 1922), *P. proliferum* (Padmanabhan, 1965), signal grass - *Brachiaria mutica* (Ramakrishnan, 1948; McIntosh, 1951; Govindarao and Koteswararao, 1958; Krishnaswamy et al., 1959; Veeraraghavan and Padmanabhan, 1965), barnyard grass - *Echinochloa crus-galli* (Kuribayashi et al., 1953; Suzuki and Hashimoto, 1953; Johnson, 1954), cup grass - *Eriochloa villosa* (Nisikado, 1926), centipede grass - *Eremochloa ophiuroides* (Johnson, 1954), St. Augustine grass - *Stenotaphrum secundatum* (Johnson, 1954; Malca and Owen, 1957), *Distichlis* sp. (Campacci, 1950), bermuda grass - *Cynodon dactylon* (Sprague, 1950), ryegrass - *Lolyum* spp. (Trevathans 1982), *Eleusine indica* (Teng, 1932), common reed - *Phragmites communis* (Kuribayashi et al., 1953 and Narita et al., 1956), giant reed - *Arundo donax* (Srinivasan and Vijayalakshmi, 1957), *Zizania latifolia* (Goto et al., 1954; Narita et al., 1956), *Oryza montana* (Steyaert, 1948), *O. minuta* (McIntosh, 1951), bottle brush - *Hystrix patula* (Sprague, 1950), *Commenlia benghalensis* (Thirumalachar et al., 1956), *Cyperus rotundus* (Luttrell, 1954), *C. compressus* (Thirumalachar et al., 1956) and *Rhyncheletum repens* (Wilson, 1962).

In the 'Index of Plant Diseases in the United States' and 'Diseases of cereals and grasses in North America' (Sprague, 1950) the following were described as hosts of *P. grisea* : *Agrostis stolonifera* (bent), *Andropogon* spp. *Digitaria* spp. *Echinochloa colonum* (Jungle rice), *Eragrostis* spp. (love grass), *Fluminea festucaceae* (river grass), *Leersia oryzoides* (rice cut-grass), *L. virginica* (white grass), *Muhlenbergia* spp. (muhly), *Panicum texanum* (Texas millet), *Paspalum plicatum* (brown—seed paspalum), *Sorghum vulgare* var. *sudanense* (Sudan grass) and *Hystrix patula.*

Other monocot families in which *Pyricularia* spp. have been described are common ginger - *Zingiber officianale* (Nisikado, 1926) wild ginger - *Z. mioga* (Nisikado, 1926), *Costus speciosus* (Sawada, 1917), yellow zedoary - *Curcuma aromatica* (Sawada, 1959), common garden cana - *Canna indica* (Roldan, 1938; Sawada, 1959), banana - *Musa sapientum* (Hoette, 1936; Wardlaw, 1940; Stehle, 1954; Meredith, 1962), sugar cane - *Saccharum officinarum* (Hastings de Gutierrez, 1955; Srinivasan and Vijayalakshmi, 1957). Many *Pyricularia* spp. on several other hosts of monocot families (Hashioka (1971, 1973; Itoi et al., 1978, 1979; Arase et al., 1984) have been reported.

Although much work has been done on the host range of the B1 pathogen the results by various authors are always contradictory, primarily because of the failure to adopt standardized procedures by various workers in different countries. Since the collateral hosts play a very important role in the life cycle of the pathogen a detailed information on host range of the B1 fungus is quite essential and needs a thorough investigation.

Life Cycle

Kuribayashi (1928) and Sueda (1928) reported that mycelia and conidia, especially the former, play an important role in hibernation and perpetuation of the fungus. Mycelia usually persist in infected straw for one year when it is kept dry and in the lesions and nodes they remain alive for two years or more. Conidia are viable on infected straw and seeds up to three years (Bremer et al. 1948) and even in adverse conditions of temperature (Abe, 1935).

Yamanaka and Kobayashi (1962) reported that the dark coloured conidia produced by *P. oryzae* under certain cultural conditions have no significant role in the life cycle. However, Yamaguchi (1977) reported, a few phases of the life cycle while Kim (1987) stressed that B1 fungus completes life cycle during leaf B1 season within a week. The life cycle of *P. oryzae* is given in Fig. 13 (Manibhushanrao, 1992)

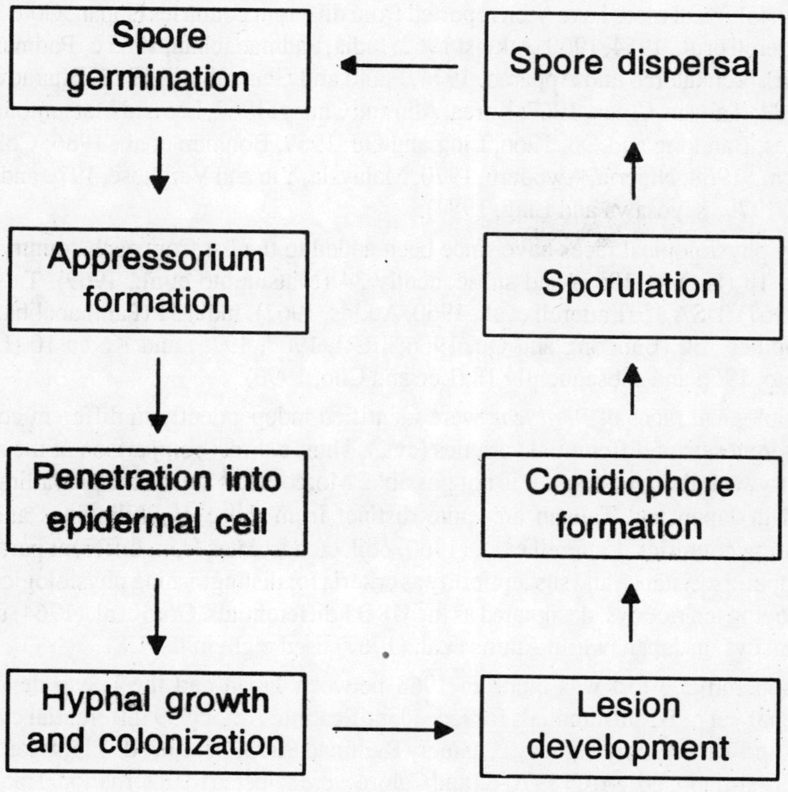

Fig. 13: Life cycle of *P. oryzae*. (Manibhushanrao,1992).

Physiological Races

Sasaki (1922) was the first to report the existence of strains of the B1 fungus which differ in their pathogenicity. Subsequently, classifications based on morphological, colour and colony characteristics (Tochinai and Shimamura, 1932), degree of sporulation in culture (Tochinai and Shimura, 1932; Konishi, 1933; Aoki, 1935) became available. These classifications were, however, arbitrary and not recognised by others.

Race distribution and differentiation

Through successful hybridization a resistant cv. Futaba was introduced, and remained resistant for ten years after which its resistance suddenly broke and it became susceptible. This led to intensive studies on physiological races in Japan around 1960. As a result, twelve cvs. were selected as differentials: two of tropical, four of Chinese and six of Japanese origin. Many physiological races have been reported from different countries (Japan, Goto, 1965a; USA, Latterell et al., 1954, 1960, Atkins, 1962; India, Padmanabhan, 1965 c, Padmanabhan et al., 1970, Venkatarao and Apparao, 1974, Padhi and Chakrabarti, 1972, Bharadwaj and Singh, 1982; Taiwan, Chien, 1967; Korea, Ahn and Chung, 1962, Lee and Matsumoto, 1966; Philippines, Bandong and Ou, 1966, Ling and Ou, 1969, Bonman et al., 1986; Columbia, Galvez et al., 1968; Nigeria, Awoderu, 1970; Malaysia, Yin and Varghese, 1970 and China, Luo et al, 1979, Kiyosawa and Ling, 1984).

Many physiological races have since been added to the list from each country. Japan registered 18 (Hirano, 1967) and subsequently 34 (Matsumoto et al., 1969), Taiwan 27 (Chien, 1967), USA 16 (Latterell et al., 1960, Atkins, 1962), India 31 (Padmanabhan, 1965 c). Philippines 250 (Bandong and Ou, 1966; IRRI, 1967, 1975) and Korea 10 (Lee and Matsumoto, 1966 and subsequently 18 (Lee and Cho, 1990).

Physiological races of *P. oryzae* were identified independently in different countries using different sets of differential varieties (cvs.). Thus, a direct comparison of the races of one country with those of another is not possible. Moreover, it seems the prevailing races identified in Japan and Taiwan are quite distinct from those of Philippines and other neighbouring countries. Latterell et al. (1960) collected isolates from different parts of the world adopting resistance and susceptibility as criteria for distinguishing physiological races and employing ten rice cvs. designated as the US B1 differentials. Goto et al. (1964) used 12 differential cvs. in Japan, while Atkins et al. (1967) used eight in the US.

A cooperative effort was made in 1963 between Japan and the US to develop an international set of B1 differentials for race identification. Among 39 differential cvs. from Japan US and Taiwan only eight cvs., namely Raminad Str. 3, Zenith, NP 125, Usen, Dular, Kanto 51, Sha-tiao-tsao (S) (CI 8970-S) and Caloro were selected for international race group differentiation. Eight race groups IA to IH were designated and the races were called international races followed by numerals to indicate race numbers (Atkins et al., 1967). Standardization of the numbering of the international races has been suggested to avoid any confusion (IRRI, 1967: Ling and Ou, 1969)

Similar cooperative effort was made in analyzing the B1 races through an IRRI- Korea collaborative project on rice research (Lee et al., 1988). The annual fluctuation of races (Lee et al., 1975; Ryu et al., 1987); and a new method for differentiating pathogenic races of *P. oryzae* (Lee et al., 1987) were reported in Korea.

Single lesion and lesion types

Reports on the recovery of many physiological races from a single lesion or from a group of lesions on one leaf are available. In India, from four lesions on the susceptible cv. Tjiremas, 184 monoconidial isolates (MCIs) were made among which 34 races were identified by testing against Philippine and international differentials (IRRI, 1967). Out of 56 and 44 MCIs from the first and second lesions,14 and 10 races respectively were identified (Ou and Ayad, 1968). Giatgong and Fredericksen (1969) reported seven races from 20 sister MCIs from testing on 4 cvs. Goto et al. (1967) also reported ten races from ten MCIs for a single lesion.

Bonman et al. (1987b) studied the reactions to an international differential set inoculated with 130 MCIs obtained from a culture. There was little variation among the MCIs from a single source, in contrast to Ou and Ayad's finding (1968). The data support the findings of Latterell (1975) that the rice B1 pathogen is relatively stable in its pathogenicity. If this is so, the quantitative resistance to B1 of rice cvs. is not due to the extreme pathogenic instability of *P. oryzae*. If the reactions of the cvs. depend on lesion types, it is clear that inoculating monoconidial subcultures obtained from susceptible-type lesions result in the production of more susceptible lesions and more resistant-type lesions from intermediate-type lesions (Hwang and Lee, 1980). Further, Lee and Cho (1990) reported that among eight isolates from a total of 213 isolates derived from nine lesions of four rice cvs. reacted differently to the Philippine differential set (Table 4).

Table 4 : Reactions of the Philippine differential set to monoconidial isolates (MCIs) of *Pyricularia oryzae* from single lesions tested in Korea and in the Philippines (Lee and Cho, 1990)

Differential Cultivar	Cultivar reaction[a]												
	Korea		Philippine										
	M 23—3	M 23—2	IR—1		IR—2	IR—3	M 23—2		M 23—3	T—1			T—2
Kataktara DA-2	R	S	R	R	R	R	R	R	R	R	R	R	R
CI 5309	R	R	R	R	S	R	R	R	R	R	R	R	R
Chokoto	R	R	R	R	S	R	R	R	R	R	R	R	R
CO 25	R	R	R	S	S	S	S	S	S	S	R	R	S
Wagwag	R	R	R	R	R	R	R	R	R	R	R	R	R
Pai kan-tao	R	R	R	R	R	R	R	R	R	R	R	R	R
Peta	S	S	S	S	S	S	S	S	S	S	S	S	S
Raminad Str. 3	R	R	R	R	R	R	R	R	R	R	R	R	R
Taiching T.C.W.C.	S	S	S	S	S	S	S	S	S	S	S	R	S
Lacrosse	S	S	S	S	S	S	S	S	S	S	S	R	S
Sha-tiao-tsao	R	R	R	R	R	R	R	R	R	R	R	R	R
Khao-tah-haeng 17	R	S	S	R	R	R	R	R	R	S	S	R	S
Philippine race no.	PG-13	PA-48	PD-46	PD-45	PB-789	PD-45	PD-45	PD-61	PD-45	PD-46	PG-14	PH-1	PD-46
No. of MCIs	19	18	22	3	25	24	24	1	25	20	3	1	28

[a] M 23 (Milyang 23), IR-I (IR 442-2—58), T-1 (Tetep)

Protoplast regenerated isolates

Han et al. (1989) studied the pathogenic variation of *P. oryzae* using the protoplast-regenerated isolates (PRIs) derived from KI-307, KI-3315b, KJ-105 and KJ-301 races. These were compared with their mother isolates in terms of race differentiation and pathogenicity. These four PRIs were inoculated to the Korean differential set, and three of them lost 13-56% of their pathogenicity. PRIs originated from KI-315b were differentiated into three races as KI-315a, KI-307 and KI-315b and PRIs from KJ-301 into KJ-201, KJ-105 and KJ-301. In general, the majority of the PRIs belonged to the same race groups, but the rest showed a different pathogenic response to the Korean differential set compared with their original races.

Race shift and evaluation

Earlier, Quamaruzzamen and Ou (1970) in the Philippines reported monthly changes of the pathogenic races of *P. oryzae* in the Bl nursery. Yamada and Iwano (1975) in Japan also found an increase or decrease in C-group races with varying rice cvs. with *Pi-k* gene. Later, Chung (1974) and Marchetti et al. (1976) built up new races with Tongil group cvs. and IC-17 race on cv. Laballe in Korea and Texas, (USA) respectively.

Rice differential cvs. which are considered susceptible to certain isolates/races may have different effects on their multiplication as well as shift in pathogenicity. For instance, race C-8 has more advantages than C-1 for multiplication (Yamada and Iwano, 1975) while, of two isolates when inoculated together, one becomes predominant after four generations (Yamada and Iwano, 1975). The multiplying potentials of race C-1 and C-3 on Kanto 51 type cvs. Tatsuminochi and Fuker 69 were compared with C-3 produced fewer spores and smaller lesions than C-1 on the former cv. but had no difference and was noticed on the latter Fuke 69 (Yaegashi and Kobayashi, 1976). Veeraraghavan and Dath (1975) identified two physiological races IC-1 and IC-17 (according to Ling and Ou, 1969 classification) after analyzing 48 isolates of *P. oryzae* collected during 1971 and 1972. Veeraraghavan and Dath (1976) analyzed a further 110 isolates during 1973 and 1974 from 19 states in India and found that only the race IC-17 was prevalent. This was the predominant race among the 31 reported by Padmanabhan et al. (1970). These authors suggested that the high frequency of a single race over a period of two years might indicate a possible host-parasite equilibrium marked by the absence of reports of breakdown of resistance in any rice cv. during that period. Further screening with 246 isolates during 1975 confirmed the exclusive presence of the race IC-17. Veeraraghavan (1975a) has further confirmed the stability, under a different method of storage, of isolates of *P. oryzae* belonging to the race IC-17.

Veeraraghavan and Dath (1975) suggested the need for reclassifying C-type spots - circular spots about 2 or 3 mm in dia. with a central ashy zone and a dark, purplish-brown margin - as a susceptible reaction, since both C and D spots possess grey centres and brown margins and reveal sporulation. Furthermore, spots falling between type C and type D do occur on some of the differentials. They pointed out that the international B1 differential Usen gave an ambiguous reaction and should be replaced by a cv. with clear-cut reaction. They also felt Caloro as a narrow-lesion cv. and Raminad Str. 3, which does not set seed or flower at high latitudes, creating a problem for seed production, should be replaced. The ambiguity in the reactions of the differntials to race IC-1 and IC-17 are presented in Table 5.

Table 5 : Pathogenicity-pattern of isolates of *P. oryzae* belonging to race IC-1 and IC-7 (Veeraraghavan and Dath, 1975)

Int. race	Raminad Str. 3	Zenith	N. P. 125	Usen	Dular	Kanto-51	CI 8970 (S)	Caloro
IC 1	R	R	S	S	S	S	S	S
IC 17	R	R	S	R	S	S	S	S

R : Resistant S : Susceptible.

Veeraraghavan (1975b) suggested a new method of classification and nomenclature of physiologic races of *P. oryzae* Cav. The 256 possible pathogenicity patterns on the eight international B1 differntials were sorted into 16 groups with the English letters A to P indicating the race group, and numbers indicating the physiologic races. Under this system of classification, all isolates of *P. oryzae* producing a susceptible reaction on all the international B1 differentials have been categorized into one race group. Similarly, those which evoke a resistant reaction on all the differntials have been accommodated in a single race group. Susceptibility or resistance of an individual differential or sets of differentials, taken in a serial order from the list of differntials, has been the criterion for classification.

While discussing the lack of uniformity in the evaluation procedure, Veeraraghavan (1975b) stressed the importance of monitoring the integrity of the genotypes of the differentials. The designation of the pathotype depends on the genotypes of the differentials used. He suggested that a culture of the pathogen with a known pathotype designation should be utilized as a control for inoculating a set of differentials used for distinguishing unknown pathotypes. Thus, three sets of differentials are essential to identify the pathotype of a culture. These will help (i) to check the correctness of the genotypes using a standard culture, (ii) monitor the unknown race and (iii) serve as an uninoculated check. He suggested that reports of pathotypes in *P. oryzae* would only be valid when the accession numbers of the differential hosts and their source are given in any communication along with the registration number of the standard type culture collection.

Veeraraghavan (1985) has proposed the term 'aggressive race' for the virulent isolate which can produce lesions on the leaves of rice cvs., namely, Carreon, Tetep and Tadukan, known to be resistant to several isolates/races. He suggested the grouping of isolates into aggressive races utilizing cvs. of rice as *P. oryzae* agrosensors, namely, Carreon (P. I. 372279), Tetep (P. I. 280682), Tadukan (P. I. 380001) and Dular (P. I. 180061). He feels classifying aggressive races is more meaningful than grouping for which proper evidence is lacking. The merits of these suggestions need discussion before implementation.

In China, physiological races were identified from the rice growing areas all over the country (except Taiwan) based on their virulence to seven Chinese rice differntials. They were grouped into 43 races in seven groups (ZA, ZB, ZC, ZD, ZE, ZF and ZG) of which ZG-1 constitute more than 43% of the total isolates (ACCR, 1980). In 1987, the number increased to 85 races in eight groups. Kiyosawa and Ling (1984) expressed difficulties in differentiating the Chinese races as most Chinese rice differentials have multiple-resistant genes. Further, distribution and changes in race pattern and virulence, their population in Yunnan Province have been reported (Iwano, 1990; Iwano et al. 1990).

Race ZF-1 was the predominant race during the 1970s in North China, *japonica* rice-growing region, while during the 1980s race ZE-3 became predominant. Similarly, race group ZB increased rapidly instead of ZG in Guangdon province, ever since the hybrid rice Xianyu series was widely grown (Sun et al. 1987). Studies also showed that the extensive growing of cvs. with specific resistant genes changed the virulence pattern of the fungus (Jin et al. 1986; Chen et al., 1987 and Feng et al., 1989). A few investigators (Luo et al. 1979 and Sun et al., 1981) reported the existence of a few relatively stable races despite reports on large variation in the virulence pattern among the fungus races. This is in agreement with the observations made on the stability of race IC-17 in India (Veeraraghavan, 1975a).

Kiyosawa et al. (1981) emphasized the need for an accurate method for testing reliability of identification of races. They suggested that the main requirement for each of the differential cvs. should be sensitivity to a particular type of fungal race and not the wide range of resistance controlled by two or more genes. For instance, He et al. (1989) interpreted resistance of upland cv. Haonaihuan using F_3 lines of three combinations of Toyonishiki (Japanese cv.) X Haonaihuan (Chinese upland cv.), Todoriki Waste (Japanese cv.) X Haonaihuan and Lijanxintuanhegu (Yunnan cv.) X Haonaihuan against three Japanese fungus isolates contributed by two major and two minor genes. One of the major genes effective only in the case of one Japanese isolate is *Pi - i* or an unknown gene while another major gene is effective only for another Japanese isolate. The minor genes have different ability to make plants resistant. The result suggested that the two minor genes responsible for resistance to all fungus isolates of Kunmin area in China may be used for breeding stable resistant cvs. since the two major genes were effective only for the two Japanese fungus isolates. This gets to the heart of what physiological races mean and why they are important to rice breeders. However, most plant breeders today are only concerned with finding out the resistant genes that the B1 pathogen isolates can overcome.

Discrepancies

Three sources of error are discernible in identifying physiological races. The first is through differences in interpretation of variation in lesion characteristics on a single leaf; these may be due to physiological and developmental differences between the different parts of the leaf or to differences in micro climate. For instance, in the Standard Evaluation System (SES) of IRRI (1988), tiny brown spots classified as the 13th type lesion are considered a resistant reaction. 'Slightly elongated to roundish with necrotic grey spot and brown margin' (Goto et al., 1964) is classified as a susceptible reaction for race identification. Lesions with ashygrey centres and brown margins treated as equivalent to 'C' types (Padmanabhan et al., 1967, 1970) are also considered susceptible. Secondly, the methods adopted by different workers are another source of discrepancy. Manibhushanrao and Day (1972) described that differences in response of a rice cv. to different temperature treatments prior to inoculation also affect race identification profoundly. The third possibility is contamination with conidia of other races and this can be gauged with uninoculated check plants.

Ou (1980) emphasized that B1 fungus is extremely variable. But others have disagreed (Kiyosawa, 1976 a; Ezuka, 1979 and Kiyosawa et al., 1981). At the other extreme, Veeraraghavan (1975a, 1986) claims that there is no race pattern involved in *P. oryzae*. Earlier, Latterell (1972, 1975) reported many physiological races but emphasized that changes in race pattern were rare. Others also stressed the pathogenic stability in *P. oryzae* (Bharadwaj and Singh, 1983; Latterell and Rossi, 1986). Latterell further pointed out that the race picture was exaggerated and led to an unfortunate controversy. This debate should be quickly resolved so that the value of breeding for specific resistance can be clearly assessed.

Ou and Ayad (1968) claimed to have identified 22 different physiological races among 56 single-conidial cultures from one typical lesion on a susceptible rice cv. in the Philippines. The authors commented "............. if more varieties were used as differentials, each conidium in the culture could probably be shown to be a distinct race". In fact, only one of the 17 differentials used in the study remained consistently resistant and only one remained susceptible. Subsequently, Ou (1980) mentioned that 'the so called race or virulence pattern appears to be only temporary' after working for several years on the instability of *P. oryzae*. However, Stakman and Harrar (1957) while discussing the physiological races in a pathogen, stated that only 'that which can be recognized with reasonable certainty and facility by physiological characters including pathogenicity can be termed as race'. Thus, a cautious approach is always necessary while working with physiological races of *P. oryzae*. In the light of contradictory views on the pathogenic variability in *P. oryzae*, Day (1973) reviewed the available work to illustrate variations in culture and discussed the probable discrepancies that may arise.

Suzuki (1965 b, 1967) proposed a theory for race identification in *P. oryzae* by application of fungal cytogenetics in addition to pathological principles. He coined a new term, 'homocaryonization' (verb : homocaryonize), a procedure to obtain the component homocaryons dissociated from their parent heterokaryons by the combined method of saltation and monospore assay. A number of conidia showed the major races S-2, C-2, SD-2 or CD-2, while S-O, C-O and C-1 revealed as genetically stable homocaryons as judged by both monospore assay and nutritive selection.

Despite the literature on variability and physiological races of the B1 fungus is extensive, several questions remain unanswered. None of the investigators who worked on physiological races of *P. oryzae* maintained the host (rice cvs. used as differentials) by vegetative (clonal) propagation. They also did not maintain on a single isolate of the B1 fungus as standard reference culture for comparison. Thus, the earlier results on evolving physiological races seems to be questionable. Application of sophisticated techniques using PRIs, and repeated DNA sequences to differentiate races may be necessary. Newer approaches through molecular biology to study variability will have to be tried in B1 research.

Varietal Resistance

Fulton (1908) reported that cv. Berton, when introduced in Italy in 1828, was almost immune of 'brusone' (Bl) but later it rapidly succumbed. Nisikado (1927) listed a number of cvs. as resistant to B1, but not immune to B1 in Japan. Further, Nakatomi (1927) found that certain cvs. from India, the Philippines, and Java were most resistant to B1 in Japan. Suzuki (1934, 1937) and Abe (1936) screened several rice cvs. against B1 pathogen for identifying the source of resistance. Apart from rice (*O. sativa*), wild *Oryza* spp. such as *O. cubensis* and *O. perennis* were found susceptible, while *O. alata*, *O. eichingeri* and *O. glaberrima* were resistant and *O. latifolia* and *O. minuta* highly resistant to *P. oryzae* (Kawamura, 1940).

In India, Sund araman (1927) and Ramiah and Ramaswamy (1936) were the first to carry out crosses between the resistant cvs. GEB 24 and CO 4 and the susceptible Korangu Samba. Subsequently, Thomas (1938, 1939) identified CO 4, CO 11, ADT 6 and 10998 were highly resistant but the next year both CO 4 and CO 11 reported breakdowns.

Cralley and Adair (1942) reported from the US that rice cvs. Zenith, Arkansas and Fortuma were resistant in Arkansas while Anderson et al. (1947) tested 19 cvs. and identified five groups, namely, very susceptible, susceptible, moderately susceptible, moderately resistant and resistant. Waldee and Roberts (1947) listed a number of resistant cvs. suitable for Japan. Okada and Maeda (1956) studied the crosses between resistant *indica* cvs. and *japonica* cvs. and concluded that *indica* cvs. were more resistant in temperate countries while the *japonica* cvs. tended to be more resistant in the tropics. Ito (1965) identified cvs. Kameji and Aikoku to be highly resistant in Japan and Asahi, Omachi and Shinriki as moderately resistant and very susceptible respectively.

Padmanabhan (1959, 1963) evolved a standard procedure for screening and evaluating the reactions of rice cvs. based on the extent of disease development. No appreciable differences were observed among the 43 cvs. tested at the Central Rice Research Institute (CRRI), Cuttack, for varietal reaction to B1. However, Chakrabarti (1976) reported Tetep, Tadukan and Zenith to be resistant to all the physiological races then identified in India. Subsequently, John (1977) reported cvs. IET 1444, RPC 2B 849, CR 12-178, RP5-2 and Jaya to be resistant among the dwarfs for cultivation in the plains.

Evaluation for varietal resistance

Evaluation of a fairly large number of cvs. against B1 is possible under field conditions by screening them either at the nursery stage as well as at neck or at the panicle emerging stage of the plants. To handle a large number of cvs. and get an accurate assessment of the degree of resistance/susceptibility, natural infection (severe disease) under uniform environmental conditions is necessary. A few workers have developed varietal testing under natural conditions in the field as well as by artificial inoculation of the test plants under green/net/ glass house conditions. A uniform evaluating system (IRRI, 1988) is necessary to evaluate the reaction types for obtaining an accurate assessment of the degree of resistance/susceptibility.

Uniform blast nursery

Ou (1965 b) developed an improved method of screening the rice cvs. in the uniform B1 nursery. This was adopted in about 45 stations by more than 25 countries participating in the International Rice Blast Nursery (IRBN) programme at IRRI. During 1962-64, 8214 cvs. were tested in IRRI B1 nurseries and 1957 of them were found resistant. About 450 remained resistant to B1 consistently at seven research stations of the Philippines and only 75 cvs. were found resistant at all the testing stations of IRBN (IRRI, 1966). Extensive studies on the inheritance of B1 resistance indicated that in most cases that it was simply inherited.

In the IRBN, the recommended application of N-fertilizer (120-160 kg N/ha), maintenance of high humidity by sprinkling water (3-4 times) depending upon the weather conditions is quite essential. Plots, square or rectangular, with short rows about 1 m long with 10 g seed for each test cv. were recommended (Ou, 1963). The use of a susceptible cv. as border rows (0.5 m length) to retain high humidity in the testing rows and to build up natural inoculum in the nursery is also essential.

Shigemura (1955), Ou and Nuque (1963) and Padmanabhan (1965) recommended more endemic or hot-spot areas for testing the reactions. The IRRI (1968) reported that about 60 Philippine races equivalent to 34 international races were detected in the uniform B1 nursery trials conducted over a period of 21 months.

Use of established races

Yamasaki and Niizeki (1963) used four races of *P. oryzae* to classify resistance while Yamasaki and Kiyosawa (1966) and Kiyosawa (1967 c) classified rice cvs. based on the resistant genes. Yamada et al. (1969) tested 1349 and 1181 local cvs. using seven (C-3, C-6, N-1, N-2, N-3, N-4 and N-5) and five (T-1, C-1, N-1, N-2 and N-4) representative races found in Japan. Ezuka et al. (1969 a) studied the reactions on 373 cvs. and Kozaka et al. (1970) on 691 cvs. obtained outside Japan against seven local and 21 isolates (of which 11 from outside Japan) respectively. Although this study identified resistance genes in each of the cvs. it failed to identify the overall resistance of the cvs. tested. Other reports on the source of resistance (Kawamura, 1940; Mello-Sampayo et al., 1954; Katsuya 1959, 1960, 1961; Lin and Lin, 1961) used established races but there is much discrepancy among them as a few cvs. identified as resistant by some authors were found susceptible by others.

Methods of inoculation

Artificial inoculation is quite necessary when testing a specific race in a set of rice cvs. either by spraying or by syringe inoculation, sheath or smear inoculation using conidial suspensions/dust/paste as required by the method used.

Spraying the spore suspension on pot plants in green house studies as well as excised leaves under *in vitro* conditions to test the varietal resistance is in practice. Young leaves of 20-30 day old rice seedlings were detached and cut into uniform lengths (about 5 cm) and floated in 50 ml solution of benzimidazole/kinetin (40 µg/ml) in petridishes, lined with blotting paper, to give adequate humidity. The floated excised leaves/potted plants were sprayed uniformly with the spore suspension using an atomizer/sprayer and were incubated for 10-12 d. To build up high humidity for potted plants, water spray or use of dew chamber is suggested. For inoculum, either dried spore dust (Andersen et al., 1947; Kahn and Libby, 1958; Panzer and Beier, 1958) or a suspension with an adhesive agent (Andersen and Henry, 1946) was used with success. The disease scoring was done on the intensity of reactions on the leaves.

Kuribayashi and Terazawa (1953), Suzuki (1967a) used injection as an artificial inoculation method in rice B1 disease. This syringe inoculation gave uniform results in screening cvs. for resistance.

Ou and Nuque (1963) and Manibhushanrao (1980) injected the first leaf sheaths of one-month old seedlings with spore suspension through the hypodermic needle of a syringe. After 4-6 days, the symptoms were discernible on the merging young leaves especially in the region of needle punctures. This technique ensured 100% infection for screening a large number of cvs.

Sheath inoculation is ideal for screening varietal resistance under *in vitro* studies (Sakamato, 1949). Leaf sheaths are cut into uniform size (7-10 cm), filled inside with spore suspension and incubated for about 40 h at 24-28° C. The cells of the inner leaf surface of the sheaths are examined under the microscope. Host penetration and the extent of colonization of the invaded cells were used as the criteria for determining the degree of resistance using a standard scale (Takahashi, 1965).

Guochang and Shuyuan (1989) developed a simple smear method for inoculating rice plants with the B1 pathogen for resistance screening. Smear inoculation is carried out by smearing the mixture $1\text{-}2 \times 10^5$ conidia/ml suspended in 1-2% carboxymethyl cellulose onto the rice leaf with a brush. The method is suitable for inoculation at several growth stages of the plants but is more efficient at the 2-4 leaf stage and requires only a small quantity of spore suspension to produce a high infection frequency. The efficacy of the smear inoculation, tested on 12 *indica* and 5 *japonica* cvs., was more as it produced a disease index higher than spray inoculation but slightly lower than syringe inoculation (Table 6). However, these workers claim that it is far superior to the spraying techniques but is on par with the injection method for evaluating neck B1 infection as in both cases the infection frequency was 100%.

Table 6 : Efficacy of smear inoculation of leaf blast (Guochang and Shuyuan, 1989)

Variety	Bl index[a]			Bl score[b]		
	Injection	Smear	Spraying	Injection	Smear	Spraying
Indica						
Chei-Tang	0	0	0	0	0	0
Nong 8506	0	0	0	0	0	0
Gui-Chao 2	0	0	0	0	0	0
Hou-Zeng-Zao	5·6	4·9	1·5	0·5	0·4	0·1
Zao-Shuang 1	14·8	14·3	4·0	1·3	1·3	0·4
Zhong 83-40	69·6	68·6	17·8	6·3	2·6	1·6
Zhong 83-49	76·4	68·6	45·5	6·9	6·2	4·1
Hong-Tu 3	86·9	71·2	48·9	7·8	6·4	4·4
B 40	83·0	80·4	45·8	7·5	7·2	4·1
8004	96·9	73·9	48·3	8·7	6·6	4·4
Fr-Jiu-Qing	95·1	85·8	53·5	8·6	7·7	4·8
Zuo 5	94·8	80·4	56·2	8·5	7·2	5·1
Japonica						
Pi-4	0	0	0	0	0	0
Kusabue	18·3	24·3	3·5	1·6	2·2	0·3
BL 1	87·2	52·4	41·0	7·8	4·7	3·7
Dong-Nong 363	71·8	68·6	42·7	6·5	6·2	3·8
Li-Jiang-Xing-Tuan-He-Guo	100·0	100·0	92·4	9·0	9·0	8·3

[a]Disease index = $\frac{\Sigma \text{ (number of plants with a given score} \times \text{value of that score)}}{\text{total number of plants} \times \text{value of the highest score}}$

[b]Based on *Standard evaluation system for rice.*

Identification of resistance source

Around 1967-68, the IRRI identified a number of cvs. as resistant through IRBN trials, taking the existing racial picture into account. In the US, cv. Saturn was found resistant to race 1 and 6 which are widely prevalent in Arkansas, Florida and Louisiana (Atkins et al., 1965). It was a replacement to Nato which became susceptible to race 1 and 6. A second cv. Dawn has been found resistant to many races. In India, Padmanabhan et al. (1974) found that cvs. Tadukan, Tetep and Zenith were resistant to most of the local races identified and these were subsequently used in the hybridization programmes for B1 resistance.

Method of scoring

The reaction types (lesions) of B1 infection on rice leaves depend on the cv., virulence of the pathogen and other environmental conditions. The lesion types are used as the main criteria for evaluating varietal reaction. The size, shape, colour and number of lesions on each leaf or unit of leaf area are the general parameters considered. Stunting of plants, due to B1 attack is also considered (Kuilman, 1940).

Hashioka (1950 a) considered five classes of reaction types—from susceptible to resistance—and designated them as A, B, C, D and E which correspond to the reversal order (Plate 13) adopted by Padmanabhan and Ganguly (1959). The former scale is almost equivalent to the types 5, 4, 3, 2, 1 and 0 of the Standard Evaluation System (SES) of IRRI. Ono (1953) classified the host reactions into four types, brown spot, white spot, acute (Sharp pointed) and chronic (long continued) types. Abumiya (1955, 1958) described the zones of lesions as white (w), blackish purple (p), brown (b) necrosis (perhaps green) (g) and yellow (y) and the combination of these alphabets such as bpg for the description of the lesions and the respective capitals to indicate greater intensity of the colour. Padmanabhan and Ganguly (1959) added I, II, III to each group to indicate the number of lesions and 1, 2, 4, 9 and 16 to each type and 2, 5 and 10 to each lesion number, group respectively. The product of the two scores was considered as the final reading. Podhradszyky (1961) suggested four groups of the lesions based on the size and development. Goto (1960), Latterell et al. (1960) and Hung and Chein (1961) adopted lesion types 1 and 2 as resistance, type 3 as intermediate reaction and types 4 and 5 as susceptible in studying pathogenic races in Japan, the US and Taiwan respectively. Ou (1961) used modified scales 1 to 4 based mainly on the lesion types and 5 to 8 based on the percentage of leaf area killed, that is 25-100% respectively. IRRI (1965) and Ou (1965) further suggested a scale of seven units (1-7) for the uniform B1 nursery. Goto et al. (1967) working with a US-Japan cooperative research programme on international races classified lesions into seven types (A-G) resistant (A,B), moderate (C, D) and susceptible (E, F & G) reactions (Plate 14). Subsequently the International Rice Testing Programme (IRTP) of IRRI (1979) modified the SES scale to 10 classes- 0, 1-9 to make it more suitable for computer programming.

Ahn et al. (1988a) developed a modified coding system to take into account all the major lesion types as 0, 1, 3, 5, 7 and 9 from resistant (no lesions) to highly susceptible (rapidly coalescing small, whitish, greenish or blueish lesions without distinct margins) reactions. Lesion type 1 indicated the incompatible reaction of the host to a virulent isolate while type

3 considered also an incompatible reaction controlled by major genes as seen in many *japonica* cvs. inoculated with *indica* isolates producing lesions with yellowish centre. Lesion types 5, 7 and 9 were considered typical susceptible reactions, produced by compatible isolates.

Other methods for scoring B1 for accurate assessment included use of leaf sheaths (Suzuki, 1967a) as well as cytological responses of host cells for resistance against invasion of the B1 fungus using whole-leaf clearing and staining techniques (Koga and Kobayasi, 1980).

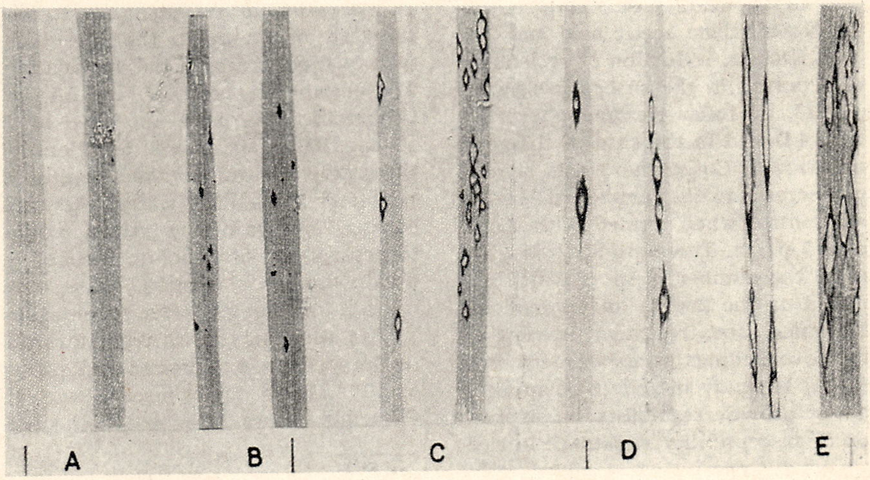

Plate 13: Lesion types of blast disease of rice. A,B: resistant; C : moderately resistant; D, E: susceptible (Padmanabhan and Ganguly, 1959).

Plate 14: Reaction types of rice seedlings induced by *P. oryzae*. A: resistant C, D: moderately resistant E,F,G : susceptible (Goto et al., 1967).

Physiology of Host Plants

It is becoming clear that the physiological and biochemical changes in the host under pathogenesis should be documented in order to elucidate the mechanism(s) of resistance. Any changes in the N, constituents of the host under pathogenesis have also been implicated to resistance. Besides N phosphorus and potassium, silica, growth regulators and trace metals are also implicated in the resistance mechanism(s) of rice.

Nitrogen

High N—fertilizer application always induces heavy incidence of blast (Hori, 1898; Miyazaki, 1938, 1930; Bokura, 1930; Murata et al., 1933; Ikata et al., 1938; Krishnaswamy, 1952; Otani, 1952a; Padmanabhan, 1953; Ganguly et al., 1954; Apparao, 1956; Suryanarayanan, 1958d). The severity of disease varies with the air temperature (Hashioka, 1943a) soil and climatic conditions such as temperature (Hashioka, 1943b), level and method of applying the N-fertilizer (Thomas, 1938; Sawada, 1940), the cv. used and the age of the host (Hashioka), 1943c). However, Hashioka (1944a, b) considered that the relation between resistance and the environmental temperature has to be ascribed mainly to the N-metabolism of the host as this influence of fertilizers on host resistance is stronger at low temperature.

Heavy application of green manures is believed to cause high B1 incidence (Nishikado, 1926 and Murata et al. 1933). A close correlation was reported between B1 incidence and accumulation of soluble N and total protein N (Otani, 1948, 1950) and increased N content (Amin and Venkatarao, 1979). Suryanarayanan (1958a) reported that besides altering the N metabolism of the host, the low temperatures favoured amide synthesis, especially glutamine. Apparao (1964) reported that manuring increases the total N content but greater accumulation of N is noticeable in susceptible cvs. than in resistant ones. The influence of manuring on N content was greatest during the tillering and ear-head stages when the rice plant is more susceptible to the disease.

A few workers reported that accumulation of N lowers the silicated epidermal cells in rice leaves, and reduces resistance (Ikeda, 1933; Suzuki, 1934; Abumiya and Kobayashi, 1955 and Wakimoto and Yoshii, 1958). Matsuyama (1975) reported that greater accumulation of N decrease hemicellulose and lignin in the host tissue and thereby reduces the mechanical resistance to B1.

Several workers (Ohata et al., 1966; Otani, 1952 a; Tokunaga, 1959; Tokunaga et al., 1966) reported a close relationship between the soluble N content and the susceptibility of rice. Sadasivan et al. (1965) correlated between the greater accumulation of free amino acids at low night temperature with the susceptibility of the host tissues.

Phosphorus

Unlike N, phosphorus (P) accumulation has no significant effect on B1 disease incidence

(Krishnaswamy, 1952; Ganguly et al., 1954). However, Ikata et al. (1938) and Okamoto (1950) reported that application of large quantities of P fertilizers reduced the disease severity. Kozaka (1965) found that in P deficient soils the supply of phosphate reduces the disease until normal growth occurs while further application increases the disease severity.

Potassium

Reports of reduced infection following the application of potassium (K) fertilizers are available (Miyazaki, 1928, 1930 and Ito and Hayashi, 1931). However, Murata et al. (1933) and Chiba and Yamashita (1957) showed that large scale K fertilizer application did not reduce the disease in plants receiving high N. Application of K alone increased the B1 incidence but the disease could be reduced by adding Mg (Sako and Takamori, 1958). K also influences resistance of rice tissues to B1 infection (Moscobriofrio and Moralesabanto, 1975).

Silica

It is well established by many workers that silicification of epidermal cells has a direct correlation with the resistance of rice plants to B1 (Onodera, 1917; Nagai and Imamura, 1931; Miyake and Ikeda, 1932; Hemmi, 1933 and Suzuki, 1934 c). It has been reported that the silicated epidermal layer blocks or prevents the physical penetration by the infection peg of the pathogen (Hemmi, 1933, Suzuki, 1934c and Ito and Sakamoto, 1939). Ishizuka and Hayakawa (1951) ascribed a role for silicic acid (SiO_2), as high incidence of B1 was reported in its absence. Akai (1953) demonstrated that application of SiO_2 in the soil increased the resistance of panicles and spikelets of rice to B1.

Miyake and Adachi (1922) reported that the cv. Bozer resistant to B1 contained heavy deposits of silicon on the surface. Large and older leaves had more silicified cells than younger leaves. Similarly, Adyanathaya and Rangaswamy (1952) found more silicated epidermal cells in the resistant cv. than in the susceptible one. Volk et al. (1958) noted that resistance to rice seedlings could be increased by the addition of silicon in the nutrient medium. Yosida et al. (1962 a, b, c d) reported that 90% or more silicon accumulation occurred in alternate layers with the cellulose layer in the epidermis, vascular layer as well as sclerenchymatous cells. Tasugi and Yoshida (1958) showed the accumulation and desposition of silica around injured cells as this served as a barrier against pathogen invasion.

Growth regulators

Subramanian et al. (1965 a, b) found an increase in the number of lesions in the B1 susceptible cv. CO 13 on treatment with indole acetic acid (IAA) and 2,4-D. In the case of resistant CO 29, similar results were obtained consequent to application of naphthyl acetic acid (NAA), 2,4-D and Kinetin. However, such a trend was not noticed when spraying IAA on the intact plants (Veeraraghavan and Padmanabhan, 1976) nor did floating the detached leaves with indolebutrytic acid (IBA) (Suryanarayanan, 1958a) influence resistance. Admittedly, the effect of auxin production in B1 disease development has not been critically investigated.

Amino acids

Venkatachalam et al. (1957) evaluated the role of free amino acids in the reaction of CO 4 and ADT 10 susceptible and resistance cvs. to B1 respectively. The concentration of all the amino acids, in general glycine, aspartic acid, glutamic acid, threonine, alanine and valine in particular was more in ADT 10 than in CO 4.

Host-Pathogen Interaction

Modern research aims to unravel the physico-chemical processes of the diseased plant. This helps in determining the mechanism of fungi toxicity, toxins and their mode of action and the role of hormones and other antimicrobial substances. The importance of extracellular and oxidative enzymes in the etiology of the disease is also emphasized.

B1 fungus - rice interaction studies have been reported by several workers (Yoshii, 1936; Kawamura and Ono, 1948; Suzuki, 1965; Hashioka et al., 1968), especially with different host cvs. and fungal isolates/races on diverse parameters, adopting their own concepts of resistance and methods of investigation. Interaction of rice plants with the B1 pathogen without symptoms has been reported by Veeraraghavan (1988).

Genotype-nyctotemperature

The effect of low nyctotemperature (NT) on B1 severity and disease development has been reported by many workers (Suryanaryanan, 1958a; Ohata et al., 1966; Manibhushanrao and Day, 1972; Kato and Sasakii, 1974; Kimura and Kobayashi, 1980). Suryanarayanan (1958 a) emphasized that the resistance of *indica* cvs. of rice to B1 was governed by genotypes-NT interactions. Most cvs. tested exhibited resistance when grown with an NT above 25°C. However, a few cvs. showed typical susceptibility and resistance at 20° C while at 15° C, breakdown of their resistance was observed. This was later confirmed by getting a good correlation between low NT and disease incidence in the hilly areas of Tamil Nadu (Govindaswamy, 1964) and in the plains of India (Padmanabhan, 1965).

Ramakrishnan (1966a) and Subramanian (1967) reported that Low NT affect the concentration of individual free amino acids in rice leaves; greater quantities of glutamine, β- aminobutyric acid and glutamic acid are present in contrast to the low quantities of phenylalanine and tyrosine. Blast infected rice leaves, in general, shown an increase in the content of phenylalanine and glutamine accumulates in low NT grown plants. The temp. of N metabolism is higher in a low NT regime which favours accumulation of soluble N, particularly the amides, and restricts protein breakdown and turnover.

Ramakrishnan (1966b) studied the various carbohydrate fractions and total phenols in both healthy and B1-infected rice leaves grown at low NT. The quantity of reducing sugars in both susceptible and resistant cvs. decreased as infection progressed while the quantity of phenols increased. The effect of low NT on the infection phase in B1 disease of rice with reference to resistance/susceptibility was studied by Ramakrishnan (1966c). It has been further established that low NTs favoured greater nitrate reductase activity in *indica* rice cvs. (Fig.14) (Subramanian and Sadasivan, 1964) and accelerated N metabolism of the host (Sadasivan et al., 1965). Subsequently, Sadasivan (1968) proposed that genotypes-NT interactions affect the C/N balance of the host and related this to the B1 resistances/ susceptibility.

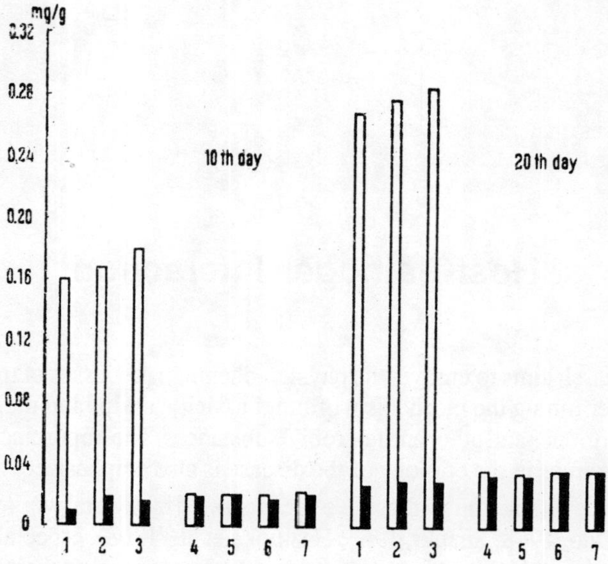

Fig. 14 : Nitrate reductase activity in india rice varities. 1-3 : susceptible varieties; 4-7; resistant varieties. Ordinate : Nitrate reductase activity (mg nitrite nitrogen formed per g fresh weight). White columns; 20°C nyctotemperature; black columns; 30°C nyctotemperature. (Subramanian and Sadasivan, 1964)

The effect of three NTs - 15, 20 and 30°C - with a day temperature of 30° C on the lesion development in rice plants susceptible and resistant to *P. oryzae* was studied by Subramanian (1967). There was an increase in infection with a fall in NT while an actual breakdown of resistance occurred at an NT of 15° C (Table 7).

Table 7 : Effect of varying night temperature on number and size of lesions (Subramanian, 1967)

Cultivar	15°C		20°C		30°C	
	Mean number of lesions/leaf	Mean size of of lesions mm length × breadth	Mean number of lesions/leaf	Mean size of lesions mm length× breadth	Mean number of lesions/leaf	Mean size of lesions mm length × breadth
CO 13 (S)	20	9 × 5	12	8 × 4	4	2 × 2
GEB 24 (S)	23	8 × 5	10	8 × 3	3	3 × 1
ADT 10 (S)	21	9 × 6	10	8 × 4	4	2 × 1
CO 4 (R)	29	6 × 4	1	3 × 1	0	0
CO 25 (R)	24	8 × 5	2	2 × 1	0	0
CO 29 (R)	21	6 × 6	2	3 × 1	0	0
CO 30 (R)	22	6 × 5	3	3 × 2	0	0

(S) = Susceptible (R) = Resistant

Manibhushanrao and Day (1972) investigated the effect of different periods of low NT on the susceptibility of different rice cvs. The low NT (15° C) treatment varied from a total of 16 h (4 h per day on four consecutive days) to a total of 144 h (12 h on 12 days); seedlings were also grown at 30° C constant. It was of interest to note that three cvs. Kanto 51, Lacrosse and Caloro, were resistant in all the treatments. The cv. CO 13 showed a susceptible reaction only after 16 h of low NT treatment (Table 8). These results confirm that low NTs lead to a partial breakdown of resistance. Further, plants that received maximum low NT treatment (144 h) were 10 to 15% shorter than those grown at a constant 30° C. Intermediate treatments had varying effects between these extremes as shown by Ohata et al. (1966). Whether low NT altered the physiological age of the host tissue or affected the host metabolism in some other way is not known. It is also not clear whether the effect of low NT is independent of the darkness; perhaps low day temperatures, if they prevailed in nature, may have similar effects.

Table 8: Blast development on rice cultivars after varying periods of low night temperature (15° C) (Manibhushanrao and Day, 1972)

Rice cultivar	Reaction types[a]									
	Treatment periods[b]									
	Control	12 days			8 days			4 days		
	Constant 30° C	12	8	4 h	12	8	4 h	12	8	4 h
Raminad Str. 3	R	S	R	R	R	R	R	R	R	R
NP-125	R	S	S	S	S	S	S	S	S	R
Usen	R	S	S	S	S	S	S	S	R	R
Kanto 51	—	R	R	—	R	—	—	R	—	—
Lacrosse	—	R	R	—	R	—	—	R	—	—
Caloro	R	R	R	R	R	R	R	R	R	R
CO 13	R	S	S	S	S	S	S	S	S	S

[a] R = Resistant. S = Susceptible. — = No symptom.
[b] Treatments began after 14 days at 30 °C

Veeraraghavan and Dath (1975) found variations in the lesion development in cv. Usen, perhaps sensitive to low NT, and suggested replacing the differential cv. with the susceptible CO 13. Further, Manibhushanrao (1979) made an attempt to confirm the differences in the reaction types of the international rice differentials as due to low temperature (15° C) treatment as well as as to major international races of *P. oryzae*. The race identification on the basis of lesion size and development on eight international differentials did not vary significantly when plants were grown at 30° C, while at low temperatures two cvs. NP 125 and Usen succumbed to the disease by producing typical susceptible 'E' spots, although these are resistant to race 1E-1 and 1F-1 when grown at 30° C. This resulted in a shift in the identification of race group IE and IF to ID and IE respectively. Similarly, race ID-1 was identified under the IC group since the race could not infect NP 125 under low temperature conditions (Table 9).

Table 9: Reactions of international differential cvs. to physiological races of *Pyricularia oryzae* at low night temperature 15°C. (Manibhushanrao, 1979)

Cultivar	30°C									15°C								
	International race numbers									International race numbers								
	IA 101	IB 37	IC 1	IC 17	ID 1	ID 13	ID 14	IE 1	IF 1	IA 109	IB 33	IC 1	IC 17	ID 1	ID 13	ID 14	IE 1	IF 1
Raminad str 3	+	+								+								
														IC			ID	IE
Zenith		+									+							
NP 125			+	+							+	+	+	+			+	
Usen	+	+	+		+	+	+			+	+	+		+	+	+	+	+
Dular		+	+	+	+			+			+	+	+	+			+	
Kanto 51		+	+	+	+			+	+		+	+	+	+			+	+
Sha Tiao Tsao (S)	+	+	+	+	+	+	+	+	+	+	+	+	+	+	+	+	+	+
Caloro	+	+	+	+	+	+	+		+	+	+	+	+	+	+		+	+
Aichi Asahi	+	+	+	+	+	+	+	+	+	+	+	+	+	+	+	+	+	+

+ Reaction

Nature of Resistance

Resistance of rice plants to B1 disease is characterized by fewer lesions of small size and intense tissue browning associated with limitation of both growth and sporulation of the pathogen. The nature of resistance in rice plants to B1 infection has been extensively investigated by Suzuki (1965c). Earlier investigators tried to correlate the accumulation of silicon content (silicic acid) with resistance and the various nitrogenous compounds including amino acids, amides or soluble N with susceptibility. Suzuki (1933 a, b, c) observed that growing rice under dry conditions favoured B1 infection while those grown in a submerged seed-bed or in flooded conditions were more resistant (Suzuki, 1934; Akai, 1939).

Resistance of rice plants to *P. oryzae* is influenced by both constituent factors and a variety of physiological and biochemical responses that culminate in the prevention of infection by the pathogen. In addition to phytoalexins, which are produced in response to infection, an array of phenolics such as chlorogenic acid (Suzuki et al., 1953), *p*— hydroxybenzoic, vanillic, *p*-coumaric and ferulic acids (Kuwatsuka and Oshima, 1961), salicylic acid (Ishi et al., 1962), protocatechuic, cinnamic, *p*-coumaric and caffeic acids (Varga, 1970) are known to be present in rice leaf tissues. Several of the preformed substances are prohibitins, toxic to *P. oryzae*, and may act as the first chemical barrier in host defence. The pathogen races vary in their sensitivity to these compounds. There is evidence that at later stages, after penetration, rice plants also dynamically respond to infection. Most of the reports are based, however, on correlation studies.

Another group of 'non-fungitoxic' compounds that control rice blast are the inhibitors of melanin biosynthesis in the pathogen (Tokousbalides and Siser, 1978; Woloshuk et al., 1980). This group includes compounds tricyclazole, pyroquilon, pthalide, coumarin, 2,3,4,5,6-pentachlorobenzyl alcohol; 4,5-dichloro-4 methyltetrazole (1,5-a) quinazolin-5-one, *N*-methyl-2 quinolone and S-triazole (4,3-a) quinolone (Woloshuk et al. 1981; Woloshuk et al., 1983).

Suryanarayanan and Ou (1978) implicated a role for epicuticular waxes in B1 resistance while Kato et al. (1988) ascribed a role to unsaturated fatty acids for B1 resistance. Manibhushanrao and Zuber (1978) reviewed the role of antifungal substances in disease resistance in cereals.

Host surface interaction

Although *P. oryzae* spore germinates well and germ tubes form appressoria on the leaf surface of both susceptible and resistant plants (Griffiths, 1969) they also form under different temperature conditions (Sadasivan et al., 1965). A quantitative difference has also been reported depending on host resistance and the age of leaves (Abe and Yasuda, 1962; Kumar and Sridhar, 1987). Since about 60% of spores formed appressorium on the rice leaf surface even when grown under high night temperature, a situation not conducive for B1 infection. Sadasivan et al. (1965) came to conclude that the factors for susceptibility and resistance may not operate at the pre-penetration stage. However, the quantitative difference in appressorium formation under different conditions may not be totally without any significance. Though spores germinate equally on the leaf surface of both susceptible and resistant rice cvs., germ tube growth is soon arrested on the former and many appressoria are formed as compared to the longer and much branched germ tubes with fewer appressoria on the latter (Kumar and Sridhar, 1987). Suzuki (1951) reported earlier that more appressoria are formed on rice plants grown at low soil temperatures associated with greater penetration than on those grown at higher temperatures. Stoppage of germ tube growth and appressorium formation are important responses of morphogenesis regarded as essential pre-requisites for successful host penetration. Kumar and Sridhar (1987) observed that epicuticular wax extracted from the leaves of susceptible cv. Karuna inhibited germ tube growth and stimulated appressorium formation, thus aiding larger number of infections in susceptible plants. While the differential actions of fungal germ tube on susceptible and resistant (cv. IR 8) leaves do not ultimately determine their disease response, these appear to contribute to it, as there is reduced appressoria formation on resistant host surface, thus keeping the number of successful infections restricted and limiting the intensity of pathogen attack. This is consistent with the fact that resistant cv. develops only fewer spots of smaller size.

Subbarao and Suryanarayanan (1957) investigated the nature of leaf exudates from two B1 susceptible and resistant rice cvs. Though these had four amino acids in common, the susceptible cv. had glycine and the resistant one succinic acid. Later, Suryanarayanan (1958d) reported that susceptible cvs. exude more glutamine than resistant cvs. while lack of α-aminobutyric acid (Zsoldos, 1959) was also considered to be related to B1 susceptibility. Increase in the number as well as amount of amino acids (Kozaka and Sonku, 1963), especially that of glutamine (Suryanarayanan, 1958d; Hashioka and Ikegami, 1961) in dew or on leaves as crystals (Suryanarayanan, 1967) has been suggested to be related to B1 susceptibility of rice plants under low temperature conditions. However, susceptible plants also exude large amounts of glutamine under high night temperature conditions, suggesting doubtful involvement of glutamine in host susceptibility to *P. oryzae*.

Though Ito and Sakamoto (1940) concluded that resistance of rice to B1 pathogen is correlated with resistance of leaf epidermis to mechanical penetration, no unequivocal evidence is available.

Miyake and Adachi (1922), Suzuki (1937), Akai (1939) and Adyanthaya and Rangaswamy (1952) correlated resistance with greater silicon content of leaves and the presence of more silicated epidermal cells and attributed a role for such morphological and anatomical features in the B1 resistance. Nagai and Imamura (1931), Ishizuka and Hayakawa (1951) and Otani (1952b) reported increased susceptibility to B1 due to heavy application of N to be correlated with a reduction in the number of silicated epidermal cells and reduced accumulation of silica. Older leaves with more resistant reaction have more silicated cells in the epidermis than younger leaves. Rice plants grown in soils with high moisture level suffer much less from B1 infection than those grown in soil with low moisture, under upland conditions, and such resistance is associated with higher levels of silicification in leaves (Suzuki, 1934a). Later, Kim et al., (1985) reported that leaves of susceptible plants from wet land flooded plots contained significantly more silica and that when upland plots were flooded, the level of silica in leaf tissue increase greatly and when the flooded plots were drained, the level decreased, though not equally in all cvs. tested.

This has been confirmed by Hori et. al. 1957 that doubling of N application decreased the number of silicified cells in the uppermost leaves (Fig. 15a) and the number of silicified cells is larger in older leaves (heading stage) which are also more resistant to B1 than younger leaves (uppermost leaves) with less silicified cells (Fig. 15b). However, Volk et al. (1958) found an inverse correlation between silicon and susceptibility. Although it appears attractive, the silicon hypothesis does not explain all the facts of this problem as there are always a certain number of unsilicified epidermal cells in the resistant cvs. and there is no reason why the fungus should not penetrate these cells. Akimoto (1939) studied the varietal difference in absorption of N and silica in relation to rice B1. Further, in some resistant cvs., the level of silicification may even be lower than in susceptible cvs. (Kawamura and Ono, 1948). So, it should seem that though correlation between B1 resistance and the degree of silicification in epidermal cells is generally good, the nature of B1 resistance is much more than simple resistance to mechanical penetration. Nevertheless, the greater accumulation of silica in leaf epidermis is expected to reduce the thrust of initial attack by reducing the number of (successful) penetrations. Suzuki (1951) correlated changes in anatomical characters of the rice panicle and a low ratio of potassium to ammonia in the epidermal cells at low temperatures with susceptibility to B1.

Suryanarayanan (1958b) and Sadasivan et al. (1965) reported that low NTs of 20°C or below during the growth phase of rice seedlings rendered the host more susceptible to B1 infection. This was attributed to the accumulation of glutamine in particular in the leaves of susceptible cvs. (Suryanarayanan, 1958a). Although glutamine stimulated conidial germination of *P. oryzae in vitro*, this amide is not required for sporulation whereas glycine is essential (Manibhushanrao, 1971). Ohata et al. (1966) reported that seedlings treated under low temperature conditions became very susceptible because of a decrease in peroxidase activity.

Puncture resistance

Ito and Sakamoto (1939) studied the relationship between puncture resistance (measured by the puncture of leaf epidermis using a needle tip of 30μ from beneath a torsion balance) of leaf epidermis and B1 resistance. The degree of resistance increased in proportion to the amount of silicate applied as well as plant age. Again, Yoshida et al. (1962.a, b, c, d) pointed out that the role of silicon in rice plants was mechanical and the higher the puncture resistance, the higher the resistance to B1. Ito and Sakamoto (1940) had earlier pointed out that the puncture resistance to B1 could not be attributed solely to the silicification of rice leaves; rather, it must be attributed mainly to the nature of the protoplasm. This has led to the study of the role of chemical constituents on the basis of host-parasite interaction.

Constituent chemical compounds

Against B1 infection, resistance to initial host penetration by the pathogen seems to be less important than resistance to its spread in the host tissues. A close correlation has been reported between B1 susceptibility in rice and the presence of soluble nitrogenous substances, particularly dibasic amino acids (Otani, 1952a) or the ratio between soluble N and total N. Correlation is also evident between low NT induced susceptibility and increased accumulation of amino acids (Sadasivan et al., 1965) or total N (Ohata et al., 1966). Kim et al. (1985) reported under wet land conditions for two B1 susceptible cvs. great differences

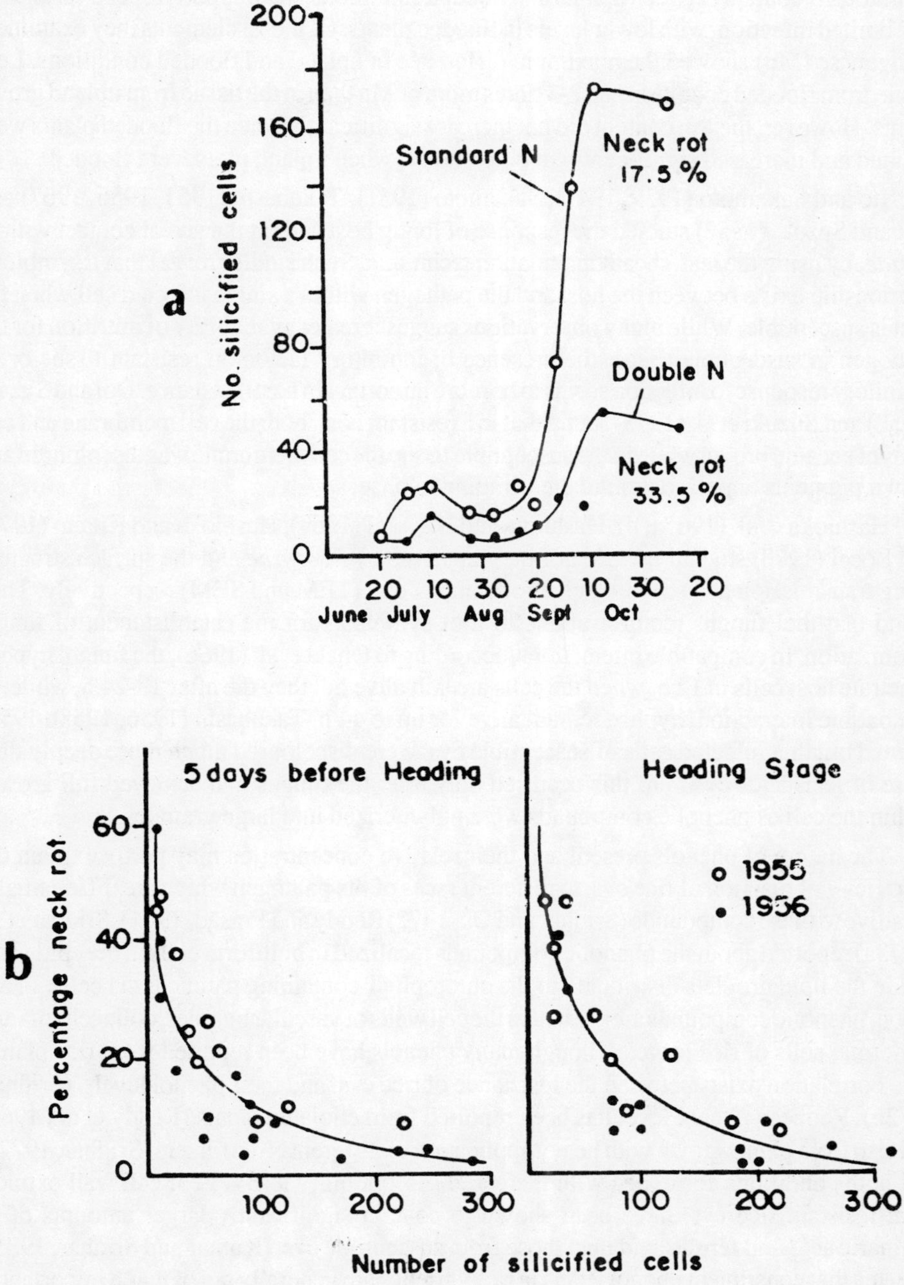

Fig. 15ab: (a) Seasonal changes in number of silicified cells in upper most leaves of rice plants treated with standard and double amounts of nitrogen fertilizer. (b) Correlation between percentage of neck rot and number of silicified cells in upper most leaves at two stages of growth (Hori, et al., 1957)

in the total N content between upland and flooded conditions, that support respectively heavy and limited infection, with lower levels in flooded plants. Of the 13 elements they examined, manganese (Mn) showed the maximum difference in upland and flooded conditions. Leaf tissue from flooded condition had 2-3 times more of Mn than in the tissue from upland grown plants. However, the Mn content did not increase significantly when the flooded plants were drained and increased significantly in only one cv. when upland plots were flooded.

Ito and Sakamoto (1939, 1941), Sakamoto (1951), Takahashi (1951, 1956, 1967) and Doi and Suzuki (1952) studied the response of living host cells at the site of contact with *P. oryzae*, by using the leaf-sheath inoculation technique. Such studies reveal that a symbiotic relationship exists between the host and the pathogen within a single infected cell when the host is susceptible. While many observations suggest greater availability of nutrition for the pathogen in susceptible tissue, the presence of inhibitory factors in resistant tissue or its inhibitory response to infection seems to be more important in host resistance. Doi and Suzuki (1952) and Suzuki et al. (1953) found that in a resistant host, both the cell membrane and cell content became brown while in the susceptible tissue the cell membrane was discoloured and brown pigments appeared around the invading hyphae.

Hashioka et al. (1967a, b), Hashioka and Ikegami (1969), Hashioka and Kitano (1971) and Locci (1970) studied invasion of the rice tissue by *P. oryzae* and the surface structure using transmission and scanning electron microscopy (TEM and SEM) respectively. They found that the fungus requires about 20 h of symbiosis for the establishment of fungal colonization. In compatible interactions, according to Ohata et al. (1966), the fungal hyphae penetrate host cells in 12 h, when the cells are still alive but they die after 18-24 h, while, in compatible interaction, hyphae remain alive for up to 44 h. Takahashi (1956, 1958b 1959) pointed out that infected cells of susceptible cvs. were discoloured much more deeply than those of resistance cvs., but this occurred only after the fungus had achieved full growth within the cell as phenolic compounds were polymerized into large granules.

The nature of phenols present and their relative concentration may partly explain the specificity of reaction of rice cvs. to different races of the pathogen which are differentially sensitive to these compounds (Sridhar and Ou, 1972; Reddy and Prasad, 1974). Sridhar et al. (1973a) reported about the phenolic compounds localized in bulliform cells in the epidermis and in the lipid droplets distributed in the chlorophyll-containing parenchyma cells. In addition, phenolic compounds are present in the cell walls of vascular bundles, collenchyma and epidermal cells of rice leaves. Though many phenols have been reported from rice plants, little correlation exists between the resistance of rice cvs. and their phenol levels (Sridhar, 1972b). Reduced phenol level has been reported from etiolated plants (Reddy et al., 1969) and also from plants grown with heavy application of different N-fertilizers (Sridhar, 1972a) and both conditions correlated with increased susceptibility to B1. Leaf cell wall extracts from resistant rice cvs. have been shown to contain significantly larger amounts of *p*-coumaric acid and ferulic acid than those from susceptible cvs. (Kumar and Sridhar, 1985). It seems that constituent phenol levels in rice varieties are generally not of much importance, but the alterations in phenol level leading to greater accumulation of phenols and their subsequent oxidation may be of some significance in the expression of B1 resistance.

Ohata et al. (1963) found that the degree of invasion of leaf sheath epidermis in rice by *P. oryzae* was closely correlated with the total flavonoid content in the corresponding tissue; ferulic, *p*-coumaric, *p*-hydroxybenzoic and vanillic acids and two other unidentified phenolic compounds increased in the tissue around the leaf lesions. Increased peroxidase activity and decreased catalase activity take part in the oxidation of phenolic compounds in the tissue around the infected area and polyphenol oxidase is hardly detectable in rice leaves (Toyoda and Suzuki, 1960).

There is another possibility of phenol oxidation because *P. oryzae* has a unique copper-containing protein capable of oxidising monophenols, *o*, *m* and *p*-diphenols (Ogasawara et al., 1957; Tamari and Kaji, 1955, 1957). Tamari and Kaji (1955) reported that the ability of the rice plants to synthesize chlorogenic acid is related to resistance. Although the production of phytoalexin in rice leaves invaded by *P. oryzae* has been reported by Uehara (1958), it is not enough to explain the arrest of hyphal growth in Ohata's repeated experiments (Suzuki, 1965). However, these results show a strong possibility that post-infectionally accumulated phenolic compounds play an important role in resistance of rice to the B1 fungus.

Toxins

The fungus produced toxins such as α-picolinic acid, piricularin (Tamari and Kaji, 1954), pyriculol and tenuazonic acid (Iwasaki et al., 1969; Umetsu et al., 1972, 1973, 1974; Narayanarao and Suryanarayanan, 1974b) that cause natural looking B1 lesions and also inhibit the growth of rice seedlings. Of these, piricularin is the most active and semi-specific as it affects B1 susceptible rice plants more than resistant plants. Further, the toxin produced *in vitro* inhibited plumule elongation and O_2 uptake in cvs. susceptible to B1 but not in those resistant to it (Shanmugam and Ramakrishnan, 1965). Tamari and Kaji (1954, 1955) reported that the B1 fungus produced a piricularin binding protein, a copper oxide which binds piricularin and destroys fungitoxicity but not its phytotoxicity. Although the role of piricularin toxin in the disease syndrome is not well understood it is known to induce an increase in polyphenols and oxidases (Togashi et al., 1969) and its toxicity is counteracted by chlorogenic or ferulic acid (Tamari and Kaji, 1955), both present in rice plants.

Tamari et al. (1965) reported that *P. oryzae* was found to excrete piricularin into host tissue even at the infection site. One of the roles assigned to piricularin is its ability to delay the hypersensitive reactions (Takahashi, 1957b; Togashi et al., 1969) and a hypersensitive response is common in B1 resistant cvs. (Ou, 1985). This property of piricularin implied in the prolongation of symbiosis appears to be negated in resistant reactions owing to the combination of the toxin with polyphenol constituents of rice plants such as chlorogenic acid and ferulic acid (Tamari and Kaji, 1957; Kuwazuka and Oshima, 1962). This view cannot, however, explain why certain resistant cvs. are resistant to some but not to other races.

Phytotoxin(s) produced during spore germination by *P. oryzae* (Arase et al. 1990a) induced the orange-brown or 'Sekiguchi-lesions'—like necrosis on the leaves of cv. Sekiguchi-asahi but not other plants. The induction of the leaf necrosis by the toxin(s) was light dependent (Arase et al. 1990b) and effect of thermal (55°C for 15 sec.) and chemical (cycloheximide) pre-treatments also suppressed leaf necrosis formation under light (Arase et al. 1993).

Phenols

Phenol metabolism in rice plants in relation to B1 resistance has been extensively investigated by Japanese workers. A large variety of phenolic compounds was reported by them. Uehara (1958) was the first to show post-infectional antifungal substances in rice. A drop diffusate technique with detached rice leaves using conidial suspension of *P. oryzae* indicated the antifungal activity of the supernatant which was shown to be due to a phenol. Blast resistant rice cvs. are reported to contain more polyphenols than susceptible cvs. (Suzuki, 1965), whereas the reverse relationship also exists (Ramakrishnan, 1966c). No direct correlation has been found between resistant cvs. and the level of phenolic compounds (Sridhar, 1972a, b). However, the significance of cell wall phenols in resistance to rice B1 was emphasized by Kumar and Sridhar (1984). It appears that the involvement of phenolics in B1 disease resistance is not beyond doubt.

Toyoda and Suzuki (1960) studied the changes in the activities of phenols and oxidative enzymes in the rice tissue. Ayyamperumal et al. (1968) and Veerraju and Prasad (1971) revealed the presence of phenolic pre-infectional antifungal substances in rice. Veerraju and Prasad (1972) screened a number of cvs. for this inhibitor and observed that CO 4 and CO 13 possessed the highest and lowest concentrations of the inhibitor. Surprisingly enough, these cvs. are highly resistant and susceptible respectively to *P. oryzae* (in most B1 races). Etiolated rice plants, which contained only traces of phenol were more susceptible to B1 than field grown plants (Reddy et al., 1969). Qualitative differences in phenol between B1 resistant and susceptible leaves have been suggested (Wakimoto and Yoshii, 1958). Significantly, more *p*- coumaric acid and ferulic acid have been reported from resistant than from susceptible cvs. (Kumar and Sridhar, 1985).

The effect of some phenolic compounds on spore germination and growth of hyphae within the invaded tissue was studied (Ohata et al.,1966). Ferulic, *p*-coumaric and vanillic acids inhibited the growth of the germ tubes as well as the invading hyphae. On the other hand, *p*-hydroxybenzoic acid slightly promoted the growth of the germ tube as well as hyphae in the host tissue while chlorogenic acid neither promoted nor inhibited them.

Wakimoto and Yoshii (1958) and Ohata et al. (1966) reported the reduction in polyphenol contents of rice plants in response to N-fertilization. Sridhar (1972b) and Prasad and Ragunathan (1972) confirmed that N-fertilization generally decreased the phenolic compounds in both resistant and susceptible rice cvs. However, phenol synthesis was enhanced in response to *P. oryzae* inoculation and it was more pronounced in the resistant cv. than in the susceptible one.

As in many other host-pathogen combinations, B1 resistant rice cvs. also accumulate phenols more readily than susceptible cvs. when infected (Sridhar, 1972b). Prasad et al. (1972) working with three differentially virulent races on two cvs., resistant and susceptible, reported that there was more accumulation of total phenols as well as *o*-dihydroxy phenols in both cvs. infected with the least virulent race M4 than with the highly virulent races M5 and AP 37 in the early stages of disease development. Similar observations were made using different physiological races on the susceptible cv. Peta (Sridhar and Ou, 1974).

The possible involvement of aromatic compounds in disease resistance is well known. Their chemical defence may be three fold. Firstly they are present in healthy plants in concentrations sufficient to inhibit growth and sporulation of the B1 fungus and these are the preformed resistance factors. Secondly, as a response to infection, their concentrations increase markedly, restricting the invading pathogen (Ohata et al., 1966). Finally, certain post-infectional products, not normally found in healthy plants, are also synthesized after infection. The data on the accumulation of phenolic compounds in rice tissue are conflicting. For instance, Sridhar and Ou (1974) reported the ubiquitous occurrence and accumulation of phenolic compounds in B1 infected rice plants. There are instances of marked phenol accumulation in the inoculated susceptible rice cv. also (Sridhar and Mahadevan, 1979).

Phytoalexins

At the second stage of infection, resistance to spread of the pathogen in rice tissue appears to depend not so much on the constituent inhibitory environment as on the dynamic host responses that create an inhibitory effect on the pathogen. A hypersensitive response is common in B1 resistant rice cvs. Quite, early,Kawamura and Ono (1948) had suggested that inactivation or death of the invading hyphae was due to some toxic substances produced in the course of host-parasite interaction. Available anatomical evidence indicates that resistant host cells respond rapidly to hyphal invasion quite early during infection and this is followed soon by the appearance of dead cells filled with resin like substances. Tissue browning and the associated biochemical processes, namely, phenol synthesis and their oxidation, toxify the cellular environment and these seem to restrict the development of the B1 fungus. Resistant rice plants synthesize inhibitors (phytoalexins) in response to infection. Pre-formed and other post-infectionally accumulated antimicrobial substances play a crucial role in the host's 'innate defence' and their involvement in its defence mechanism is widely accepted.

It is pertinent to mention that the production of phytoalexin type toxic substances in rice plants on interaction with *P. oryzae* was first reported by Uehara (1958) using the drop-diffusate method which employs conidial suspension of *P. oryzae*. Uehara (1962) reported that rice cvs. respond differently to B1 infection in respect of production of such substance, resistant cvs. producing more than susceptible ones. Ohata and Kozaka (1967) identified the presence of antifungal substances in a mixed inoculation (inocula with avirulent and virulent spores). Lesion size was inversely related to the concentration of the avirulent spores in a mixed inocula of *P. oryzae*. Two UV fluorescing phytoalexin - like substances were detected in and around the lesion area, the one with Rf value (0.89) being more inhibitory than the one with 0.69 Rf value. These substances are also produced when rice plants are infected with other pathogens or treated with $CuSO_4$. They were also not identified with any phenols or coumarins already reported from diseased rice tissues. They appear to be specific to infected rice plants but not to the B1 disease alone. Further, from time to time many researchers (Misawa and Takahashi, 1968; Takahashi et al., 1969; Izawa et al., 1970; Nishimura et al., 1970) have reported the existence of phytoalexin - like substances in rice plant infected with *P. oryzae* but the chemical characterization of the substances has not been reported.

Cartwright et al., 1977; 1981) characterized the phytoalexins isolated from *P. oryzae* infected rice leaves treated with 2, 2-dichloro -3, 3-dimethyl cyclopropane carboxylic acid (WL 28325) as well as UV irradiated rice leaves as momilactones A and B, diterpene lactones. These compounds had earlier been obtained as plant growth inhibitors from rice husk (Kato et al., 1973, 1977) as well as antimicrobial agents against *Xanthomonas oryzae and P. oryzae* (Watanabe et al., 1979). These phytoalexins are significantly different from those earlier described, which are phenols or sesquiterpenes. Production of momilactones in resistant rice plants inoculated with an avirulent race of *P. oryzae* has been reported by Bird and Callow (1982), but Matsuyama (1983) failed to detect any such compound in rice leaves and these substances in rice husk identified as phytoalexins are doubtful. Shimura et al. (1981) reported the production of four antimicrobial substances in response to *P. oryzae* infection in probenazole treated rice plants that developed resistance. Two major substances have been identified as α-linolenic acid (cis-9, cis-12, cis-15) and octadecartienoic acid while the remaining two are of minor importance (Sekizawa and Mase, 1980).

The production of momilactones increased when rice plants were pretreated with WL 28325 prior to infection with B1 pathogen and momilactone B was scarcely observed in untreated B1 leaves (Cartwright et al., 1980). Kumar and Sridhar (1984) reported phytoalexin production from rice - *P. oryzae* interaction as well as induction by wounding.

Later, Akatsuka et al. (1983; 1985) and Sekido et al. (1986) reported the novel phytoalexins, oryzalexins A, B, C and D from B1 - infected rice leaves. These were confirmed as (+) - sandracopimaradiene derivatives by chemical and MS, IH-NMR studies for orzalexins A, B, C and D as 3-oxy—7–oxy— (I), 3-oxy-7-oxy- (II), 3,7-dioxo- (+) - sandracopimaradiene (III) and 3, 7-dihydroxy-(+)- sandracopimaradiene respectively (Kono et al., 1985). Mori and Waker (1985) synthesized the oryzalexins A, B and C. The identification and semi-quantitative analysis by mass chromatography of oryzalexins A, B, C and D from leaf B1 lesions of rice cv. Koshikari were found to be at 75, 40, 180 and 200 µg/3g fresh weight respectively, while they were not detected from non-lesion leaf parts (Sekido et al., 1987). Further, Sekido and Akatsuka (1987) studied the inhibitory effect of oryzalexin D against *P. oryzae* and concluded that it showed the strongest inhibition in a group of oryzalexins against spore germination. A positive correlation between its concentration and the resistance of older lower leaves has been demonstrated (Matsuyama and Wakimoto, 1985).

The production, purification and characterization of anti-B1 substance (S-1) were further reported from B1-resistant lower rice leaves and also from B1 infected and UV irradiated rice leaves which are associated secondarily in disease resistance (Matsuyama and Wakimoto, 1985, 1987). They were identified as diterpenoid compounds and one such is distinctly different from formerly reported oryzalexins (Matsuyama and Wakimoto, 1988).

While using drop diffusate technique, Kumar and Sridhar (1984) observed that resistant plants produced phytoalexin only when injured and such production was enhanced by the pressure of B1 pathogen. The response of susceptible leaves to injury alone or to injury plus pathogen was less significant. Since the extract of infection drops from uninjured resistant leaves showed only traces of fungitoxicity even 72 h after inoculation they expressed doubt as to the role of such phytoalexins in host resistance to initial establishment of the pathogen.

Recently, Kumar and Sridhar (1993) reported the influence of N-fertilization and chemicals that alter the host metabolism influenced the phytoalexin production. There was accumulation of phytoalexin with α-picolinic acid in resistance cv. alone but with probenazole treatment, both resistant and susceptible cvs. showed marked accumulation. These results suggest that probenazole does not induce phytoalexin accumulation as such but may act as a 'sensitizer' to activate it.

Enzymes

Changes in phenol content may depend on alterations in the activity of key enzymes responsible for their biosynthesis. Phenylalanine ammonialyase (PAL) is probably important in the synthesis of various phenolic constituents including flavonoids, isoflavonoids and hydroxycinnamic acids. The biosynthesis of phenylpropanoids from phenylalanine and tyrosine is initiated through the action of PAL and tyrosine ammonialyase (TAL) respectively, peroxidase (PO) functioning terminally in the process. Matsuyama and Diamond (1973) have shown lower levels of PAL (32%) and TAL (22%) activities in rice plants grown with high levels of N-fertilizers (Table 10). Manibhusanrao et al. (1988) summarized the role of phenolics as well as PAL, TAL and other oxidative enzymes in host resistance against potential pathogens including B1 pathogen.

Table 10 : Activity of PAL and TAL in rice plants with N-fertilized rice plants (Matsuyama and Diamond, 1973)

Treatment	Expt. 1	Expt. 2	Expt. 3
		PAL activity*	
N-fertilized	0·185	0·150	0·165
Control	0·235	0·235	0·267
		TAL activity #	
N-fertilized	0.141	0.134	0.094
Control	0·176	0·157	0·141

* OD of *t*-cinnamic acid at 268 nm
\# OD of *p*-coumaric acid at 333 nm

The presence of phenolic compounds in higher plants, their oxidation following injury, either mechanical or due to infection, and the relatively high toxicity of the oxidation products have drawn the attention of scientists for many years. The possible relationship of these factors to hypersensitivity has prompted many workers to ascribe a role to polyphenoloxidase or DOPA - oxidase (PPO) in host resistance. The intense browning characteristically associated with B1 resistance reaction in rice plants obviously suggests an involvement of PPO. Reports on PPO production by B1 pathogen (Neufeld et al., 1958; Tamari et al., 1965) as well as germinating conidia (Sridhar and Ou, 1972) are available. However, only traces of PPO have been reported from healthy rice plants (Oku, 1967) and B1 infected rice leaves (Toyoda and Suzuki, 1960) and any possible role for it in B1 resistance has even been discounted (Sridhar, 1972a). Nevertheless, Muralidharan (1974) has related PPO to B1 resistance. Recent studies have shown that suseceptible rice cv., substantially protected from B1 under field conditions by seed treatment with cupric chloride, sodium selenite and IAA, responds to infection with significantly greater increases in both total phenol level, peroxidase (PO) and PPO activity, the increases being closely correlated with the levels of resistance induced (A.K. Sinha - personal communication).

Activation of PO has been a general response to infection and has often been implicated with host resistance (Farkas and Király, 1958, 1962). Enhanced PO activity has been reported from B1 infected rice plants (Toyoda and Suzuki, 1960; Sridhar, 1978). Much greater PO activity in cv. Peta was reported when infected with a mildly virulent race of *P.oryzae* than when infected with its virulent race (Sridhar and Ou, 1974). Increased susceptibility of rice plants to high doses of N-fertilizers has been linked with reduction in PO activity (Sridhar, 1972a). This may be due to an increase in PO isoenzymes in B1 infected rice leaves due to *de novo* synthesis (Fig. 16) as reported by Sridhar (1978).Matsuyama et al. (1977) reported striking differences in the PO activities of *Pyricularia* spp. from various host species.

Root application of probenazole, a systemic compound, provided rice plants with lasting protection from B1 (Watanabe et al., 1979), and the protected plants responded to *P. oryzae* infection hypersensitively, the arrested hyphal growth being associated with an increase in activity of certain enzymes involved in lignin synthesis including PO (Iwata et al., 1980). Further, Wan et al. (1986) and He et al. (1989) found that resistant cvs. had one more enzyme band in the isoenzyme spectrum than in susceptible ones. The browning reaction is attributed to increased PO activity in the infected tissue but contradictory views are also available. Thus, sound proof for a PO role in the resistance of rice to B1 is still lacking.

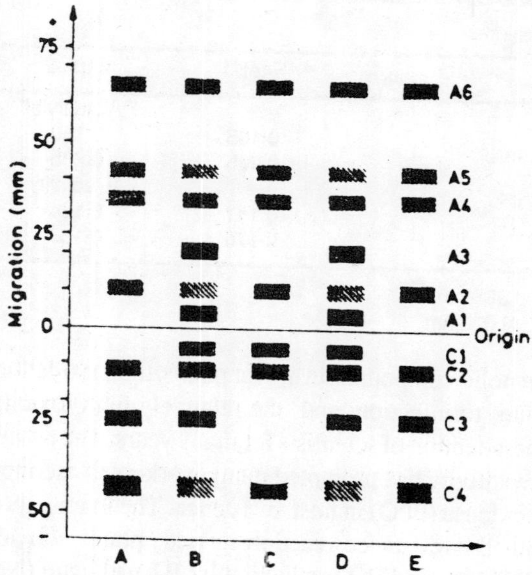

Fig. 16: Electrophoretic pattern of peroxidase isozymes. Leaf extracts from healthy (A), and inoculated (B), intact plants and healthy water treated (C), inoculated water treated (D), and inoculated actinomycin- D treated (E) leaves of cv. Khao-teh-haeng 17 (Sridhar, 1978).

Increased application of N-fertilizers to rice plants augmented the level of free ascorbic acid and decreased the ascorbic acid oxidase activity of leaves (Sridhar, 1972 a). Further, Sridhar and Ou (1974) reported that the depletion of ascorbic acid was high along with an increase in ascorbic acid oxidase in the cv. Peta infected by the less virulent races (P 26 and P 149) of *P. oryzae,* hence the resistant response. In contrast, depletion of ascorbic acid and an inhibition of ascorbic acid oxidase during pathogenesis characterised the interaction between the cv. Peta and virulent races (P 38 and P 150), to which the cv. was susceptible.

Ramaswamy (1979) studied the role of phenolics and lignin in rice B1 disease and suggested that the defence mechanism mediated through phenolics and lignin appeared earlier in resistant than in susceptible reactions. Increased lignin accumulation at the site of infection has been implicated in the expression of probenazole induced resistance in rice to B1 (Iwata et al., 1980). Treated plants developed restricted lesion like the resistant plants. Earlier, Sridhar et al., (1973a, b) reported that lipid bodies in rice leaves discoloured before the appearance of visible browning during B1 infection.

Permeability changes

Increased permeability damage to the host cell membrane is indicated by the leakage of ions as large quantities of electrolytes are released. At the same time, the cell loses the ability to accumulate mineral salts. Earlier, Ito and Sakamoto (1939; 1940) and Sakamoto (1948) showed that ammonium ions increased the water permeability of epidermal cells of rice which decreased their puncture resistance to invasion by the B1 fungus *P. oryzae.*

The leakage of ions has been detected at the initiation of B1 infection (Dube,1971). Ambalanathan (1981) showed that $(NH_4)_2SO_4$ and NH_4Cl increased tissue permeability to a greater extent in the B1 susceptible cv. HR 12 than in the resistant cv. IR 8 and this effect was pronounced even with a very low concentration (N/10,000). He, further, reported that in light α - picolinic acid induced a pronounced increase in electrolyte leakage from the susceptible cv. (HR 12) but not from the resistant (IR 8) leaves. In darkness, however, high rates of electrolyte, phenol and amino acid leakage were evident with the toxin treated resistant, but not with susceptible leaf tissues. Not only ammonium salts but also the toxin α- picolinic acid and the enzyme lipoxygenase (ECI. 13.1.13) caused differential permeability changes in the leaves of B1 resistant and susceptible cvs. However, the pattern of electrolyte leakage induced by lipoxygenase in the leaf tissue of these two cvs. appears to be reversed in the presence of ammonia.

Lokeswari (1987) observed changes in permeability of resistant and susceptible rice leaf tissues also when treated with two more toxins of *P. oryzae*, namely, tenuazonic acid and pyriculol besides α- picolinic acid. Although increasing N nutrition increased the permeability of the resistant IR 8 leaves but decreased it in the susceptible cv. HR 12, contrary effect was observed in toxin treated tissues. This anomaly has been explained on the basis of making of electrolyte levels in the ambient solution by the host factor leaking out faster and in larger quantities from toxin-treated resistant than from susceptible tissues.

Sekido and Akatsuka (1987) studied the effect of oryzalexin D on the cell membrane permeability in the B1 fungus. It scarcely inhibited the respiration, caused leakage ot K and inhibited the uptake of glutamate by mycelial cells. These results suggest that interference with the cell membrane function is respcnsible for the primary mode of action of oryzalexin D against B1 pathogen.

Disease Resistance

Disease resistance in rice plants generally depends on their ability to tolerate the B1 pathogen or restrict its development. Here, the role of preformed toxic materials, oxidative and biosynthetic enzymes that appear to confer resistance was discussed. Changes in disease reaction from susceptible to resistant or *vice-versa* can be induced by pretreating plants with a variety of physical and chemical agents, depending on the rice-B1 pathogen combination involved. Such induced resistance has attracted much attention in recent years.

Cross-protection

Ohata and Kozaka (1967) and Kiyosawa and Fujimaki (1967) reported that a mixture of avirulent and virulent strains of *P. oryzae* produced smaller lesions on certain rice cvs. as compared to the virulent strain alone. Hilda and Suryanarayanan (1976) reported cross-protection in the B1 disease of *Panicum repens* using incompatible isolates. The protection achieved was explained in terms of competition between the isolates for infection sites and the interaction between the host and the first organism. Further, the involvement of preformed and post-infectional antifungal substances in the compatibility of *Pyricularia* spp. with *P. repens* was studied (Hilda and Suryanarayanan, 1977).

The sectoring in *P. oryzae* colonies and the association of mycovirus (Ferault et al., 1971) may yield avirulent races of *P. oryzae* which can be used for cross-protection/ immunization of rice plants against virulent races of *P. oryzae*.

Phyllosphere microflora

Most phyllosphere microorganisms on rice plants are saprophytes and some may influence the outcome of B1 infection. A yellow gram-positive bacterium, often associated with *P. oryzae*, reduced the incidence of the disease but not the size of the leaf spot (Kitani, 1941). However, the contribution of phyllosphere microflora to B1 disease development has never been critically investigated.

Induced resistance

Any innovative approach to disease management must depend upon an understanding of the damage the B1 fungus causes and how the rice plants limit the damage. Resistance to *P. oryzae* in rice is expressed both during pre-penetration and post-penetration stages of pathogen development and rice plants have both constitutive and inducible resistance mechanisms (Sridhar et al., 1988). Induced resistance in the host surface, although incomplete, may have an epidemiological significance in reducing the amount of potential inoculum, consequently limiting the disease spread. Activation of the innate defence mechanisms of the plants using biotic (microbial) and abiotic (non-toxic exogenous chemicals and elicitors of microbial and host origin) agents will protect them against invading pathogen. The mechanisms reported to be associated with induced resistance are phytoalexins, enzymes such as chitinases, β-1-3- glucanases, proteases, PO, non-enzymatic antimicrobial proteins, peptides, lignin, callose, hydroxyproline - rice glycoproteins and pathogenesis- related (PR) proteins, etc.

A few reports are available on the biological induction of B1 resistance in rice by prior inoculation with a biotic agent or treatment with fungal metabolites. Root application with mycelial extract of *Cephalothecium* sp. (Yoshii, 1950) and seed treatment with culture extract of *P. oryzae* (Kumamato, 1948) developed in rice plants had appreciable resistance to B1. Prior inoculation with the less virulent race of *P. oryzae* was also effective in developing the more virulent race (Ohata and Kozaka, 1967).

Iwano (1987) induced resistance by pre-inoculating rice plants with a spore suspension of incompatible race before the natural infection. This suppressed remarkably the leaf panicle B1 infection. The same effect was also observed in field trials. Although leaf B1 was suppressed by pre-inoculating any incompatible B1 races, there were differences in the degree of leaf B1 suppression among the isolates used. The maximum leaf B1 suppression was observed when isolates inciting the very small eye spots with brown margins were used as the inducer agent.

Further, spraying foliage or treating seeds with a wide range of chemicals such as heavy metals, metabolic inhibitors, amino acids, plant growth regulators and miscellaneous organic compounds has been shown to induce resistance to B1. Effective protection of rice from B1 disease was achieved by treating grain with compounds of the dichlorocyclopropane group and was associated with higher PO, IAA- oxidase, and catalase activity of infected plants leading to more rapid formation of melanin (Langcake and Wickens, 1975) as well as tenuazoic acid (Lebrun et al., 1984). Inhibition of melanin biosynthetic reactions in *P. oryzae* by certain compounds that prevent the rice B1 was reported (Wheeler and Greenblatt, 1988).

Similarly, susceptible rice seedlings raised from seeds treated with dodecyl DL-alanine hydrochloride (Arimoto et al., 1976), 2, 2, dichloro 3, 3 -dimethyl cyclopropane (DDCC) (Langcake and Wickins, 1975), seedlings treated with systemic fungicidal cyclo-propane derivatives (Cartwright et al., 1980) or a systemic fungicide compound probenazole (Watanabe, 1977; Watanabe et al., 1979; Shimura et al., 1981; Kumar and Sridhar, 1990) also become resistant to the B1 fungus. In field experiments under upland conditions, substantial control of B1 disease was achieved by treating seeds with a group of phytoalexin inducer chemicals. Protective effects recorded with cupric chloride, ferric chloride, cysteine, cycloheximide, sodium malonate and *p*.-chloromercuric-benzoate were particularly strong and persistent (Sinha and Sengupta, 1986).

Pathogenicity and virulence

Two terms *pathogenicity* and *virulence* have been used in the literature for many years and are often confused. Pathogenicity is the ability to cause disease. It is often defined as the property of a pathogen that allows it to incite disease. Virulence is the property of the pathogen that determines the severity of the disease or disease-inducing capacity. In other words, host resistance and pathogen virulence are complementary aspects of the host-pathogen interactions and each must be assessed in relation to the other. For instance, if isolate 'A' of a pathogen induces severe disease in a host and isolate 'B' induces negligible disease, the host can be said to possess low resistance to 'A' and high resistance to 'B'. Conversely, if the hosts are genotypically identical it may be said that isolate 'A' is virulent and isolate 'B' is non-virulent or avirulent. Virulence is quantified by differences in disease severity and generally expressed as disease severity index (DSI).

Geovirulence and its implication

Earlier Matsumoto et al. (1969) indicated the type of distribution of virulence of

P. oryzae from the tests conducted from the multilocation IRBN nurseries located at different geographical area. Subsequently, Ahn et al. (1988b) proposed the term "geovirulence" to refer to the overall virulence patterns of rice B1 population in a testing site shown by qualitative reactions of a set of diverse groups of cvs. included in the IRBN tests. Four major distinct geovirulence patterns of rice B1 population were found based on reactions at 24 locations in 14 countries (located in both the hemispheres) through I and IV, common in East Asia, Nepal and Italy. These have narrow ranges of virulence; patterns III and IV, found in South and South-East Asia, Latin America, and some parts of China, have wider ranges (Fig. 17). Further, where mostly *japonica* cvs. are grown pattern I and II are prevalent, while regions where *indica* cvs. are grown represent pattern III and IV (Ahn and Sehu 1990). A co-evolution among the B1 pathogen population and rice genotypes over wide geographical locations with the geovirulence pattern was correlated.

Resistance phenomenon

It is important to get a clear picture of the meaning of the term "resistance". Resistance does not necessarily mean freedom from disease, rather it is a quantitative phenomenon, applied on the basis of the magnitude of disease present on a plant or cv. and is the inverse of susceptibility. Thus, it can be high, moderate or low which correspond respectively with low, moderate and high susceptibility. If the inoculated plants are totally free from disease development they can be regarded as totally resistant or immune.

Some important relevant definitions may be considered here. Resistance is defined as the ability of the host plant to suppress or restrict the growth of the pathogen. It is also expressed as the ability of a pathogen to withstand or oppose the operation of or to lessen or overcome the effects of an injurious pathogenic factor such as a fungicide. Most plant pathologists often describe resistance in their own terms but generally these do not correspond precisely. Generally, resistance may be described in terms related to the mode of inheritance. More details are given under horizonal/vertical resistance.

Crill and Khush (1979) described a susceptible rice plant or cv. as one which develops lesions (diseased symptoms) when exposed to a pathogenic race of the B1 fungus within the environment suitable for disease development and a resistant plant or cv. as one which does not develop these symptoms under the same conditions. In this definition, resistance is considered absolute as measured by symptom development and the plants with typical symptoms are considered susceptible and those without them as resistant. The expression of resistance in rice to B1 is complex as it involves the host, the pathogen *(P. oryzae)* and environment ideal for infection and disease development.

Inheritance of resistance

Studies on the inheritance of resistance to leaf B1 began in Japan in 1917 by Sasakii (1923) followed by Takahashi (1965) and Yamasaki and Kiyosawa (1966). Most workers (Nakatomi, 1926; Hashioka, 1950a; Okado and Maeda, 1956; Abemiya, 1959) employed different methods and different strains of the fungus and no comparison was possible between their experimental results. Subsequently, most of the systemic work was carried out to analyze genetically world rice cvs. for B1 resistance by several workers (Kiyosawa, 1967a, b; 1968; 1969a, b; 1970; 1972a, b; 1974a, b; 1976b; 1981; Kiyosawa and Cho 1980;

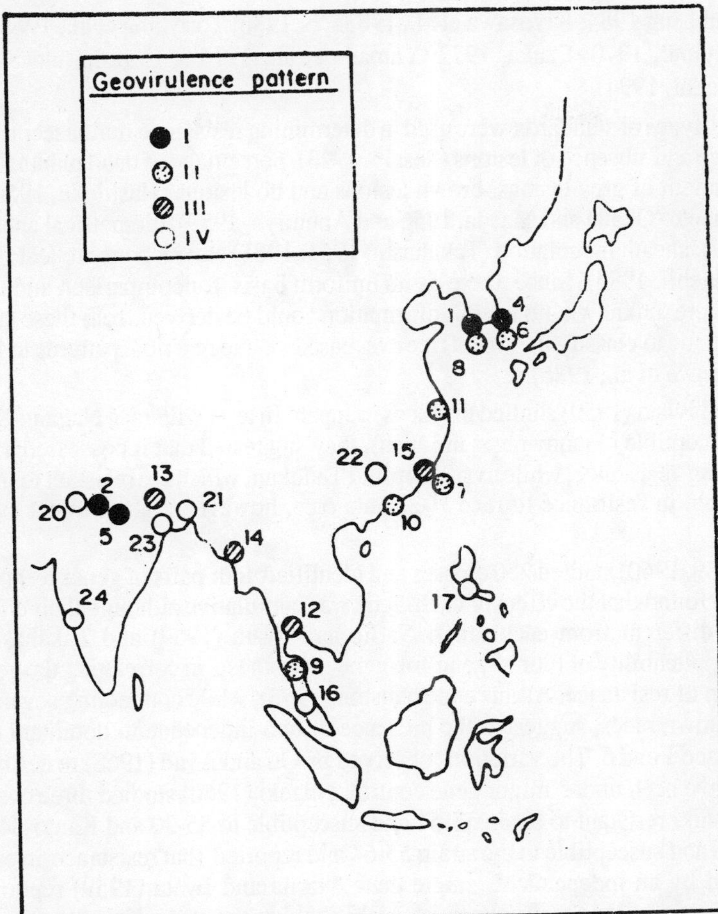

Fig. 17: Classification of geovirulence patterns of 24 test locations based on Ward's cluster analysis of leaf B 1 reaction.

Location :

Pattern I

1. Suweon, Korea
2. Parwanlpur, Nepal
3. Moralara, Italy*
4. Icheon, Korea
5. Kankal, Nepal

Pattern II

6. Mllyang, Korea
7. Chia- YI, China
8. Irl, Korea
9. Kuangut, Thailano
10. Guangzhou, China
11. Hangzhou, China

Pattern III

12. Bangkhen, Thailand
14. Yezin, Burma
13. Bhur, Bhutan
15. Fuzhou, China

Pattern IV

16. Bumbong LIma, Malaysia
17. Los Baños, Philippines
18. Golanla, Brazil*
19. VIllavicenclo, Colombia*
20. Almora, India
21. Shillong, India
22. Changsha, China
23. Joydebpur, Bangladesh
24. Lonavala, India

* Not shown in the map (Ahn et al. 1988b).

Kiyosawa and Ling 1984; Kiyosawa et al., 1983a, b; 1986; Toriyama et al., 1968; Shinoda et al., 1971; Goto, 1970; Ezuka, 1972; Yamada et al., 1976 a; Higashi and Kushibuchi, 1978: Purba et al. 1994).

Different types of standards were used in determining resistance and susceptibility such as the presence and absence of lesions (Sasaki, 1923), percentage of dead plants (Nakatomi, 1926), assessment of grey lesions, brown lesions and no lesions (Hashioka, 1950a), lesion types y, b, g, p etc (Okada and Maeda, 1956 and Abumiya, 1959), biometrical analysis (Oka and Lin, 1957), sheath inoculation (Takahashi, 1951; 1965) and response to leaf protoplasts (Nomura and Ishii, 1985). Since there is no uniform basis for comparison and the fungus strains used were unknown, no useful information could be derived from these studies. An attempt was made to classify the world rice cvs. based on the reaction patterns to B1 fungus strains (Kiyosawa et al., 1986).

Narita and Iwata (1959) studied the races in Japan. In tests with race Nagano-89, against cv. Asahi (susceptible to many races in Japan), they suggested that it possessed a dominant gene controlling resistance, while cvs. Tetep and Tadukan, which are resistant to most races, are intermediate in resistance to race H-2. This race, however, cannot infect cv. Ishikari Shiroki.

Goto (1959, 1960) studied 250 crosses and identified four pairs of genes responsible for resistance and found that the effect of each gene was cumulative although their effects were more or less different from each other. Narita and Iwata (1959) and Takahashi (1965) reported the applicability of Flor's 'gene-for-gene' hypothesis in correlating the gene action and expression of resistance. Atkins and Johnston (1965), while conducting several experiments with known races, suggested the presence of two independent, dominant genes for resistance to race 1 and 6. The variations observed by Ou and Ayad (1968) to define various races might have been under minor gene control. Niizeki (1960) studied three cvs. - Aichi-asahi, Ayanishiki resistant to strain 55-54 and susceptible to 55-20 and Kanto-54 resistant to strain 55-20 and susceptible to the strain 55-64 and reported that resistance in each strain was controlled by an independent single gene. Narita and Iwata (1959) reported that a dominant gene controlled the resistance of Aichi-asahi to the strain Nagano -89.

Yamasaki and Kiyosawa (1966) studied in detail the inheritance of resistance to B1. They grouped B1 resistant Japanese and Chinese cvs. into five types, namely, Shin 2, Aichi-asahi, Kanto 51, Isikari -shiroki and To-to - based on the reactions of these cvs. to seven isolates representing seven of the 14 races found in Japan. Kiyosawa (1967 a, b) identified two genes *(Pi-a* and *Pi-ta)* in cv. Pi No. 1, another two (*Pi-a* and *Pi-z*) and one *(Pi-ta*$^{2)}$ in Pi No. 4. Further, Kiyosawa (1968 a) studied the genotype of the fungus isolates from the results of genetic analysis of rice cvs. belonging to Aichi-asahi, Kanto 51 and Ishikari-shiroki types based on the gene-for-gene hypothesis and designated them as *Pi-a*, *Pi-k* and *Pi-i* respectively. Genes *Pi-k* and *Pi-i* showed variations while gene *Pi-a* showed complete dominance and all the three behaved independently. A number of genes were identified; *Pi-m* in cv. Minehikari, *Pi-a* and *Pi-k* in To-to group of cvs. (Kiyosawa, 1968 b), *Pi-k*^{s} in Shinriki (Kiyosawa, 1969 a), *Pi-a*, *Pi-k* and *Pi-k*^{p} in cv. Pushur (Kiyosawa, 1969 b) and *Pi-k*^{h} in HR-22 (Kiyosawa and Murty, 1969).

Kiyosawa and Yakoo (1970a, b) and Asaga et al. (1975) reported that resistance of Toride-1 was controlled by a dominant gene *Pi-z*^{t}, while resistance of Toride-2 to five fungus isolates (P-2b, Ken 52-33, Hoku-1, Ken 54-20 and Ken 54-04) was again controlled by *Pi-z*^{t}. Shinoda et al. (1971) reported the linkage group of a newly found B1 resistance gene *Pi-s*. Yamada et al. (1976) indentified a single resistant gene for rice B1 fungus in nine differential cvs. They used the Gilmour's octal system for numbering the races as Shin *(Pi-k^5)*, Aichi-asahi *(Pi-a)*, Iskikari-shiroke *(Pi-i)*, Kanto 50 *(Pi-k)*, Tsuyuake *(Pi-m)*, Fukunishiki *(Pi-z)*, Yashiromochi *(Pi-ta)*, PI 4 *(Pi-$ta^{2)}$* and Toride 1 *(Pi-$z^{t)}$*. The inheritance of resistance was studied in crosses Sensho X/H (Goto, 1970) and cv. Reiho (Inbe and Matsumoto (1980) and the upland cv. Sensho (Higashi and Saito, 1985) to the B1 fungus.

Kiyosawa, 1972a, 1981, Kiyosawa and Cho, 1980 Kiyosawa and Ling 1984, Kiyosawa et al., 1986, Marchetti et al., 1989, He et al., 1989 and Purba et al. 1994 in a series of experiments identified several genes in exotic cvs. from various countries and the respective genes identified are summarized in Table. 11. To date, 14 genes *(Pi-a, Pi-i, Pi-k^s, Pi-k, Pi-z, Pi-ta, Pi-ta^2, Pi-z^t, Pi-k^p, Pi-k^m, Pi-k^n, Pi-b, Pi-t* and *Pi-?)* have been indentified (Kiyosawa, 1974a, 1976b, 1978; Inbe and Matsumoto, 1985; He et al., 1989). Resistance in upland cvs. could not be explained by these 14 identified genes indicating that additional resistance genes may be present in them (He et al., 1989). Minor genes have been reported in upland cv. Norin Mochin 4 (Abe et al., 1974), the US cv. Dawn (Kiyosawa, 1974b) and IRRI culture IR 1905-81-3-1 (Kiyosawa et al., 1983b) and upland cv. Haunaihuhn (He et al., 1989).

In the US, Jennings (1963) identified monogenic resistance in a cross (Cheianung 242 (R*) × Tjere Mas (S**) against the IRRI isolate 1-146 with one or more modifying genes. Further, Atkins and Johnston (1965) designated the resistance gene in Northrose (also Nato) as *Pi-1* and that in Zenith and Gulfrose as *Pi-6*. They have also studied the crosses Northrose × Zenith, Northrose × Gulfrose and Northrose (Nato) × Gulfrose inoculated with US race 1 and 6 found that Zenith and Gulfrose were susceptible to race 1 and resistant to race 6 while the reverse was the case with Northrose (Nato). Jodon and Atkins (1966) reported duplicate genes for resistance (CI 9541 × PI 264817) to the US race 6 while Rosero (1967a) reported single dominant major gene to races 2, 6, 16 with minor effect on cvs. Colusa and Zenith. Earlier, Tepora and Jennings (1966) identified five genes *(Pi-a, Pi-b, Pi-c, Pi-d, Pi-e)* for leaf B1 resistance.

Eroutor et al., 1979 reported that resistance to races IB- 54 and IG -1 was controlled by a single dominant gene in the cv. Lebonnette. Further, Marchetti et al., 1986 studied inheritance in the US rice cvs., identified and characterised genes for resistance to local rice pathotypes of *P. oryzae*. The new recessive gene *Pi-n* as well as *Pi-k^s* were identified in Brazos (resistant to race IB-54); *Pi-g* in Gulfrose (IG -1), and *Pi-d* in Lebonett (IB -1). The *Pi-z* gene was also identified in Gulfrose (IC -17, IG -1 and IH -1), *Pi-k^s* in Vista, and *Pi-k^h* (IB-45, IB-54, IG-1, and IH- 1-0?).Two unnamed dominant genes *Pi-?* (IH -1) independent of *Pi-z* and *Pi-k^h* (IB -49) were identified on cvs. PI 331581 and Usen respectively.

In India earlier, Ramiah and Ramasamy (1936) studied the inheritance of resistance in cvs. GEB 24, CO 4 (R) and Korangu Samba (S). In the cross with GEB 24, inheritance appeared to be simple while in the cross with CO 4 it was more complex. Rath and Padmanabhan (1972, 1973) and Padmanabhan et al. (1974) reported that resistance in the three cvs. Zenith, Tetep and Tadukan was controlled by three dominant genes and the presence of any two of these conferred the resistance. While crossing Zenith and Tetep with cv. PI 180061 (S) or Yakeioko (S), Rath and Padmanabhan (1972) found it to possess one dominant gene conferring resistance to race IA -11 and ID -1 biotype C. The gene for resistance present in Zenith was independent of that of Tetep. Later, Padmanabhan et al. (1973) studied the inheritance of B1 resistance in the F_1 and F_2 generations of 10 crosses involving Zenith, Tadukan and Tetep (R) S 67 (R), S 67 (MS)+ and CO 13 and CI 5390 (S),

---

* R = resistant, ** S= susceptible to blast, +MS = moderately susceptible to blast.

Table 11 : Resistance genes identified in some exotic cultivars (Kiyosawa, 1972, 1981 ; Kiosawa and Ling 1984 ; Marchetti et al. 1986 ; He et al. 1989; Kiyosawa and Cho 1980; Kiyosawa et al. 1986 and Purba et al. 1994)

Origin	Cultivar	Genes identified
Korea	Dozai chall	Pi-i
	Jae keum	Pi-a
	Pal tal	Pi-a
	Tongil	Pi-a,Pi-b
	Myliang 23,	Pi-b
	Packyeong	Pi-b
	Iri 333,	Pi-b
China	Usen	Pi-a, others
	Yakei-ko	Pi-k
	Reishiko	Pi-k
	To-to (short grains)	Pi-k, Pi-a
	Choko-to	Pi-k, Pi-a
	Hokushi Tami	Pi-k, Pi-a, Pi-m
	Pe Bi Hun	Pi-a
	To-to (long grains)	Pi-k^{s}
	Taichung 65	Pi-k^{s}
	Sha-tiao-tsao	Pi-k^{s}
	Oka-ine	Pi-ta
	Pai-kan-tao	Pi-ta, others
	Nan 56	Pi-z, Pi-k
	Zong Hua-4	Pi-z
	Zong Hua-78883	Pi-z
	Zong Hua-7829	Pi-i
	Dong nong-363	Pi-a, Pi-k
	Hejiang-18	Pi-a
	Haunaihuhn-A	Pi-i
	Kanto-51	Pi-k
	Du dao	Pi-a
	7801	Pi-b
	C4-63	Pi-b
	Sifeng	Pi-t
	Zhaitang	Pi-t
Philippines	Tadukan	Pi-ta and/or Pi-ta^{2}
IRRI	IR 24	Pi-b
	IR 36	Pi-b
India	HR 22	Pi-k^{h}, others
	CO 25	Pi-z^{t}, Pi-a, others
	TKM 1	Pi-z^{t}, others
	Charnack	Pi-k*, others
	CO 4	Pt-z^{t}
	CR 44	Pi-b
Pakistan	Dular	Pi-k*, others
	Pusur	Pi-k^{p}, Pi-a, others
Vietnam	Te-tep	Pi-k*, others
	Morak Sepilai	Pi-z^{t}
	Kontor	Pi-z^{t}
Thailand	Leuang Tawing 77-12-5	Pi-z^{t}
	Chao Leuang 11	Pi-z^{t}
USA	Zenith	Pi-z, Pi-a
	Caloro	Pi-k^{s}
	Lacrosse	Pi-k^{s}
	Blue Bonnet	Pi-a
	Brazos	Pi-n, Pi-k^{s}
	Gulfrose	Pi-g. Pi-z
	Lobennet	Pi-d, Pi-k^{h}
	Vista	Pi-k^{s}
	PI-331581	Pi-z
	Dawn	Pi-k^{h}
USSR	Roshia No. 33	Pi-k *
Indonesia	Tjina	Pi-b, others
	Tjahaja	Pi-b, Pi-t, others
	Bengawan	Pi-b, others
Malaysia	Milek Kuning	Pi-b, others
	Kontor	Pi-z^{t}
France	MM 10	Pi-z^{t}

*Any allele at the pi-k locus

using race ID-1 of *P. oryzae*. Kaur et al. (1976) developed a high yielding cv. Ratna (R) a cross (IR 8× TKM 9).

In China, inheritance studies in B1 resistance to rice began relatively late in 1980 with a tentative set of local differential cvs. Lin et al. (1980) studied 16 crosses between R and S parents and found that 12 crosses with Toride 1, Toride 2 and Pi 5 as the R donors segregated into a ratio of 3:1, indicating that the resistance is controlled by a pair of dominant genes. Four other crosses with Pi 1, Kungyi and Hunger as R donors had the F_2 segregation into a ratio of 9:7, indicating that two pairs of complementary genes are involved. Wan et al. (1986) crossed ten R cvs. with five S ones and the majority of the F_2 population indicated non-cytoplasmic inheritance. Shen et al. (1986) reported that resistant cvs. such as Gaitang, Hongjianzhan, Tetep and 745461 BG had one pair of dominant genes and were closely linked and all allelic, while Zhang and Shen (1986) using three R cvs. (IR 2061, IR 29 Gemali) and a Chinese race ZB 15, reported that resistance was governed by non allelic duplicate dominant genes. Duan et al. (1987) worked with 21 crosses and found F_2 populations of 12 crosses segregated into a 3:1 ratio, those of two into a 15:1 and those of seven crosses into 9:7, indicating that their resistance was governed by one pair of dominant genes, two pairs of duplicate dominant genes and two pairs of complementary genes respectively.The progress in B1 resistance research in China was reviewed recently (Zhuge et al. 1990).

Kiyosawa and Ling (1984) reported resistance genes in some Chinese differentials using Japanese fungus isolates and identified *Pi-a* and *Pi-k* in Dongnong 363, *Pi-k* in Kanto 51 and *Pi-a* and *Pi-i* with Hejiang 18. Duan et al. (1987) identified the gene constitutions of B1 resistance in 36 indigenous rice cvs. in Yunnan province using Japanese isolates and found that 24 cvs. had *Pi-z*^{t}, five *Pi-a* and *Pi-k*, three *Pi-a* and four *Pi-k*^{s} *isolates Pi-a*, *Pi-i* and *Pi-k* genes.

In Taiwan, Ou and Lin (1958) reported that resistance to leaf B1 was dependent on one or more incompletely dominant genes. Wo (1965) studied eight crosses with four Tiawanese belonging to three physiological races and reported that three genes conditioned B1 resistance. Hsiesh and Chien (1967) and Hsiesh et al. (1967) identified four genes for resistance in many cvs. and designated them as *Pi-4*, *Pi-13*, *Pi-22* and *Pi-25* as per the reactions obtained with the respective races. Ezuka (1973) indentified one *Pi-i* gene each in cvs. Akishino-mochi and Takasago-mochi instead of *Pi-a* alone as was reported earlier.

Maximoo (1969) and Azizi (1972) working in Egypt studied 19 crosses made between some local and introduced cvs. and reported three pairs of dominant genes governing resistance to B1. The genctic study on the resistance of B1 to cv. Tongil was studied in Korea (Kowon et al., 1975). Other genetic studies in B1 resistance include those by Rosero (1967b) and Eruotor et al. (1978).

At IRRI, Mackill et al. (1985ab) studied the genes for resistance to the Philippine isolates of *P. oryzae* as well as some *indica* rice cvs.

The inheritance of neck B1 resistance has been studied by several workers (Hashioka, 1950b; Abumiya, 1959; Ou and Nuque, 1963; Padmanabhan, 1965a; Chang et al., 1965; Samoto and Ouchi, 1968; Azizi, 1972). Most workers concluded that resistance was dominant in some crosses and recessive in others. Genes controlling leaf B1 resistance were independent of those controlling neck B1. However, there is always a correlation between leaf and neck B1.

Field resistance to B1 is controlled by minor genes (Hsieh and Chen, 1967; Hirano, 1967), major genes (Kiyosawa et al., 1967) and the penetration and establishment phase by polygenes (Rath and Padmanabhan, 1973) while resistance to the spread of infection was again controlled by major genes (Kaur, 1972).

Much superior genotypes for various traits inducing to B1 resistance have been identified and genetic analysis has been made to explain the nature of inheritance of resistance. However, precise knowledge on the hereditary genetic components in rice cvs. through quantitative analysis is lacking. Perhaps more literature is available on B1 inheritance of resistance than on other diseases of rice. Still many challenges remain.

Gene action and expression

Takahashi (1965) proposed a working hypothesis for explaining the relationship between gene action and expression of resistance for rice B1 disease (Fig. 18). It was assumed that host cv. I is most resistant and IV highly susceptible, while II and III are intermediate. Each of the *P. oryzae* isolates A, B, C, and D has a different degree of virulence controlled by resistance genes in the host. It was further assumed that A has several pairs of genes controlling B1 resistance in rice cvs., B indicate there are several races of pathogen specific in their host reaction, C shows that the higher the degree of resistance in a host cv. the fewer the fungus races attacking it while D indicates that a susceptible host cell permits symbiotic mycelial growth for a fairly long period whereas the resistant one inhibits it by a hypersensitive reaction.

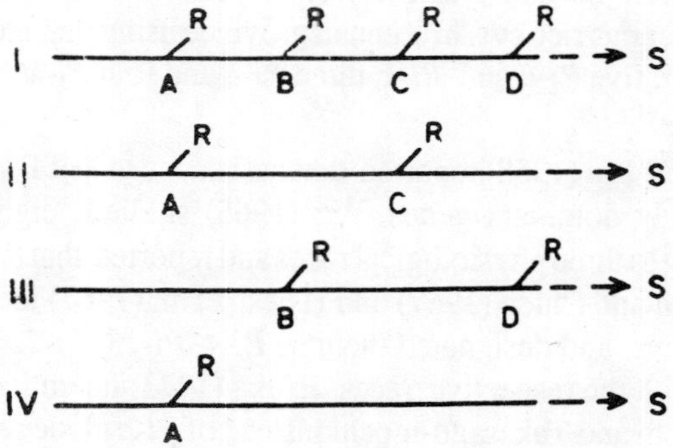

Fig. 18: Simplified model showing relation between gene action and expression of resistance (Takahashi, 1965).

Since the chemical control is expensive as well as ineffective in some cases where the isolates develop fungicidal resistance, growing resistant cvs. has become imperative for minimizing the losses caused by B1. Although superior genotypes for various traits including B1 resistance have been identified and genetic analysis has been made to explain the nature of inheritance of resistance, precise knowledge on the heritable genetic components in rice cvs. through quantitative analysis is lacking.

Near-isogenic lines

Mackill et al. (1988) and Mackill and Bonman (1990) developed a set of near-isogenic lines (NILs), each of which possesses only one B1 resistance gene. *Indica* cv. CO 39

(Amaravathi) was used as one of the parents and through the recurrent cycles of selection and back crossing, resistance genes from Tetep, LAC 23, 5173 and Pai-KanTao were incorporated. The NILs were classified on the basis of their reactions to a diverse set of isolates of *P. oryzae*. Seven groups were identified, with group I having gene *Pi-1*, group II having *Pi-2* and V having *Pi-3* while groups II and VI have alleles of the same gene *Pi-4*[a] and *Pi-4*[b]. He and Shen (1990) also identified NILs in *indica* rice cvs. with B1 resistant genes. Inukai et al. (1992a, b) further identified B1 resistant genes *Pi-2(t)* and *Pi-2* on NILs.

Molecular markers

McCouch et al. (1988) used restriction fragment length polymorphisms (RFLPs) as molecular markers for studying disease resistance genes, particularly with NILs. Using this analysis Yu et al. (1990) located the gene in cvs. C 101A51, C102A51, C103A51 and C105A51 *(Pi-2)* on chromosome 3, about 70 cM from the waxy locus (Fig. 19). Similarly, another (unnamed) gene was located in NIL C105TTP1 on chromosome 6.

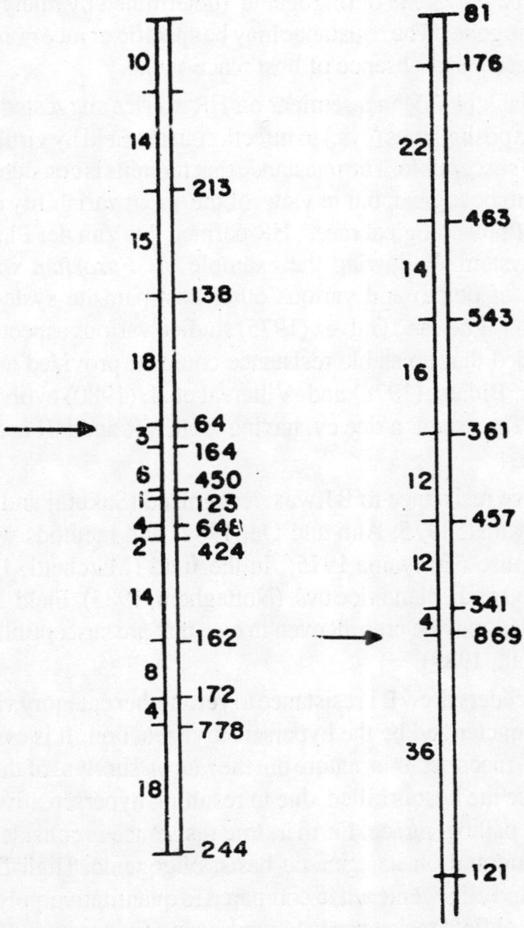

Fig. 19: Probe RG64 on chromosome 3 is closely linked to *Pi-2* and RG869 is closely linked to the resistance gene in C105 TTP-1 on chromosome 6 (Yu et al., 1990).

When some of the NILs were inoculated with a matching pathogen race, they were found to have 'residual' resistance relative to the recurring parent. Further, NIL C101LAC showed fewer and smaller lesions and NIL C103TTP had fewer lesions than CO39. No information is available on whether the residual resistance was due to the resistance gene locus or other loci in the NIL carried through back crossing. Presently, the NILs are used in fundamental studies on B1 resistance and may also be useful as improved B1 differentials for the tropics.

Horizontal/vertical resistance

A new concept of horizontal resistance (HR) or field resistance and vertical resistance (VR) or monogenic resistance was put forward by Van der Plank (1968). These terms were coined in an epidemiological context and have been widely used. Resistance may be either *horizontal* (uniform) when it is evenly spread against all races of a pathogen or *vertical* (differential) when it is completely effective against some races of a pathogen but not against others. Resistance may be *polygenic* or oligogenic (determined by many genes both major minor)or *monogenic* (one gene). The resistance may be specific or race non-specific or active or passive (denoting presence or absence of host reactions).

Further, Van der Plank (1975) in a seminar on HR to rice suggested 'determining HR as field resistance', by exposing lines (cvs.) to infection in the field by virulent races to which the lines are (vertically) susceptible. The resistance that remains is considered HR. The ideas expressed by him seem to be logical, but in view of the great variability of the fungus, and the possible existence of physiological races, HR defined by Van der Plank probably does not fit in the rice-B1 system. Following the example of *Fusarium* wilt of tomato and watermelon, late blight of potato and various other host-parasite systems, Yorinori and Thurstron (1975) and Rodriguez and Galvez (1975) studied various aspects of HR in the rice -B1 system and concluded that no stable resistance could be provided even under limited variability of the fungus. Bidaux (1978) and Villareal et al. (1980) evolved some rice cvs. possessing HR for B1. The use of a rice cv. having both VR and HR is the most effective means of controlling B1.

Field or quantitative resistance to B1 was recognized (Sakurai and Toriyama, 1967; Kozaka, 1975 and Toriyama, 1975; Ahn and Ou, 1982) and methods were developed to measure in the green house (Toriyama,1975), in the field (Marchetti, 1983; Notteghem, 1985), as well as irrigated and upland rice cvs. (Notteghem, 1985). Field resistance to neck B1 was also identified but may be present even in cvs. that are susceptible to leaf B1 (Koh et al., 1987; Bonman et al., 1989).

The japanese rice breeders drew B1 resistance to yet another category viz. *true resistance* which is specifically characterized by the hypersensitive reaction. It is evident that 'highly virulent'/'super virulent' races occur in nature but the rate of survival of their progenies will be extremely low because the host is killed due to resulting hypersensitivity reducing their capability to support the pathogen/race. Further, true resistance is considered qualitative in terms of its gene function and on its genetic basis, oligogenic. High field resistance is considered quantitative and oligogenic when compared to quantitative polygenic of the field resistance. Identification of field resistance was made using field tests in B1 nurseries using artificial inoculation (Ezuka et al., 1969b; Yunoki et al., 1970b). The relationship between qualitative and quantitative resistance to B1 was studied by Ou et al. (1975).

Ezuka (1972) further reviewed the work and concluded that field resistance to rice B1 was not always non-specific to the strain of the causal organism. However, under environmental conditions suitable for epidemic disease development, HR could not provide adequate protection against B1 in Korea in the late 1970's. Thus scepticism arose regarding HR as a long-term rice B1 control measure and the trend shifted towards monogenic resistance was not effectively utilized to combat B1 disease. However, monogenic resistance to B1 was incorporated in Tongil and the subsequent cvs. developed with *indica, japonica* as one of the parents controlled B1 for six years (1972 to 1977) in Korea until the 'new' race was identified in winter 1978 and an epidemic occurred in 1979. Despite this setback, monogenic resistance is considered the most desirable alternative for long-term rice B1 control in Korea.

Earlier, the Institute de Recherches Agronomiques Tropicales et des Cultures Vicrieres (IRAT) in the Ivory Coast studied the field resistance by measuring the rate of disease development (*r*-value) with leaf and neck B1 stages (IRAT, 1977; Bidaux, 1978). Further, Kiyosawa et al. (1977) tested field resistance by measuring infection rate while the International Institute of Tropical Agriculture (IITA)(1979) studied a number of rice lines and cvs. selected under green house conditions but the results were inconclusive.

High field resistance of upland rice cv. Kuroka is governed by more than three major genes including complementary genes (Shinoda et al., 1971) while Higashi and Saito (1985) reported field resistance of upland rice cv. Sensho, controlled by more than five additive genes located in five linkage groups II,VIII, X and XI + XII (Table 12). Incidentally, Goto (1970) reported that it was governed by three additive genes.

Durable resistance

Lee et al. (1987) reported that a few cvs. possess durable resistance to rice B1 while Bonman and Mackill (1988) recognized that durability is specific to the rice growing environment. It is effective against one population of a pathogen in the region when the cv. possessing the resistance is grown commercially on a large scale for a long period, may be 10 years or more. For example, IR 36 has been durably resistant to B1 in a particular environment such as tropical, lowland, rice growing areas (Yeh and Bonman, 1986). However, this resistance will probably not prove useful in highly B1-conducive environments such as many of the world's upland rice areas. Here, the resistance types based on(a) inheritance monogenic and polygenic and (b) magnitude of the effect (complete or incomplete) are to be employed.

There is great diversity in the rice-growing environments of India, many of which are endemic. Therefore, it is necessary to search for cvs. of *indica* types as some exhibit durable B1 resistance. Since 1970, IR 20 is being widely grown as a late *kharif* crop in tank-irrigated areas of south India and although the reproductive phase favours disease development the crop escapes disease, and proves to be resistant while other cvs. are badly damaged. Perhaps durable B1 resistance is associated with partial resistance controlled by an undetermined number of genes (Bonmann and Mackill, 1988). Recently, Reddy and Bastawisi (1990) while testing rice genotypes to two representative *P. oryzae* isolates of Egypt found cvs. IR 36 and Rasi as of durable resistance in many rice-growing environments. On the other hand, in Korea, Lee and Cho (1990) found 6 rice cvs. (Seomjin, Daseong, Youngsan, Shinounchal, Shinkwang, Taebaeg) and six elite lines (M80, M 82, Sangju 5, Iri 377, S 341 and Sangju 6) with a high degree of durable resistance for more than seven years. These results suggest that durable resistance, if selected properly, can be used as a long-term strategy for B1 control.

Table 12 : Linkage between field resistance to leaf blast and marker loci, and effects of the resistance genes in crosses of Sensho/Testers (Higashi and Saito 1989).

Linkage group	Chromo-some	Marker gene	Linkage between[a] Marker locus and resistance	effect of resistance gene(s) linked with marker locus		Tester
				additive[b]	dominance[c]	
I	6	*wx*	ns	(0.35)		H-84
		fs-1	ns	(0.13)		H-84
		C1	ns	(–0.17)		H-337
		Ur-i	ns	(0.20)		H-232
II	II	*D-2*	*	0.62		H-79
		1g	***	2.03	±	H-79
		Ph	***	1.73	+	H-337
III	3	*1ax*	**	0.86		H-453
IV	10	*d-6*	*	0.69		H-84
		d-6	ns	(–0.02)		H-126
		Rc	ns	(0.08)		H-5
VI-IX	2	*gh-1*	ns	(0.05)		H-143
		St-2	ns	(0.01)		H-143
		ri	ns	(0.22)		H-68
		nl-1	ns	(–0.08)		H-103
		gl-1	ns	(–0.53)		H-103
VII-V	1	*Dn-1*	ns	(0.04)		H-232
		Dn-1	ns	(0.01)		H-147
		I-BF	ns	(–0.16)		A-5
VIII	9	*la*	**	1.09	+	H-79
		sp	***	0.89	+	H-452
X	8	*bl-1*	***	0.39	+	H-68
XI+XIII	5	*bc-1*	*	0.62	±	H-79
		bc-1	s	(0.55)		H-84
		d-14	***	1.79	+	H-147
		Hg	ns	(0.23)		H-126

[a] ***, **, * : Singnificant at 0.1%, 1% and 5% levels, respectively.
[b] Additive effect : Difference of the means of blast scores between tester type group and Sensho type group on the marker gene.
[c] + : Resistance is dominant. ± : No dominance. + :Resistance is recessive.

The concept of 'slow blasting' (SBlg) is of recent origin and refers to the occurrence of disease but at very low incidence. Here, resistance appears to be evenly spread against all races of the pathogen and supposedly serves to slow down an epidemic as it is in progress. Villareal et al. (1980) studied the SBl tendencies of B1 fungus by calculating the apparent infection rate *(r)* of 16 rice cvs. and several of them showed lower *r* values while Ahn (1982) also observed SBlg resistance to rice in the Philippines.

Wang et al. (1989) studied the physiological biochemical chracteristics of SBlg rice cvs. and found that these cvs. had a higher content of silicon, phenolics and flavonoids than the susceptible ones. In the infection process the hyphae were restricted in the single infected cell of all SBlg plants. However, neither hypersensitivity reaction nor necrosis occurred.

Wang and Chen (1987) also reported that the disease progress curves (value A) were significantly lower for five resistant cvs., namely, Manjing II, Dageyu, Jingang, Ou 244 and Dianyu 1 inoculated with fungus race ZB9 than for the check cvs. Lijanxintuanhegu and Zhaolouzao. The SBlg value A (Wilcoxson formula) was negatively corrected with infection efficiency (r = -0.98) and sporulation capacity (r = -0.81),but there was no correlation with the latent period. Sun et al. (1990) reported that cv. differences for two isolates in the latent period were statistically significant as compared with the susceptible check. The relative infection efficiency on the cvs. with SBlg was significantly less with smaller lesion size. Sporulation capacity on all test cvs. was so much lower than on the susceptible check.

No single rice cv. is expected to be resistant in all areas, at all times. However, certain cvs. will be useful in certain areas for some time and should be replaced before severe B1 infection results. Instead of depending on the existing vertical resistance to rice B1, other alternate strategies suggested by various workers are discussed.

Gene rotation

Crill and Khush (1979) and Crill et al. (1982) described the concept of race prediction and gene rotation in B1 which was successfully adopted in Korea. The programme was based on identifying monogenic resistant lines, screening them for resistance to specific races in greenhouse trials, evaluating these lines in the field and increasing the seeds of the potential candidates. Once a cv. is released to farmers in a particular country, the same cv. is tested in the field elsewhere (foreign site) also. The race of B1 pathogen attacking the new cv. at the foreign site will have the same gene for pathogenicity as any race capable of attacking the new cv. in the particular country. The cvs. that exhibit monogenic resistance can be grown successfully in the area until a new or highly or super virulent race of *P. oryzae* appears.When it is identified the rice breeders should replace the existing cv. with a cv. resistant to this race as was done in other systems like corn and tomato. The new cv. developed should be grown until again another new super race appears and then the whole procedure should be repeated.

In view of the obscurity of the biology and life cycle of the B1 pathogen rice, breeders at Korea opted for rotation of resistance (R) genes. The first cv. released will have only R_1, second one will have R_1 plus R_2 while the third one will have R_1 plus R_3 and so on. This system of R gene management emerged as it places extreme selection pressure on the pathogen population for about twice as long as the rotation of single R genes. The longer period of selection pressure presumably would lead to a reduction in the prolonged survival of genes for pathogenicity.

Single gene addition

In the B1 resistance breeding programme, other alternative strategies envisaged include single-gene addition and pyramiding of genes, instead of dependence on the conventional vertical resistance. The strategy of single-gene addition is based on the idea of using as few resistant genes as possible. A new resistance gene is added to an agronomically valuable germplasm only when needed. Resistance in some cases is lost by a stepwise increase of virulence in the pathogen, but careful monitoring of the pathogen population as well as correct timing for inserting a new resistance gene often help in releasing resistant cvs. which are also economical.

Gene pyramiding

Gene pyramiding for resistance to B1 pathogen in cvs. with agronomic acceptability is again to be considered. The cvs. that have resistance based on genes with major or minor effects and complex models of inheritance of resistance are best suited. These cvs. can be developed with different genes for resistance if the genes are identified. Hitherto, due to pathogenic variability of the fungus no single cv. could be made resistant in all areas or resistant for long periods in any one area. There is added advantage to use certain cvs. in specific areas for a longer period and they can be replaced before severe B1 infection results due to development of a super race.

Multiline concept

Yokoo (1974) proposed the use of multiline cvs. to prevent the breakdown of rice B1 disease resistance. As an estimate, the effective longevity of rice cvs. with various resistant genes for B1 disease is less than three years (Kiyosowa, 1982). Although Bl severity was decreased by planting mixtures of susceptible and resistant cvs. (Chin and Teng, 1976; Shindo, 1977; Shouji et al., 1982) as well as upland rice cv. mixtures (Bonman et al., 1985) the relationship between the ratio of B1 incidence and the proportion of susceptiblity to resistance in rice cvs. or lines in a mixed planting is not known.

Any individual plant in a multiline mixture is infected with a compatible race and the build up of disease on that plant will be mostly through self-infection. The rate of build up differs with the cvs. used. This means that a specific set of cvs. alone are more suitable for multiline conversion. The slowed self-infection types of resistance complement the effect of multilines by limiting self-infection and hence the epidemic builds-up in the multiline population.

Two different models have been developed and investigated and the criterion used was mainly the effect of distance on the disease spread. The first Gregory (1968) model is projected as:

$$Y = aD^{-b}$$

(or by transformation $\log_{10}{}^{Y} = \log_{10}{}^{a\text{-}b}\log_{10}{}^{D}$)

in which the predicted quantity of disease Y equals the negative power $-b$ of distance D times as scaling factor a.

Kiyosawa and Shiyomi (1972) offered a second model as:

$$Y = ae^{-Db}$$

(or by transformation if $\ell n\, Y = \text{ln}a\text{-}Db$)

MacKenzie and Villareal (1977) investigated these models using rice B1 to evaluate their goodness-of-fit to actual data. However, neither of the models could not adequately describe the first metre of B1 disease spread.

Koizumi and Kato (1987) suggested the best possible combination of Norin 29 susceptible to race 003 and Fukunishiki resistant to race 003 in mixed patterns (Fig.20a) for using rice multiline cvs. to control rice B1. Further, the relationship between r in mixed plantings and the proportion of the two cvs. Norin 29 and Toride 1 (resistant to race 003) is shown in Fig. 20b. The r value or Norin 29 was in proportion to the logarithm of the proportion of Norin 29 in the mixed planting. Using the equation $r_m = r_s + \text{c} \log_e \text{m}$, where r_m is r in mixed plantings, r_s is r in single planting of susceptible cv., m is the proportion of susceptible cv. in mixed planting and c is a constant, c = 0.043 and $R^2 = 0.953$ which was significant at 1% level. As the ratio of Norin 29 decreased, the proportion of the diseased area per plant of Norin 29 also dropped. Reduction of this value in mixed planting of less than 25%. No B1 disease was found in the plot with 100% planting with cv. Toride 1.

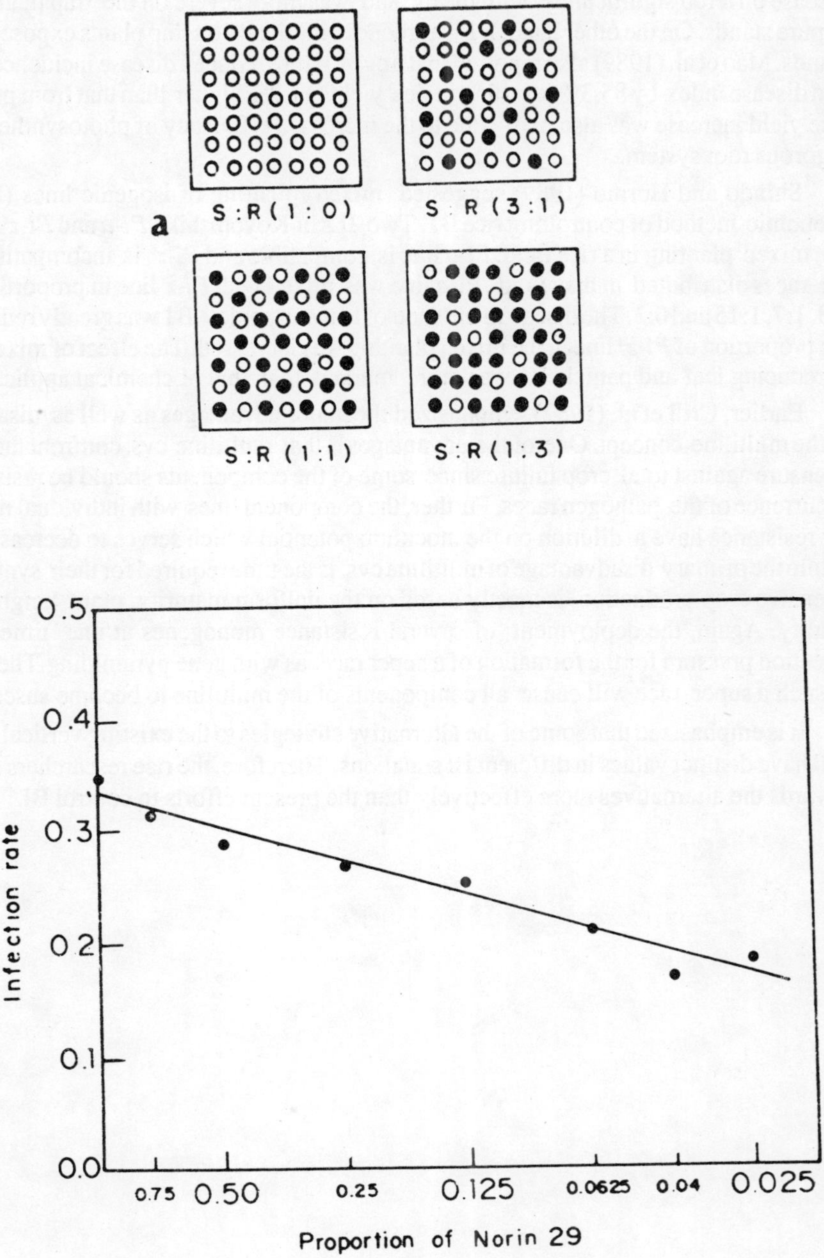

Fig. 20ab: (a) Mixed pattern of B1 susceptible cv. Norin 29 (open circles) and resistant Fukunishiki (closed circles) to B1. (b) Linear relationship between infection rate (leaf B1) and proportion of Norin 29 in a mixed plantings in the nursery (Koizumi and Kato, 1987).

Chai et al. (1989) found that mixtures of cvs. either within a hill or between hills suppressed the development of leaf and neck B1. Zhuge and Bonman (1989) observed B1 incidence in trap plants exposed to pure and mixed stands of rice cvs. They noted that the disease differed significantly with the cv. and was more severe on the trap plants exposed to pure stands. On the other hand, B1 was reduced markedly on trap plants exposed to mixed stands. Mao et al. (1989) also reported that a cv. mixture reduced disease incidence by 74.2% and disease index by 85.3% resulting in rice yields 11.8% higher than that from pure stands. The yield increase was also partly due to the increased efficiency of photosynthesis and the vigorous root system.

Shindo and Horino (1989) suggested mixed planting of isogenic lines (ILs) as the economic method of controlling rice B1. Two ILs of Koyonishiki, *Pi-a* and *Pi-z^t*, were used for mixed planting in a rice field. *Pi-a* line is compatible and *Pi-z^t*, is incompatible with all the races distributed in that field, *Pi-a* line was mixed with *Pi-z^t* line in proportions of 1:1, 1:3, 1:7, 1:15 and 0:1. The disease incidence of leaf and panicle B1 was greatly reduced when the proportion of *Pi-z^t* line in the mixed planting was increased. The effect of mixed planting in reducing leaf and panicle B1 was more marked than that of chemical application.

Earlier, Crill et al. (1982), emphasized the major advantages as well as disadvantages of the multiline concept. One of the advantages is that multiline cvs. confront the pathogen to ensure against total crop failure since some of the components should be resistant to the occurrence of the pathogen races. Further, the component lines with individual monogenes for resistance have a dilution on the inoculum potential which serves to decrease severity. While the primary disadvantage of multiline cvs. is the time required for their synthesis. The intensive crop production is mostly based on the uniform maturity, plant height and crop quality. Again, the deployment of several resistance monogenes at one time generates selection pressure for the formation of a super race, as with gene pyramiding.The evolution of such a super race will cause all components of the multiline to become susceptible.

It is emphasized that some of the alternative strategies to the existing vertical resistance will have distinct values in different Bl situations. Therefore, the rice researchers should opt towards the alternatives more effectively than the present efforts to control Bl.

The Disease

The disease is a complex phenomenon, an interaction between the host parasite and the environment. Diseased plants are distinguished by the changes in their morphological structures or physiological processes which are brought about by biotic as well as abiotic agents. Although the abiotic physical and chemical agents are considered pathogens, most pathologists avoid this term especially when deficiency of essential material is responsible for disease symptom.

Disease *incidence* and *severity* are the two related terms frequently used by the plant pathologists for measuring the intensity of plant diseases. Quantification of B1 disease is quite essential for the evaluation of the yield losses accurately. Disease assessment methods include, the estimation of *disease intensity* (expressed either as incidence or severity) or using direct measurements or indirect biochemical estimations. *Incidence* is a quantal measure expressed as the percentage of diseased plants or plant parts in a given area, while *severity* depicts the percentage of plant parts covered by the lesions/symptoms. Thus, *severity* depends on the number and size of the lesions present on the infected host and is expressed as the percentage of total disease.

Source of primary infection

The B1 fungus is reported to persist in the infected seed, diseased straw, alternative hosts, infected seedlings or even in the soil. However, the primary sources of infection are not definitely known.

Infected seeds are considered as seedling B1 infection by several workers (Kuribayashi, 1928; Suzuki, 1930; Mayee, 1974; Ranganathiah et al., 1979). However, Orton (1931) did not consider the disease to be seed-borne. Earlier, Suzuki (1930) observed the presence of the fungus within the tissues of the embryo, endosperm, glumes etc., as evidence for the seed-borne nature of the fungus.

Diseased straw also contributes as a primary source of infection (Nisikado, 1917; Sundararaman, 1925; Sueda, 1928; Dodoff and Kovacevaski, 1930; Ito, 1932; Lin and Liu, 1957) as the fungus overwinters in it. Lin and Liu (1957) reported that rice straw under water or buried in the mud for 20-d yielded no viable mycelium. On the other hand, Anikeev et al. (1985) reported the survival of mycelium on rice residues up to the new growing season.

The overwintered conidia and those found on grass hosts, namely, *Panicum repens, Brachiaria mutica, Digitaria sanguinalis* and *Leersia hexandra* (Veeraraghavan and Padmanabhan, 1965); bamboo and bamboo grass (Itoi et al., 1979); other hosts (Madden and Ellis, 1988) commonly seen on the banks of rice fields may serve as a source of primary infection. However, cross reaction studies indicated that the pathogenicity of most isolates were largely restricted to their own hosts. Thus, how far these alternative hosts help in the perennation of the fungus or serve as potential reservoirs of inoculum remains obscure.

Infected seedlings mostly serve as a source of primary inoculum and, when transplanted, two phases of disease development, namely, the multiplying phase on a single tiller and the dispersing phase among neighbouring ones (Kobayashi, 1963, 1968; Suzuki, 1969a; Kato and Sasaki, 1974) are commonly observed. The disease gradient from the plant constituting the inoculum source can be described as log N = log a - b log X, where N = number of lesions, X = distance from the inoculum source, and a and b are constants (Kiyosawa and Shiyomi, 1972; Kato and Sasaki, 1974). Regarding the field to field spread of infection, Suzuki (1969a) also reported the fitness of the above empirical formula.

Apparao (1956a, 1964) presented evidence of the poor competitive saprophytic ability of the fungus amidst a host of highly antagonistic bacteria and actinomycetes under optimum moisture level of the soil. These results indicated that the diseased host debris are unlikely to serve as the principal source of primary infection. Despite the adverse condition of fungistasis and lysis, survival of *P. oryzae* in the soil has been reported (Bhattacharya and Samadar, 1977).

Predisposing factors favouring infection

Hashioka (1965) reported that several factors affect the development of the B1. The major ones are: (i) humidity, (ii) temperature, (iii) excessive N-fertilizers, and (iv) host susceptibility. Ishiguro et al. (1988) studied the temporal dynamics of thalide on rice leaves and its efficacy in preventing rice leaf B1.

Hemmi and Abe (1932) reported that continuous wetting of the leaves was necessary to ensure infection and that the duration of wetting varied with temperature. Abe (1933) found that below 92% RH, infection failed to take place owing to the inability of spores to germinate. The invasion of the host by the B1 fungus is possible only at 92 - 96% RH.

The air and soil humidity predispose the host leading to B1 susceptibility. Tochinai and Komiya (1939) reported that the rice seedlings were unaffected with respect to B1 susceptibility by the short treatment (24-74 h) under saturated air prior to inoculation. However, when the seedlings are grown under different relative humidities of air from 75 - 100% for more prolonged periods, say, ten days, susceptibility and growth of the seedlings are considerably enhanced even at 84 - 95% RH. But the infection is less when the air is either extremely damp or dry (Hashioka, 1950b).

Rice plants are predisposed to B1 more strikingly by soil humidity than by air humidity (Hemmi and Abe, 1932) and dry soil (Kahn and Libby, 1958). Rice leaves resist physical penetration due to the dryness of the soil which has been ascribed a role in resistance but this seems to be caused by turgescense of the leaf tissues which is correlated with soil humidity rather than with the mechanical stability of the epidermal tissues (Ito and Sakamoto, 1940 and Sakamoto, 1941).

Temperatures at 21-28°C (Hemmi and Abe, 1931) are favoured at the initial stages of infection. The plants were most and least susceptible at low soil temperature (20° C) and higher soil/air temperature (28-32°C) respectively (Kuribayashi and Ichikawa, 1952). The effect of low night temperature (15-20°C) on B1 development and severity has also been reported (Suryanarayanan, 1958a; Ohata et al., 1966; Manibhushanrao and Day, 1972).

Nitrogen fertilizers have a remarkable effect on the susceptibility of rice to B1 infection.

A linear increase in infection in susceptible types with an increase in the level of N-fertilizer has been reported (Krishnaswamy, 1952; Ganguly et al., 1954; Padmanabhan, 1953; Amin and Venkatarao, 1979; De Faria et al., 1982). However, there are also reports which indicate that decreasing N level increases resistance (Suzuki, 1975).

Rice plants especially the upper (younger) leaves, are considered more susceptible than the middle or lower leaves (Miyazaki, 1930) while Abe (1936) and Hashioka (1943b) reported plants are susceptible in the seedling stage than in the tillering and booting stage. Further, Hashioka (1950a), Padmanabhan and Ganguly (1954), Podhradszky and Sudi (1957); Kahn and Libby (1958) and Ito (1965) studied the relationship between age of the host and susceptibility to B1. They recognized three distinct stages of growth, namely, seedling, rapid tillering (foliar infection) and ear-emergence stages (nodal and neck infection) when rice plants are highly susceptible to infection. This was also confirmed by Willis et al. (1968) as differential susceptibility of rice leaves and panicles to B1 pathogen was recognized. Other plant growth stages like flower emergence or booting (Padmanabhan, 1967) and panicle B1 after heading (Kato,1974) have also been found to be B1 prone stages of rice. These observations have led to the belief that resistances to leaf and neck B1 are not correlated and that different genes are involved. However, the host susceptibility and the question of reaction to B1 at different parts of the plant was not fully understood (Takahashi, 1965). Kim (1982) correlated the relationship between panicle B1 incidence and infected leaves at the booting stage as follows :

$$Y = 2.84\,x + 19.3 \qquad (r = 0.8)$$

where, Y is % of panicle B1 incidence and x is the number of B1 infected leaves at the booting stage. These relationships are similar to the results reported by Hori (1963).

Mode of infection and spread

Penetration of the host by germ-tube may be through the stomata or directly through the epidermis (Sueda, 1928) while appressoria are formed from the germ-tubes from which slender penetrating hyphae are extruded. These pierce the outer layer of the epidermis and enter the cell cavity (Yoshii, 1933). Further, the appressoria formed by hyphae on the rice leaves are not susceptible to attack, even on glass dishes (Shimada,1937). Inside the cell, a vesicle is formed and this gives rise to hyphae. Environmental conditions have a direct effect on spore germination and formation of appressoria (Rahnema, 1979).

Host penetration

Epidermal cells in leaves of wild rices, namely, *O. latifolia* and *O. minuta* reacted sharply against penetration by the B1 fungus by secreting a resin-like material around the invaded hypha, inhibiting its further growth. Dark brown granules produced in response to host penetration exert a fungistatic action (Takahashi, 1956). Successful hyphal penetration of the epidermal cells in plants grown under low night temperatures was reported while penetration was rare in plants grown under high night temperatures (Sadasivan et al., 1965). In the latter case, although the fungus succeeded in entering the host, its subsequent spread was restricted. The average time for penetration of appressorium was around 14 h at 18 - 24°C (Yoshino, 1979).

Water is necessary for spore germination, appressorium formation and penetration of the peg into the epidermal cells (Rapilly et al., 1975). The longer the water droplets remain on the leaf surface, the greater the penetration (Liang, 1979). Dew has a key role where host penetration occurs within 3 h after formation of the peg. The average infection time of the conidium population (Y) can be illustrated by the parabola $Y = 0.062x^2 - 2.98x + 48.21$, where, x = temperature during the incubation period (Kato, 1974).

Kim (1987a) reported that percent appressorium formation of *P. oryzae* ranged from 83.3-89.8% showing no distinct differences regardless of inoculum denşities and the nature of rice cvs. However, percent penetration ranged from zero in cv. Senho at 5.5×10^4 spores/ml to 7.2% in Peta at 22×10^4 spores/ml density. The number of leaf B1 lesions as well as percent penetration increased at high inoculum density.

Sunlight influences development of the B1 fungus in host tissue (Hemmi and Abe, 1931) and shading provides more favourable conditions for the pathogen. The optimum temperature for symptom development varies from 26-28°C (Hemmi et al., 1936); 27-29.5°C (Kahn and Libby, 1958); 22-26° C (Sekiguchi and Furuta, 1970) and 24° C (Yoshino, 1979). At lower temperatures (9 and 16° C) only a few brown specks appeared or they were completely absent (Kim and Mogi, 1986). These specks, when exposed to higher temperatures, increased in number (Yoshino and Yamaguchi, 1970). The mycelial growth was rapid from at 24 to 32° C with the optimum around 28°C (Yaegashi and Yanagita, 1972). This may be the reason why more lesions appeared at 28° C.

Kim and Mogi (1985,1986) studied the penetration, hyphal growth within the leaf epidermis and lesion formation using pre- and post disposed rice seedlings at 23/15°C, 23/15° C, 29/21°C, 29/21° C, 23/15° C, 29/21° C and 29/21° C, 23/15° C temperature regimes respectively for different periods. The day/night temperatures were altered from 9 a.m. to 5 p.m. and 5 p.m. to 9 a.m. in a phytotron with 70% RH under daylight. Typical leaf B1 lesions produced more in 29/21° C, 23/15° C plots than in other temperature treatments, but the total number of typical lesions decreased as the duration of treatments became longer. On the contrary, the total number of typical lesions increased by the longer duration of treatments in 23/15° C, 29/21° C plots. Seedlings disposed at the 29/21° C regime after completion of inoculation exhibited apparent hyphal growth since 96 h after inoculation while hyphal growth was suppressed until 120 h on the post inoculated seedlings exposed at 23/15° C regime. Thus, lesion formation on the seedlings disposed at lower temperatures after inoculation was delayed due to the suppressed activities of the B1 fungus after penetration. The above phenomenon was statistically analysed and strongly suggested that lesion formation is delayed under low temperature conditions which is mainly due to the suppression of hyphal growth. The hyphal growth value 11 at 23/15° C revealed better fitness of the model than at 29/21° C regime (Kim and Mogi, 1987).

The minimal periods required to initiate infection under dew-cabinet conditions were 10, 12 and 14 h at 27, 21, and 18° C respectively (Kahn and Libby, 1958) and the optimum was 8h at 25°C (Hashioka, 1950a; Barksdale and Jones, 1965; Sekiguchi and Furuta, 1970). Lesion growth depends on temperature along with leaf wetness and water status of the plant (Rapilly et al., 1975). The longer the period of wetness, the greater the number of lesions (IRRI, 1977). Wind speed also influences initial infection indirectly by its effect on the time length and the amount of dew formation, but there was no statistical correlation between wind speed at inoculation and number of lesions formed (Asai et al., 1967).

Cytological aspects of host response

Sridhar et al. (1973b) reported that lipid bodies in rice leaves get discoloured during pathogenesis. Kurosaki (1978) observed low water content in the brown tissues surrounding

B1 lesions and interpreted that the inhibited supply of water and water-soluble nutrients to the fungus enclosed inside the lesion perhaps deprive the fungus further growth similar to that of a hypersensitivity reaction that reflects the inability of the plant and the pathogen to co-exist.

Yaegashi et al. (1990) made observations on the specific interaction of cell reaction between rice and B1 fungus. The leaf cell reaction at host-parasite encounter sites was classified into six types A to E (Fig. 21) according to the fungal development and host

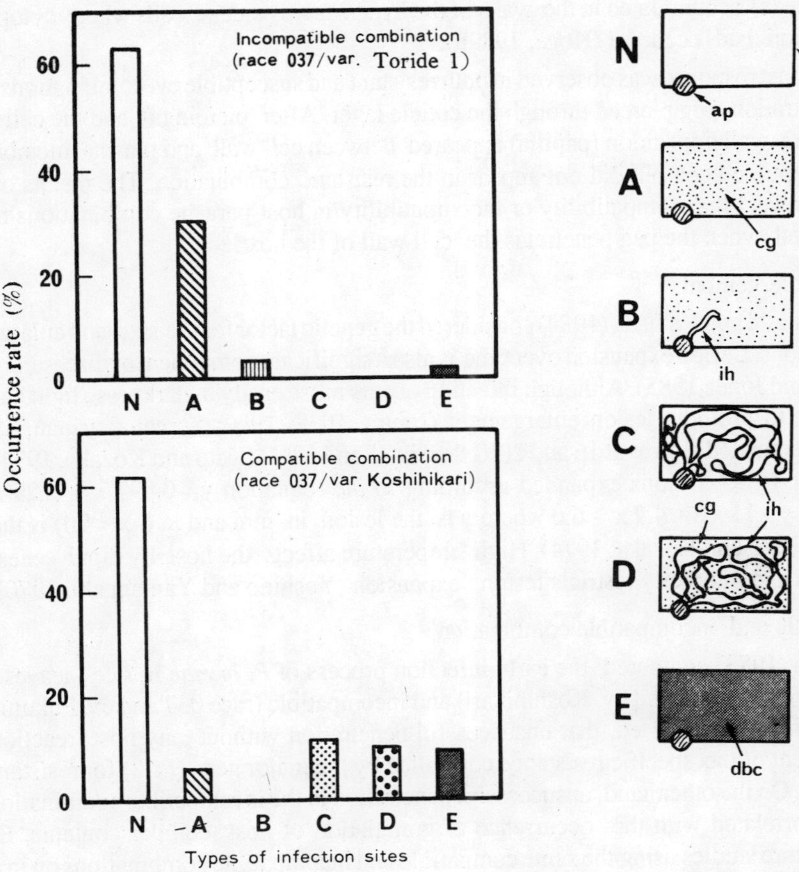

Fig. 21: Occurrence rates of each infection type (N-E) in the compatible and the incompatible combinations 72 h after inoculation with *P. oryzae*.

N: Unsuccessful penetration (no visible invaded hyphae) without any host reaction. A: Unsuccessful penetration with host cytoplasmic granules in an epidermal cell, B: A poorly-developed invaded hypha with host cytoplasmic granules in an epidermal cell, C: Well-developed invaded hyphae without host cytoplasmic granules in an epidermal cell, D: Well-developed invaded hyphae with host cytoplasmic granules in an epidermal cell, E: Deep browning of the infected epidermis and parenchyma. ap: Appressorium, cg: Granulation of the epidermal cellular contents, dbc: Epidermal cell with deep browning, ec: Epidermal cell, ih: Invaded hyphae (Koga, 1989a).

responses (Koga et al., 1986; Koga, 1989a). The necrosis (type E) may play an important role in inhibiting hyphal growth in the parenchyma but does not relate with the specificity expressed at the early infection stage. However, it was concluded that the specific occurrence of fine cytoplasmic granulation (type A and B) in resistant cvs. closely correlated with the expression of race-specific resistance (Tometa and Yamanaka, 1983; Koga et al., 1986; Koga, 1989a). Recently, Peng and Shishiyama (1988, 1989) reported that the expression of race-specific resistance in rice B1 was associated with the light brown coloration and aniline blue staining of epidermal cell walls that occurred prior to cytoplasmic granulation. Cross sections of the leaf sheath showed that the antifluorescent substance (s) accumulated in the walls of the hypersensitive dead cells where cytoplasmic granulation had occurred (Koga, 1989b).

No host response was observed in both resistant and susceptible cv. combinations before the penetration peg pierced through the cuticle layer. After piercing it and the cell wall, a collar-like wall opposition (papilla) appeared between cell wall and plasma membrane of the susceptible host, but did not appear in the resistant combination. The results of these studies reveal that compatibility or incompatibility in host-parasite combinations may be recognized when the peg penetrates the cell wall of the host.

Lesion enlargement

Yaegashi and Yamada (1984) considered the genetic factor for the size and enlargement of B1 lesions. Lesion expansion over time is also a significant component of disease progress (Berger and Jones,1985). Although infection occurs most easily in darkness, light has been reported to promote lesion enlargement (Imura, 1938, 1940; Suryanarayanan, 1958a). Lesions expanded slower at 16 and 20°C than at 25 and 32° C (Kato and Kozaka, 1974; Kato 1976). At 16° C, lesions expanded according to the equation $y = 0.897\,x + 2.29$ and at 25 x°C $y = 0.15\,x^2 + 4.9\,x - 6.0$ where y is the lesion in mm and × ($x < 20$) is the time in days (Kato and Kozaka, 1974). High temperature affects the host by rapid senescence of leaves and indirectly restricts lesion expansion (Yoshino and Yamaguchi, 1970).

Compatible and incompatible combination

Koga (1983) compared the early infection process of *P. oryzae* in rice leaves using compatible (race 037 and cv. Koshihikari) and incompatible (race 037 and cv. Fukunishiki) combination and reported that unsuccessful penetration without any host reaction was independent of the specific resistance controlled by the major gene (*Pi-2*) for resistance in rice to B1. On the other hand, unsuccessful penetration in the incompatible combination was closely correlated with the occurrence of granulation of host cellular contents. Earlier ultrastructure studies using the same compatible and incompatible combinations on invaded pathogen (race 037) on rice leaves were made (Koga et al., 1982). In the former, the hyphae, rich in lipid, densely penetrated into the host epidermis, parenchymatous cells as well as vessels and the cell organelles were destroyed by the B1 fungus. In the latter, the invaded hyphae were also found in the epidermal cells, necrotic parenchymatous cells and stomatal chambers. In the epidermal cells, both host and fungus were found moribund while in necrotic parenchymatous cells, the hyphae developed more vacuoles with less lipid cells. In the substomatal chambers, the hyphae were rich in cytoplasmic matrix and appeared to be viable. A comparison of the intercellular organelles between compatible and incompatible combinations of rice blast disease revealed that no change in the diameter of the invaded hyphae as well as the size of the nucleus in both the combinations (Koga et al., 1982). However, in the former both the mitochondria and vacuole sizes were reduced considerably (Table 13).

Table 13: Size of intercellular organelles between compatible and incompatible combinations of rice B1 disease (Koga et al., 1982).

Intercellular organelles	Diameter/size[1] (μm)	
	Compatible[2]	Incompatible[3]
Invaded hypae[4]	2.6	2.6
Osmophilic plate-let	0.75	0.20
Nucleus	2.1 × 1.7	2.0 × 1.7
Vacuole	0.84 × 0.63	2.0 × 1.5

(*1*) Mean value from 30 sites. (*2*) in the parenchymatous cells on cv. Koshihikari. (*3*) in the necrotic parencymatous cells on cv. Fukunishki. (*4*) race 307 (ken 60-90)

Recent studies in some compatible and incompatible combinations of rice cvs. and B1 fungus have suggested that specific resistance of rice cvs. to B1 is not expressed before fungal penetration into rice leaves (Koga et al., 1986; Peng and Shishiyama, 1988; Koga, 1989a and Yaegashi et al., 1990).

Epidemiology

Studies concerned with the complex effects of the environmental conditions on disease development are of prime importance for predicting any epidemic. The intensity of the epidemic is determined mainly by the abundance of the conidia, the influence of environmental conditions on infection processes, and the resistance of the host plants. These three components are influenced not only by meteorological factors, but also by the host cvs. and cultural practices used. Again, meteorological factors influence the host as well as the pathogen. However, the inter-relationship is further complicated by the introduction of new fungicides that influence both the parasite and the rice plant.

Earlier studies on epidemic outbreak and fundamental forecasting of B1 disease (Inoue, 1943; Bapreddy and Vaheeduddin, 1956; Takahashi, 1957; Chetia, 1964) showed that soil flora contributed to airflora including *P. oryzae* conidia. Asai et al. (1967), Ou (1972) and Suzuki (1975) found that the B1 epidemic will be mainly governed by the disease proneness of the host, environmental conditions such as temperature (Hashioka 1950a) and duration of leaf surface wetness, relative humidity (RH) and duration of dew (El Refaei 1977). However, Chakrabarthi (1971), Hashiguchi et al. (1973), Yoshino (1974), Padhi and Chakrabarthi (1982), Kobayashi(1984), Venkatarao and Muralidharan (1982a) and Chien et al. (1984) studied the effect of meteorological conditions on the influence and progress of B1 on rice while Friesland and Schrodter (1988) reported that high temperatures during pathogenesis and inoculum production restrain the fungus from causing epidemic.

The B1 fungus survives at varying inoculum threshold levels. Vulnerable cv. and conducive weather at the appropriate time trigger the epidemic (Suzuki, 1975). The B1 epidemic growth can be studied using either analytical or simulation techniques. Note worthy advances have been made in epidemiological studies on the pathogen and its behaviour. However, the importance of research on the host plant to understand the true dynamics of the plant disease is lacking (Seem,1988). Scholarly reviews have appeared from time to time on the epidemiology of rice B1 (Kato, 1974,1976; Suzuki, 1975 and Hashimoto et al., 1984).Later, Manibhushanrao et al. (1990) reviewed the rice B1 epidemiology and application of its techniques to disease management.

Evaluation of epidemics

The main object here is to understand the course of epidemics and to comprehend the interlocking effect of the different parts of the infection chain and the influence of external factors, mainly environmental and human. One can achieve a precise correlation between B1 disease proneness and climatic factors (Kato, 1974). This can be achieved with *in vitro* (growth chambers) experiments by analysing the concurrent and recurrent phases of the pathogen's life cycle and the host (Rotem and Cohen,1974).

Causes of epidemics

While a virulent pathogen, a susceptible host and a favourable environment are the main causes of an epidemic, there are other secondary causes for the outbreak. For instance, when a major part of the pathogen population becomes insensitive to fungicides, disease control may fail, leading to an epidemic.

The outbreak of panicle B1 epidemic is closely correlated with the extent of leaf B1 (Kuribayashi and Ichikawa, 1952; Rangaswami and Subramanian, 1958), lesions on leaves (Hori, 1963), the collar auricle and ligule (Ono and Suzuki,1960) supplying conidia for the infection of panicles. Weather influences the epidemic development either by interacting with the pathogen and/or altering the host physiology. Blast develops at temperatures varying from 19 to 30 °C (Suzuki, 1975). Three h of rain (1.5 mm) brought down eight times as many conidia as were deposited during the same time in a similar area exposed to wind (Suzuki, 1965). Higher wind velocity causes severe outbreaks of leaf and panicle B1 through injury (Hirano and Goto, 1963).

Biometeorological factors

High and low temperatures with long periods of rain (drizzling), lack of sunshine, slow winds, high humidity and drought not only increase the disease proneness of the host but also influence the life cycle of the pathogen (Suzuki, 1975).

Persistent drizzle has greater influence than heavy rain on the disease spread while, in cloudy and rainy weather, both growth and resistance of plants get reduced (Kuribayashi and Ichikawa, 1952). Severe disease incidence is reported with more than 22 d of rain. Since most of the conidia release and disperse during rainy/ cloudy weather, these may be the key factors for B1 epidemic (Kim and Min, 1989; Kim et al., 1989). Rice cvs. in non-flooded soils become more susceptible to B1 infection than in flooded soils (Hemmi, 1930; Kahn and Libby, 1958). A three day exposure to drought prior to inoculation increases the infection (Tochinai and Komuja, 1939) and if prolonged it may lead to a B1 epidemic (DeFaria et al., 1982). In upland and flooded plots disease severity of 58% and 1% respectively was reported (Kim et al., 1985). Besides, shade also influences disease development (Abe, 1931; Seem, 1988).

Spore production

Spore production and/or dissemination are important in the B1 disease cycle (Panzer et al., 1957, Kato, 1974) besides the sporulation potential (Kato and Kozaka, 1974). The effect of temperature on the sporulation potential is essential for evaluation of the role of lesions as inoculum sources. Spore formation occurs at 15 to 30°C with the optimum at 25-28° C (Atkins et al., 1967). The maximum was observed at 20°C (6500 conidia/lesion) which is double the increase at 25 or 32°C, and seven times that at 16° C (Kato and Kozaka, 1974). Sporulation was good at RH 93-100% (Hashioka, 1950b), while there was no sporulation at 93% and below 88% (Kuribayashi and Ichikawa, 1952). In natural conditions and under high humidity, the unmatured conidiophores and conidia become turgid and active (Kato, 1974). Lesions kept continuously in high RH produced spores two to several hundred times more than in dry conditions (Barksdale and Asai, 1961).

Spore production improved with increasing light intensity (Suzuki and Yoshimura, 1963); 6-8 h of darkness followed by light favours sporulation. The importance of light in stimulating sporulation was stressed by Ohmari and Nakajima, (1970); Ahn and Chung, (1974); Yamanaka et al. (1976). Cultures exposed to a period of constant darkness followed by continuous light produced an abundance of spores (Atkins et al., 1967). Continuous radiation with near-UV light (366 nm), shortened the time required to reach maximum sporulation (Ohmori and Nakajima, 1970; Ahn and Chung, 1974) but the rate slowed down remarkably under darkness (Namai et al, 1977).

Under natural conditions, many conidia could be detected at temperatures higher than 90° C (Mogi and Yanagita, 1970) and in the dry season (20.6-25° C) after a drizzle (Ou et al., 1974). The release of conidia was rapid at 95-100% RH at 11-26° C (Ono, 1966; Yamanaka et al., 1976). The requirement of a drop of water for take-off is suggested by Ono and Suzuki (1959) and the process is triggered during the day by rain showers (Ou et al., 1974; Kim and Min, 1989) or by splashing (Kato, 1974).

Alternate changes of light and darkness under saturated humidity are essential for the release of conidia (Asai, 1960; Barksdale and Asai, 1961). Wind is the strongest factor affecting the take-off of spores (Rotem, 1983). Abundant conidia are released at higher wind velocities (5 m/sec) under 80 to 100% RH (Ono and Suzuki, 1959; Misawa and Matsuyama, 1960). However, many spores are trapped within a canopy of plants at less than 1 m/sec wind velocity 2 m above the ground.

The greatest number of conidia on a single lesion was found on the ninth day after inoculation (Barksdale and Asai, 1961; Villareal et al., 1980). However, Kato et al., (1970) reported that maximum sporulation occurred 7 to 12-d after inoculation although sporulation continued for 60-d (Suzuki, 1975). Hashioka (1943 b) also demonstrated that conidia on the B1 infected panicles disseminate only in the air at more than 90% RH. Toyoda and Suzuki (1952) reported on spore production from different types of lesions in a moist chamber and found that small lesions produced fewer spores and took longer than large lesions.

Substantial variation is observed in the liberation of spores from lesions of the same size, but differing in colour. Maximum spore production is seen in grey centred lesions with dark purple to brown margin (Kato et al., 1970; Kato and Kozaka, 1974). Villareal et al.(1980) estimated that the total number of conidia per lesion varied from 4800 on cv. IRAT 13 to 33,000 on IR 442-2-58. The total number of spores liberated increased with lesion size in a colour series but the highest spore density was observed from 3 mm² lesions purple grey (pg) and yellow brown grey (ybg) and from 5-6 mm² lesions of yellow purple grey (ypg) and brown grey (bg). The spore density went up with increased lesion area (Fig.22). Broad B1 lesions, excluding purple grey, had transparent centres which were devoid of spores (Dhua, 1989).

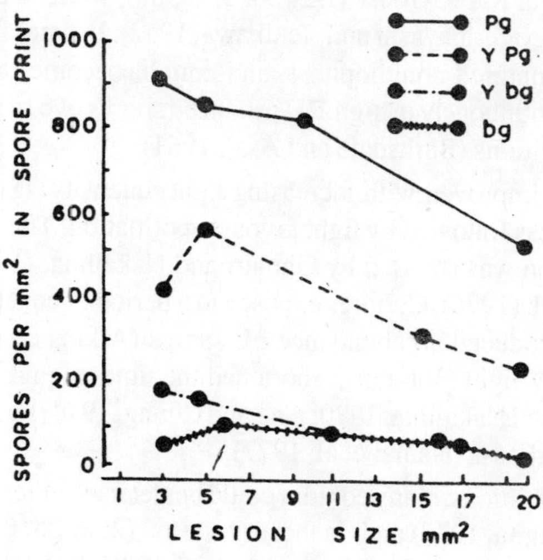

Fig. 22: Relation between size, colour and spore density in spore prints of rice B1 lesions. pg: purple gray, ypg: Yellow purple gray, Ybg: yellow brown grey and bg: brown grey. (Dhua, 1989).

Recently, Krishnan et.al (1992) have studied the spore releasing capacity of various types of B1 lesions using Spore Releasing Monitor (SRM) against weather parameters (Fig. 23). The SRM found to be ideal for collecting spore print from lesions on the coated micro slide under field conditions at rice canopy level. Purple grey lesions produced highest number of spores while spore density was least with brown coloured lesions.

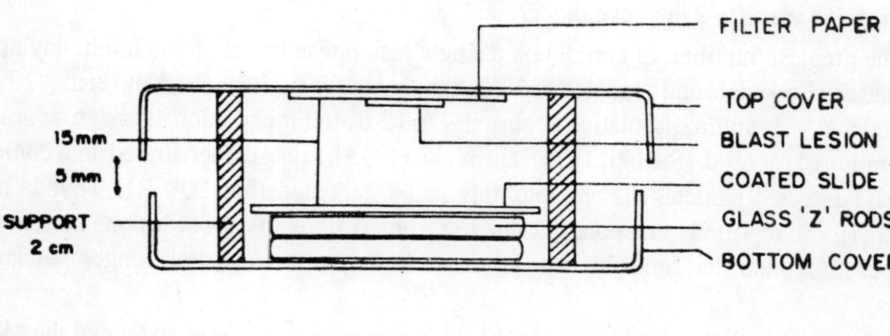

Fig. 23: Spore release monitor (SRM) (Krishnan and Manibhushanrao, 1992).

The preinfection water stress may determine the sporulation potential as the lesions on plants that had severe leaf roll produced up to 3.5 times more conidia than did those on unstressed plants (Gill and Bonman, 1988). Similarly, the spore production potential is related to the level of partial resistance of the host (Yeh and Bonman, 1986; Castano et al., (1989).

Spore discharge

Although sporulation of the B1 fungus on diseased nodes and culms occurs between 10 and 35° C, with the optimum between 18 and 30° C. However, the release of conidia from the conidiophores does not occur in this range. Kuribayashi (1928) and Kato and Kozaka (1974) studied the effect of temperature on lesion enlargement as well as the sporulation, potential of B1 fungus. Exposure of plants to 32°C causes the lesions to expand rapidly on the first eight days and levelled off shortly thereafter (Fig. 24 a). The potential reached a peak value earlier at high temperatures than at medium and lower temperatures (Fig.24b).

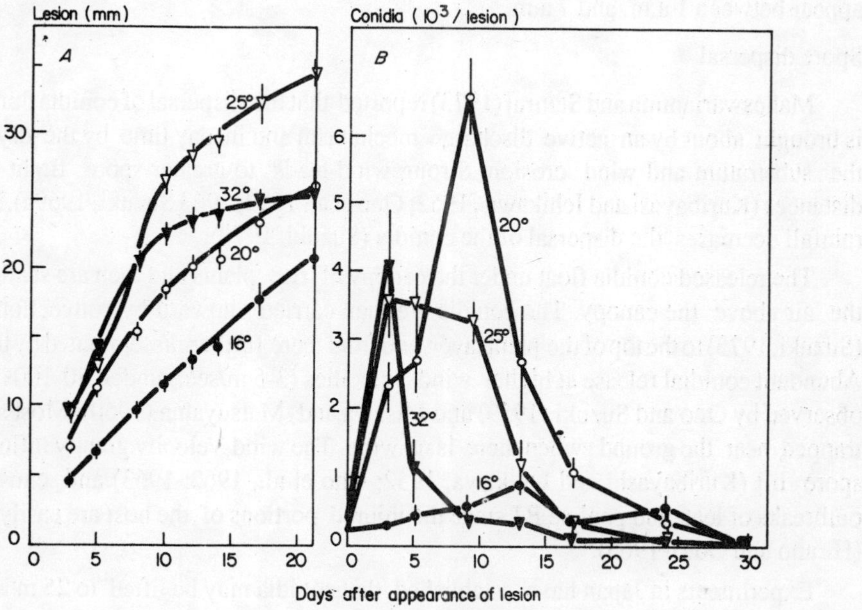

Fig. 24ab: Effect of temperature on enlargement of leaf B1 lesions (a) and on sporulation potential (b) of *P. oryzae* (Kato and Kozaka, 1974).

Light has no appreciable effect on the dissemination as most conidia disseminate at night (Kuribayashi and Ichikawa, 1952). The process begins after midnight (2 a.m.) increases till day-break (6 a.m.), and ceases at sunrise. This phenomenon seems to be more related to the humidity of the atmosphere, especially under dew conditions, rather than to the light factor (Hashioka, 1948).

A diurnal periodicity of *P. oryzae* conidia was observed in air under controlled conditions (Barksdale and Asai, 1961; Suzuki, 1969a) and in the field (Ramaligam, 1966; Ou et al., 1974). The dew period or the presence of water for spore release was demonstrated by the correlation between the duration of dew and the amount of spore released (IRRI, 1976; El Refaei, 1977) and *in vitro* conditions (Leach, 1980).

The presence of water droplets is quite essential for spore discharge and takes place within 2 min and especially during the initial 30 sec. Ingold (1964) reported that the conidium is violently discharged to a short distance by the bursting of the minute stalk-cell attached to the conidiophore. A typical lesion produces as many as 2000-6000 conidia each day and conidial production continues for 10-14-d under laboratory conditions. It is estimated that in extreme cases, 15 million conidia may be deposited in a square metre of rice field a day (Ou, 1985).

Kim et al. (1990) studied the conidia release phase critically on three different aged lesions (3 and 7-d old after lesion appearance and 4-d old but pre-existing conidia). These were removed and measured at one hour intervals for 48 h to pursue the successive conidial release under natural conditions. All the lesions exhibited 2-3 peaks on the 1st and 2nd day respectively, but the 7-d old lesions revealed only one peak on the first day. The studies revealed that the conidial release pattern seems to have a regular rhythm and most peaks appear between 1 a.m. and 7 a.m.

Spore dispersal

Maheswariamma and Samraj (1973) reported that the dispersal of conidia during night is brought about by an active discharge mechanism and in day time by the drying up of the substratum and wind erosion. Strong wind leads to greater spore flight to longer distances (Kuribayasi and Ichikawa, 1952; Ono et al. 1962, 1963 Suzuki, 1969a).However, rainfall decreases the dispersal of the conidia (Suzuki, 1970).

The released conidia float under the canopy of rice plants and then are scattered into the air above the canopy. The conidia are then carried upward by convection currents (Suzuki, 1975) to the top of the plant layer, and from there further disseminated by the winds. Abundant conidial release at higher wind velocities (3-5 m/sec.) under 80-100% RH was observed by Ono and Suzuki (1959) and Misawa and Matsuyama (1960). Most spores are trapped near the ground when there is no wind. The wind velocity greatly influences the spore lift (Kuribayashi and Ichikawa, 1952; Ono et al., 1962, 1963) and causes severe outbreaks of leaf and panicle B1 since the injured portions of the host are easily infected (Hirano and Goto, 1963).

Experiments in Japan have established that conidia may be lifted to 25 m above the ground and carried horizontally for about 20 km. It is known that conidia have also been trapped on slides in aircraft flying at over 2100 m (Ou, 1985).

Kato (1974) and Kato and Sasaki (1974) calculated the conidial dispersal in plant canopy by horizontal as well as vertical distribution using Schrödter's theory. Rain fall (1.5 mm) for 3 h brought down eight times as many conidia as were deposited during the same time in a similar area exposed to wind (Suzuki, 1965). Kim et al. (1989) also reported that conidial release was more under rainy conditions. The amount of conidia dispersed varied with wind velocities and the relation between the number of conidia and the distance from inoculum source (Fig.25a) and the height of sampling (Fig. 25b) was contemplated (Suzuki, 1969 b).

Spore deposition

The position of the leaf, that is the leaf angle, influences the number of spores deposited

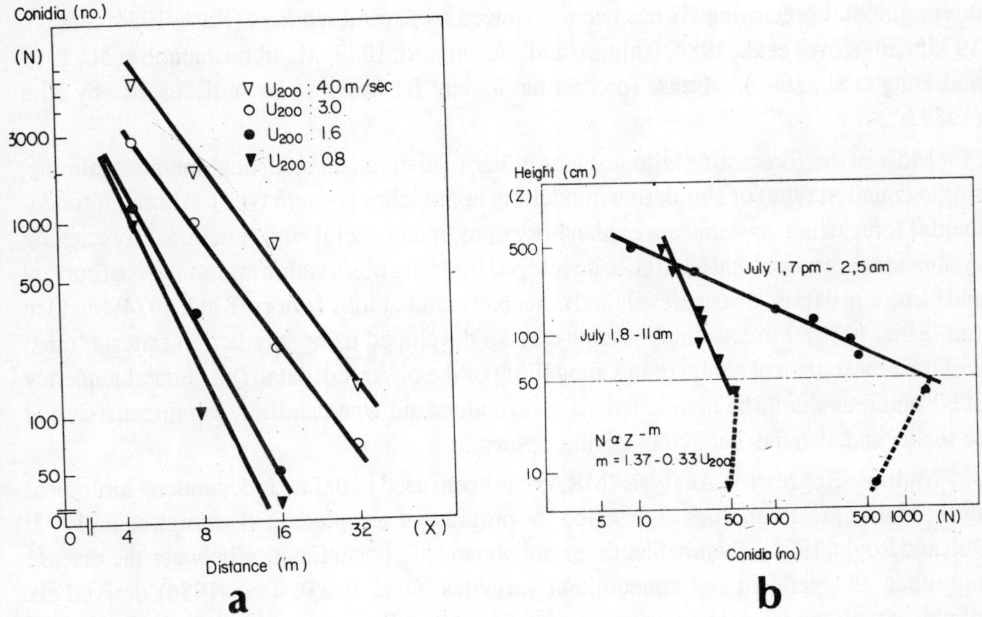

Fig. 25ab: (a) Relation between the number of sampled *P. oryzae* conidia and distance from an inoculum source under different wind velocities. This relation is given by the following formula: $N = AX^{-n}$, where n is from 0 to 1.9 and $A = 155.80\ (Q/U_{200})^{0.3}$, U_{200} shows the wind speed 2 m above the ground and Q means the number of conidia in an inoculum source. (b) Relation between height of sampling and number of sampled conidia under different wind velocities. The canopy of rice plant was 50 cm high (Suzuki, 1964).

on it. In general, the spores fall perpendicularly in the atmosphere with a weak wind and are deposited more on horizontal than on non-horizontal plant surfaces (Suzuki, 1965; 1969a). Kuribayashi and Ichikawa (1952) calculated the correlation coefficients for the spore number deposited on the slide and the severity of the disease. The number of conidia deposited (N) can be estimated by the following equation: N = number of conidia sampled × revised spore ratio by height using Suzuki's formula × coefficient of deposition × (leaf area index/area of the sampler). The coefficient of deposition was empirically determined as 1/200 (Kim et al., 1975).

Forecasting

Japan has developed an elaborate and effective forecasting system for determining the optimum time for application of fungicides to derive maximum control of the B1 disease. This will enable farmers to use minimum fungicidal sprays before the spread of the disease.

Earlier, Kuribayashi and Ichikawa (1952), Hori et al. (1960), Hori (1963), Ono (1965), Padmanabhan (1965), Ramalingam (1966), Yamaguchi (1970) and Kato (1976) studied the forecasting methods based on information available on the fungus, the host as well as the environment. Forecasting B1 has been reviewed by several workers (Kato, 1974; Suzuki, 1975; Kingslover et al., 1984, Ishiguro and Hashimoto, 1989; Manibhushanrao et al., 1989 and Teng et al., 1990). Disease forecasting in leaf B1 has also been discussed by Kim (1987 b).

Most of the forecasting is done using either mathematical/statistical methods (simple, single equation type) or simulation modelling approaches (system type). Although fundamental forecasting systems are explanatory ones, many useful and functional forecasting systems are empirical and have been developed from the observation and analysis of current and historical data on disease levels and other biotic and abiotic factors (Kato, 1974; Madden and Ellis, 1988). Forecasting systems can be developed using prediction criteria 'rabs' without any statistical analysis and modelling of the observed data. The general tendency of B1 outbreak should be mentioned at first to understand when and how the forecast should be made and also the behaviour of the spores.

Multiple Regression Analysis (MRA) has been used to define independent biological and climatological variables to describe the progress of an epidemic (Burleigh et al., 1972; Butt and Royle, 1974). This enables the establishment of a linear function between the disease and other independent and concomitant variables (Ono, 1965). Tsai (1986) derived six reliable equations ($r^2 = 0.89$ to 0.96) by using MRA between meteorological variables (average RH, h of RH over 95%, rainfall and rainy days) and the percentage of leaf area infected by *P. oryzae*.

Blast forecasting has been attempted by many workers through prediction of the disease progress, disease severity, disease incidence and time of its outbreak based on host plant criteria, disease features, spore catch, weather parameters, multiple criteria and N-fertilizer application. The available literature is reviewed here.

Host plant criteria

Takahashi (1958a) attempted host predisposition to B1 pathogen, using leaf sheath inoculation under experimental conditions. Hori (1963) studied forecasting equations that used the number of silicated cells in flag leaves as follows :

$$y = 55.5 - 22.1 \log (X)$$

where, y=% neck rot, X = number of silicated cells/10 microscopic fields (150X). This equation was valid only for cv. Asalis 1 in western Japan. Kim (1982) improved this with modifications. Further, Hori (1965) attempted other forecasting equations based on starch accumulation in leaf sheath. One such equation is given by as,

$$y = 53.57 - 0.81 X$$

where, y = % leaf B1 at the most severe stage, x = ratio of length of leaf sheath that is dyed by potassium iodide as a measure for starch density in the leaf sheath ($r = 0.78 - 0.91$). The limitation of the equation is that the starch accumulation ratio varies with seasons.

Disease feature

Hori (1963) and Kim (1982) attempted the forecasting of neck or panicle B1 based on the number of diseased leaves/hill respectively.

$$y = 1.34\ x + 3.4 \text{ (neck B1)}$$

$$y = 2.84\ x + 19.3 \text{ (panicle B1)}$$

where, y = % neck or panicle B1, x = no. of diseased leaves/hill at booting stage. Both equations were highly significant ($r = 0.8$).

Shiyomi (1970) developed a model using the lesion age as,

$$r_p(s,t) = [a/T(p)]\ [(t - s)^{p-1} \exp. (t - s)]$$

where p and a are parameters, T= gamma function, $r_p(s,t)/a$ = density function of gamma distribution.

Kiyosawa (1972c) created a model based on disease severity by computing the number of lesions as,

$$dy/dt = y\ [1 - (t/T)]$$

where, y = number of lesions and dy/dt = 0 at time T, and this equation can be transformed by integration as,

$$y = K \exp. [t - t^2/2T]$$

Kiyosawa and Shiyomi (1972) developed a model based on disease gradient, $y = \beta e^{-\alpha d} = a\gamma e^{-ad}$ where, a = number of initial susceptible type lesions on the initial source plant, γ = the degree of increase of susceptible type lesions on the initial source after the first infection generation, α = a dispersal gradient and d= distance between the initial and other plants.

The equation $y = e^{2.64} + 0.007$ (r=0.985) allows the yield loss to be estimated when the total value of the area under the rice B1 disease curve (leaf, neck and panicle) is known. The regression equation $y = 4873.43 - 43.68 x_1 - 77.84 x_2$ ($R^2 = 0.73$) expresses the yield loss caused by rice B1 disease in the leaves and panicles (Montoya et al., 1984).

Chiba et al. (1972) studied the infection rate to determine the final lesion number while Sasaki and Kato (1972) predicted B1 by counting the number of diseased spikelets plotted against time and extrapolating the curve for the next 6-d.

Prabhu and De Faria (1982) have reported a nonlinear relationship between leaf B1 index at booting stage (X) and % infected panicles at dough stage (y),

$$y = 6.36 + 9.25\ X - 25367\ X^2$$

in which X ranged from o (no lesions) to 4 (more than 25 lesions).

Spore catch

In Japan, Kuribayashi and Ichikawa (1952) computed regression equations to relate the percent neck B1 incidence (y) to the number of spores caught during 5-d periods from July to September (X). The B1 equation was :

$$y = 0.4\ X + 10.81\ (r = 0.97)$$

for spore catches of the third 5-d period of August. Subsequently, Ono (1965) suggested a simple linear regression equation to relate the outbreak area of neck B1 (y) to the number of spores caught during the 5-d period of August.

Kiyosawa (1972c) analysed the daily catch data for 12 years. He developed an equation for forecasting the number of lesions or the cumulative spore collected. $Y = y_o e^{rt}$, where, t= time in days, Y = the number of lesions or cumulative spores, r = rate of disease increase or multiplication of rate of pathogens and Y_0 = number of lesions or cumulative spore number at the time of infection.

In India, Padmanabhan and Chandwani (1960) and Padmanabhan et al. (1971) correlated the spore counts or air spore and the prevailing meteorological stages as well as growth stages of the rice plant. Maheswariamma and Samraj (1973) also correlated the atmospheric spore content and B1 incidence. Muralidharan and Venkatarao (1980) found that leaf B1 occurrence in the field could be predicted about 14-d ahead of time when plants consisting of young, susceptible rice seedlings were exposed and then examined for lesions.

Weather parameters

Padmanabhan (1965) determined several regression equations for seed bed infection, leaf and neck B1 infection followed by weather conditions such as temperature and RH. Ono (1965) used several regression equations based on external variables. e.g.

$$y1 = 12703.73 - 257.34\,x\ (r = 0.076)$$

$$y2 = 144.8 - 5.35\,a\ (r=-0.69)\ \text{or}$$

$$y2 = 0.18\,b - 5.6\ (r = 0.88)$$

$$z = 0.4\,x + 6.67\,y + 29.06\ (r = 0.96)$$

where, y1 & y2 = area of neck and leaf B1 outbreak respectively, z = area of disease outbreak (in 100 ha)x = mean % sunshine in June and July, a = mean temperature from 20 to 25 June, b = precipitation from 26 to 30 June, and x = sum of sunshine days in July, y = mixture index. The equations/methods had coefficients which varied across districts.

Ramalingam (1970) has reported that the temperature range 25-27°C RH of 86-90%, mild wind and darkness (midnight to sunrise) are critical for the development of an epiphytotic.

A method to forecast leaf B1 outbreaks was developed by Koshimizu (1982) based on weather conditions for infection, daily weather data on rainfall, temperature, sunshine hour and wind speed. Kingslover et al. (1984) described the minimum dew period required for infection as a function of temperature

$$1/D = 0.265 - 4.28/T\ (r = 0.83)$$

where, D = dew period in h, T = temperature in °C. This could be used to predict the occurrence of infection. In Taiwan, Tsai and Su (1984) and Tsai (1986) obtained data on disease progress and weather during planting of cv. Tainung 67 from 1979-84. Vehara (1985) used multivariate statistical analysis techniques such as principal components and cluster analysis to classify regions according to the degree of occurrence of leaf B1 and panicle B1. MacKenzie (1981) studied the initial occurrence and subsequent increase of the disease incidence by evaluating the microclimate. Kim et al. (1987, 1988) have developed an algorithm to index weather with respect to B1 susceptibility.

Weather and spore density

Kim et al. (1975) showed the relation between the number of B1 lesions and the number of spore density as well as the wetting period of leaves. El Refaei(1977) correlated the number of lesions with the dew period and the number of airborne spores. He developed several linear regression equations that related to weather variables and airborne inoculum spore density separately. His equation forecasts leaf B1 5-d ahead of the time.

$$y= 2.9 - 0.945d - 0.0098s + 0.1520d^2 + 0.004ds - 0.000000002d^2 s^2$$

where, y = no. of lesions/seedlings, d = dew period in h, s = no. of spores/2.8 1 air. This work is limited to B1 nursery experiments only. Other workers (Chien et al. 1984; Tsai and Su, 1985) used the above approach, while Tsai (1986) and Tsai and Su (1984) included spore catch data on several cvs.

Spore dispersal and deposition

To forecast occurrence of rice B1, several computer simulation models have been developed (Gunther, 1986; Hashimoto et al., 1984; Ishiguro, 1986; Oota, 1982; Takai et al., 1985). However, these models cannot sufficiently quantify the dispersion and deposition of B1 fungus spores in the rice canopy since quantitative data on the distribution of susceptible/non-susceptible tissue and inoculum are still lacking (Gunther, 1986). In view of this, (Koizumi and Kato (1985, 1986, 1991); Koizumi (1988) quantified the spore distribution and deposition in the rice canopy and analysed the distribution of B1 lesions on susceptible rice tissues in addition to studying the microclimate under the rice canopy.

Wind and turbulence

Earlier, Suzuki (1969a) studied the effects of wind speed and turbulent diffusion on liberation, dispersion and deposition of *P. oryzae* spores in the rice canopy. Further, the experimental data of Koizumi (1988) and the equations of Uchijima (1962), Inoue (1963), and Hore (1981) are given as follows :

The wind velocity (u) profile above a rice canopy can be expressed as :

$$u(Z)= (V_*) / k\ln ((Z-d) / Z_o)$$

where Z is height, V_* is friction velocity, d is zero plane displacement, Z_o is surface roughness length, and k is von Karman's constant.

$$d=0.2395\ Z_H^{\ 1.146305}\ (R^2 = 0.694^{**})$$

$$Z_O = 0.28875\ Z_H^{\ 0.82135}\ (R^2 = 0.510^{**})$$

$$V_* = 0.152537\ u_{180}\ (R^2 = 0.831^{**})$$

where Z_H is rice plant height and u_{180} is wind velocity at 180 cm above the ground in the rice field.

$$u(Z) = u_o \exp [-A(1 - Z / Z_H)]$$

where u_o is wind velocity at the level of plant height and A is the extinction coefficient of wind velocity in the rice canopy.

$$A = 0.4887\ L_{AI} + 0.9038\ (r = 0.597^*)$$

where LAI is leaf area index.

The turbulent diffusion coefficient (K) above a rice canopy (Uchijima, 1962) can be expressed as : $k(Z) = KV_* (Z\text{-}d)$

$$k(Z) = k\ (1 - d/Z_H)\ V_*\ Z\ (1\text{-}0.5 \int_2^{Z_H} (Z)^{H})\ dz/(V_*^{\ 2})]$$

where $c = 1.6\ (u(Z)^{-1/2})LAD^{(Z)}$ (data from Hori,1981)

(LAD : Leaf area density).

To avoid zero values of K and V_* in the model, value was changed to 2.0 cm/sec u_{180} was 0.0 cm/sec.

Sporulation index

From the data of Kato and Sasaki (1974) and Iwano (1983), the maximum number of conidiophores formed on a unit leaf lesion area was determined to be 333/mm^2 and the sporulation index was given as follows: 1.0 and 0.6 for lesions on the second youngest (n-1) and third youngest (n-2) leaves, respectively 0.3 and 0.2 for acute type lesions and chronic type lesions on the fourth youngest (n-3) leaves respectively; and 0.05 for lesions on the fifth youngest (n-4) leaves.

In the model, conidiophores grow and spores are formed when RH is more than 93% (Hemmi and Imura, 1939) and temperature range 15 - 32 at 24°C (Kato and Sasaki, 1974; Arai and Yoshino, 1987) were maintained.

The rate of conidiophore formation of *P. oryzae* was assumed to fit a normal distribution curve (with a standard deviation of 1.0). The relationship between temperature (t) and average hours (h) required for mature conidiophore formation was determined based on the experimental data of Arai and Yoshino (1987):

$$h = -0.25\,t + 14.35 \quad (15 - 27\ ^\circ C)$$
$$h = 0.59\,t - 8.5 \quad (\text{above } 27\ ^\circ C)$$

Spores of *P. oryzae* formed on each mature conidiophore. The relationship between temperature and time required for the formation of the first to the sixth mature spores on each mature conidiophore was determined by the following equations using the data of Arai and Yoshino (1987):

$$h_{s1} = 0.0275\,(t - 25.6)^2 + 0.78$$
$$h_{s2} = 0.0274\,(t - 26.4)^2 + 1.28$$

where h_{s1} and h_{s2} are h required for the formation of the first spore and the second to sixth mature spores on each conidiophore.

Spore liberation

Mature spores formed on the lesions are liberated from conidiophores by wind, rain, dew and high humidity and the liberation rates were obtained from Suzuki's (1969a) results. When wind velocity is 0.0-3.0 m/sec. the liberation rate (RI) of mature spores from conidiophores was decided to be 0.002. When wind speed is greater than 3.0 m/sec. RI was calculated as follows:

$$RI = 0.39\,u(Z) - 1.15$$

RI due to rain was set at 0.8 and RI due to high humidity (RH, above 93%) and dew was set at 0.5. When total RI was above 1.0, RI was changed to 1.0.

Spore concentration

MacCarthey and Fitt's (1985) simplified the Gradient Transter Theory to calculate the spore dispersal in the rice canopy:

$$d/dZ\,(K_Z dC/dz) + v_s\, dC/dZ = -S(Z)$$

where C is spore concentration, v_s = the fall speed of the spores (0.007) m/sec by Shimizu, 1980) and S= the source or sink term (because of particle loss, this is always negative). The equation assumes that the source emits spores continuously every h and therefore, the average concentration over time (1 h) can be computed by the equation. This equation was solved by considering the concentration C_n at many discrete heights and expressing the vertical concentration gradient in terms of definite differences as was done by Legg and Powell (1979):

$$[(-K_{n-1}(C_n - C_{n-1}) / D_Z) - (-K_n(C_{n+1} - C_n / D_Z)] / D_Z + (v_s(C_n + C_{n+1})/2 - v_s(C_{n-1} + C_n)/2)/D_z = B_n - DS_n$$

where D_z is 0.1 m in the model, B = loss of spores by the deposition, and D_s= the number of spores dispersed from the lesions.

The deposition of spores per unit volume can be expressed by the sum of impaction and sedimentation (Legg and Powell, 1979) as follows:

$$B = E_I{}^{u}CL_{AD}{}^{f}{}_{z} + E_S v_s{}^{CL}{}_{AD}\, f_x$$

where E_I is the efficiency of impaction, f_z and f_x are the ratios of leaf area projected into vertical and horizontal directions to the total leaf area, and E_s is the sedimentation efficiency (in the model, E_s is 1.0 [Legg and Powell, 1979]). Aylor (1982) gives the following equation to compute E_I:

$$E_I = 0.86 / (1 + 0.442(uv_s/gL)^{(-1.967)}$$

where L is a characteristic length of the plant element (like width of leaves) and g is the acceleration due to gravity. This equation was used to estimate E_I:

Suzuki (1969a) reported that rain decreased the spore number in the air and he evaluated the following formula to compute the same as:

$$y = 0.417484\, R_A{}^{-1.42149}$$

where, y is the rate of decrease in spore concentration and R_A, the rainfall (mm/h).

Hashimoto et al. (1984) indicated the relationship between rainfall and washing off of spores due to rain by an empirical equation. Kozumi and Kato (1990) incorporated their equation into the model. The time required for appressorium formation and its penetration of *P. oryzae* was revealed by Suzuki (1969 b) and Yoshino (1979) respectively. It is considered that the spores that formed appressoria which in turn produced penetrating pegs in the rice leaves are hardly removed by rain (Chiba et al., 1973, Yoshino, 1979). To compute the number of residual spores on the leaves, it was assumed that spores were not removed from the leaf surface by rain three or more hours after germination.

The spore concentration over the rice canopy was computed by the equation of Inoue (1963) in Suzuki (1969a).

$$C_Z = C_o (Z/Z_o) (-v_s / (kV_*))$$

where, C_o is the initial spore concentration and Z_o is the initial height.

Multiple criteria

Ono (1965) predicted neck B1 disease (y) at Yamanashi prefecture, Japan, using the formula:

$$y = 12703.73 - 257.34\times \quad r = -0.761$$

In Yamaguchi prefecture, to forecast the areas of leaf B1 outbreak (y_1) and the areas of rice fields damaged by neck B1 (y_2), the following formulae were used.

$y_1 = 144.8 - 5.35a$ $\quad r = -0.69$

$y_1 = 31.3 - 0.366b$ $\quad r = -0.79$

$y_1 = 0.18c - 5.6$ $\quad r = 0.88$

$y_2 = 75590.7 - 2886.96d$ $\quad r = -0.65$

where x = average percentage of sunshine in June and July, a =average temperature in June 20-25, b= maximum temperature in June 26-30, c = precipitation in June 26-30, d= average temperature in July 1-5, and *r* = correlation coefficient.

Kim et al., (1975) developed forecasting for leaf B1 based on the number of trapped spores and the wetting period of leaves by the following equation :

$$PSN = DSN \times RA \times MIP \times RI$$

where DSN = CSN × C × LA/GA; PSN = number of spores presumed to penetrate into the leaves; DSN = number of spores presumed to deposit on leaves; RA = ratio of spores which formed appressoria; MIP = maximum ratio of appressoria that succeeded in penetration under favourable conditions; RI = ratio of penetrated appressoria to MIP each day (calculated using the wetting period of leaves and mean temperature of that period); CSN = number of trapped spores corresponding to wind velocity and plant height; C = correction coefficient ; LA = leaf area; GA= cover glass area. The number of lesions estimated by this formula coincided with the actual number of lesions that occurred in the field. Thus, the degree of leaf B1 severity can be forecasted one week ahead of lesion appearance.

Forecasting of panicle B1 has been developed by statistical methods. By the relevant equation, the number of spikelets likely to be blasted in the next six days can be predicted. (Sasaki and Kato, 1972). To forecast panicle B1 outbreak, it is essential to compare the number of conidia deposited, the curve of daily conidium collection, the level of leaf B1 outbreak and proneness of rice and meteorological factors (Kato, 1974). There is a close correlation between leaf B1 incidence immediately before heading and panicle B1 occurrence. Further, Ishiguro and Hashimoto (1988) estimated yield loss by leaf and panicle B1 by simulators, called deterministic and stochastic respectively.

Other workers (Murumatu and Koyanagi, 1977; Kono, 1977; Shimizu, 1980; Hashiguchi and Kato, 1983) have also tried to develop statistical methods for quantitative forecasting, such as multiple regression analysis. However, the predictions from most of these workers based on regression analysis cannot be extrapolated across cvs., regions, planting periods and methods of cultivation.

Kingslover et al., (1984) presented multiple regression equations for predicting the number of leaf B1 lesions for different inoculum densities as well as different cvs. e.g.

$$y = 2.36 + 0.01\ x_1 + 0.05\ x_2 + 0.02\ x_3$$

where, y = mean number of lesions at zero inoculum density, x_1 = amount of dew, x_2 = h of dew, x_3 = leaf area at inoculation time; or

$$y = 2.28 + 0.09x_4 + 0.02\ x_5$$

where, y = mean number of lesions on a cv. Colusa, x_4 = h of dew and x_5 leaf area.

Manibhushanrao and Krishnan (1991) developed a computer-based forecasting system, to simulate the rice leaf B1 based on multiple criteria. Using stepwise regression analysis, they evaluated the best fitting model to estimate the number of *P. oryzae* spores over the paddy fields as :

$$Y_1 = 123.068 - 3.08\ x_1 + 0.321x_2 - 0.37\ x_3$$

where, Y_1 estimated number of spores; x_1 = daily maximum temperature; x_2 = daily maximum RH; x_3 = daily minimum RH. Further, they created models to predict the B1 disease progress on the cvs. IR 50 (Y_2) and IR 20 (Y_3) by correlating the disease incidence with weather parameters as well as the estimated number of spores. The predicted values were close to the observed values (Fig. 26).

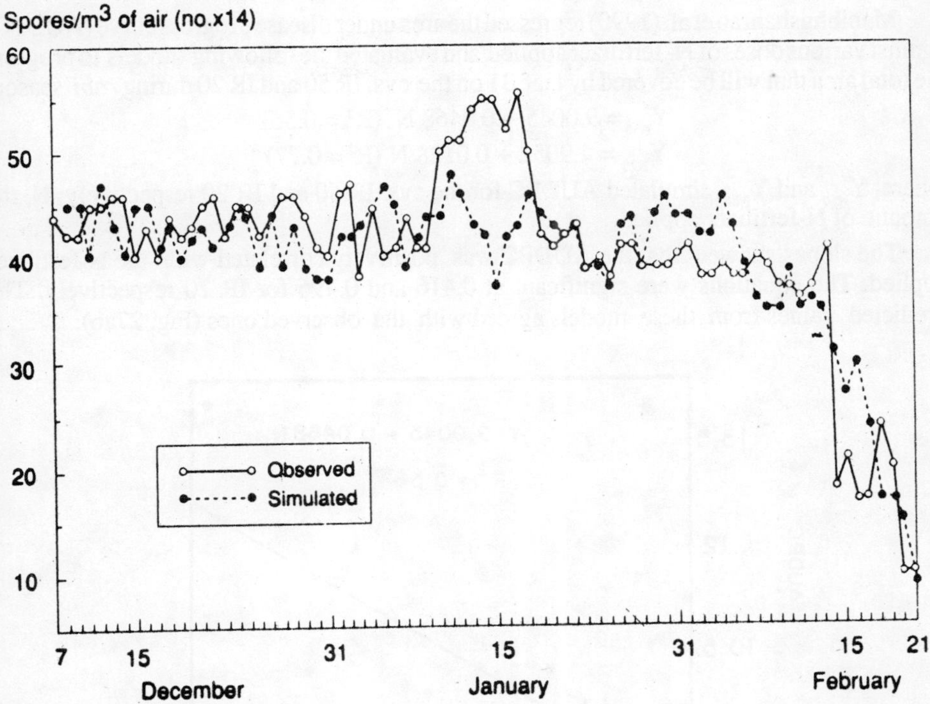

Fig. 26: Observed and simulated number of *P. oryzae* conidia m^{-3} of air. (Manibhushanrao and Krishnan, 1991).

$$Y_2 = 0.5561 - 0.0011\, x_1 + 0.0017\, x_2 + 0.2493\, x_3$$

where, x_1 = total number of spores summed for a 7-d period; x_2 and x_3 = average minimum temperature and average amount of dew for the preceding seven days from the date of disease observation respectively.

$$Y_3 = 1.0428 - 0.00007\, x_1 - 0.0102 x_2 + 0.0659 x_3$$

where, x_1 = total number of spores summed for a 7-d period; x_2 and x_3 = average minimum RH and amount of dew for the preceding 7-d from the date of disease observation respectively.

Models may predict the disease progress and help decide whether a fungicide application is needed. Moreover, specific models simulating disease development over several seasons may detect strategies to delay fungicide resistance or race adaptation. Further to simulate an epidemic, it is important to develop a model by analysing each phase of the life cycle and hence for the *P. oryzae* - rice system, the preceding sequences have to be worked out.

Forecast weather charts are useful tools in prediction of disease-weather model (Bourke, 1970). Prediction of the zero date time is more reliable. Differences in growth phase can be observed between resistant and susceptible rice plants and this fact can also be used in forecasting (Ono, 1975). Predictions based on nutrient content in leaf tissues using simple or multiple regression equations have been developed (Hori et al., 1957). However, they are not available on timely, regular and localised basis (Kim et al., 1988).

N-fertilizer application

Manibhushanrao et al. (1990) regressed the area under disease progress curve (AUDPC) against various doses of N- fertilizer applied and evaluated the following models to progress the total area that will be covered by leaf B1 on the cvs. IR 50 and IR 20 during *rabi* season.

$$Y_{N50} = 9.0045 + 0.0468\ N\ \ (R^2 = 0.56)$$

$$Y_{N20} = 1.9052 + 0.0128\ N\ (R^2 = 0.77)$$

where, Y_{N50} and Y_{N20}, simulated AUDPC for the cvs. IR 50 and IR 20 respectively; N, the amount of N-fertilizer applied.

The slopes showed that the AUDPC was positively correlated with the N-fertilizer applied. The equations were significant at 0.416 and 0.496 for IR 20 respectively. The predicted values from these models agreed with the observed ones (Fig. 27ab).

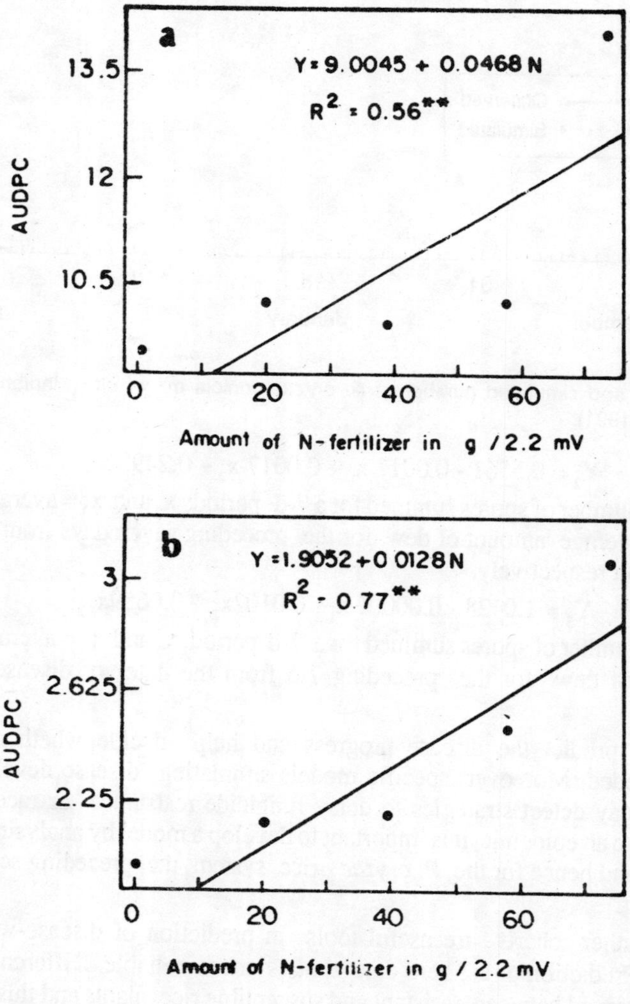

Fig. 27ab: Relationship of N-fertilizer to area under disease progress curve (AUDPC) in simulated epidemics of *P. oryzae* on the cvs. IR 50 (A) and IR 20 (B) during *rabi* season (1989-90). (Manibhushanrao and Krishnan, 1991).

Weather recorders and spore traps

The precision of weather monitors and spore traps is very important to get accurate data for quantification and assessment of disease. Traditionally, non-electronic sensors were used to monitor weather data from the fields (Madden and Ellis, 1988). Surface moisture from dew, rain, fog, irrigation or guttation drops affects foliar pathogens and is a key component of most disease forecasters. Special sensors are available to measure the duration of leaf wetness (Schnelle et al., 1963 and Hackel, 1984), thermopile pyranometers for solar radiation (Rosenberg et. al., 1983), thermistor or thermocouple for soil temperature (Sutton et al., 1984), thermoelectric, sonic and ionic anemometers for wind speed (Rosenberg et al., 1983), a tipping- bucket gauge for rain intensity (Sutton et al., 1988) and micropod mini meteorological station for all climatological parameters.

Spore dissemination studies were made earlier by using horizontal rest slides (Kuribayashi and Ichikawa, 1952). Subsequently, Ono et al, (1963) and Suzuki (1969a, 1974) used rotary samplers and correlated the B1 incidence with the number of spores collected in the trap. They could not yield either quantitative or qualitative data of air spore and the method proved unsatisfactory. Kim and Yoshino (KY) spore trap was used to estimate the amount of spores released from leaf B1 lesion under natural conditions (Kim and Yoshino, 1987). At IRRI Kramer's (Kramer et al., 1976) a 7-d spore sampler was used to study the diurnal and nocturnal periodicity of conidia released during the dry season. A few sophisticated automatic power operated volumetric spore traps were introduced. Hirst (1952) developed a suction trap with uniform efficiency which was further modified by Burkard manufacturing Co., UK. The latter was successfully used to trap the conidia of *P. oryzae* by Manibhushanrao and Krishnan (1991). Panzer et al. (1957) devised a 24 h suction-type collector for sampling quantitative and qualitative assessment which is more efficient than the Hirst trap. The efficacy of various traps has been discussed by Sutton and Jones (1976) and Wili (1985).

Simulation models

The computer based simulation modelling is a recent concept in modern sciences. Various techniques of simulation are being used currently to have a better understanding of host-pathogen complex. The resultant models enable to explore the factors that govern disease epidemics and to design the control system that minimize the yield loss. A brief description of the models that have been reported in the B1 literature, emphasizing their level of detail and utility in enhancing the B1 management is given here.

Leaf blast model

In Japan, the first leaf B1 model, BLASTL, a system analytic model that simulates the epidemic was developed by Hashimoto et al., (1982, 1984). This model, has been verified under field experiments for disease progress and use to predict efficient time for the fungicide application. A polycyclic model, PYRICULARIA, was developed by Gunther (1986) using information from the literature. PYRICULARIA was modified by Tastra et al. (1987) for upland rice farming in Indonesia and was subsequently called PYRNEW. The BLASTAM system (Koshimizu, 1983 1988; Hayshi and Koshimizu, 1988 and Sakai et al. 1990) predicted weather conditions favourable for infection using weather data of the 'automated meteorological data acquisition system' (AMeDAS) via telephone in real time. BLASTAM can be used to predict the onset of leaf B1 epidemic and the time for the first fungicidal application.

Later, LEAFBLAST was developed (Choi, 1987; Choi et al. 1987, 1988) using data from the growth chamber experiments and the literature. This model does not consider ontogenetic changes caused by environmental conditions but computes spore germination, infection, latent period, lesion growth and spore production. The model developed by Torres (1986) increases the complexity of the logistic growth function and the components considered chiefly are the pathogen's life cycle such as sporulation, spore dispersal, spore landing and infection. This model was further refined and validated at IRRI. Recently, EPIBLA (EPIdemiology of BLAst) was developed (Manibhushanrao et. al. 1990, Manibhushanrao and Krishnan, 1991) to simulate the leaf B1 incidence in tropics and the parameters used are from meteorological data, N-fertilizer application and susceptible/resistant cvs. The disease severity of leaf B1 on photosynthesis and crop growth of rice crop was modelled by Bastiaans (1991).

Panicle blast model

The first developed model in Japan for panicle B1 was from Takasahi (1982). He treated a spore deposition and penetration as a stochastic process and subdivided each panicle into small infection site units. This model does not account for secondary infections. The most comprehensive stochastic simulation model-panicle B1(PBLAST) was developed and verified by Ishiguro (1986) and Ishiguro and Hashimoto (1988, 1990) AMeDAS weather data wetness, data on host development, time of cv. cultivation practices, leaf and the number of spores formed on leaf lesion were used as input for PBLAST. It provides useful information for understanding pathosystems of panicle B1 and improving the method of fungicidal application.

Blast fungus simulator

Koizumi and Kato (1990) developed the model DISBL to simulate the dispersal, deposition and washing off spores of B1 fungus inside and above the rice canopy. This model suggests the wind speed and leaf area index of rice plants that affect spore concentration and the number of spores deposited within the rice canopy.

Blast pathosystem management

Blast and rice exist in an environment in which man's actions and attitudes determine the management options. Teng (1990a) explained how modelling and forecasting assist in B1 management. He stressed the need for modellers to provide a preliminary assessment of the environmental risk from bio-engineering organisms like B1 pathogen and suggested that simulation modelling may be the ideal tool (Teng 1990b). Further, Teng et al. (1991) reviewed the pathosystem to guide modelling and forecasting. They emphasized that the pathosystem is conceptually two interrelated subsystems, namely, the leaf B1 and panicle B1 pathosystems. The alloinfection from non-rice/host in initial epidemics increases the autoinfection in rice and hence the former is considered important in B1 management. B1 pathosystem management offers exciting challenges for the modellers for feature research.

Disease Control

Attempts to control rice B1 have been directed at the use of resistant cv., chemical protection using economic spray schedules, biocontrol as well as cultural practices. Ensuring adequate water management, observing field sanitation, judicious use of N-fertilizers on high yielding cvs. are some of the prophylactic measures that will check the disease to some extent. Integration of some of these methods will normally be used to combat this serious disease.

Breeding for resistance

Blast resistance has consistently been one of the important objectives of rice breeders in most rice growing countries. Breeding for B1 resistance in various countries has been reviewed from time to time in the US (Atkins et al., 1965), Japan Ito, 1965; Takahashi, 1965; Yamasaki and Kiyosawa, 1966; Ezuka, 1979), India (Padmanabhan,1965; 1979), Taiwan (Chang et al., 1965), Thailand (Dasananda, 1965); the Philippines (Ou and Jennings, 1969; Ou, 1979; Ikehashi and Khush ,1979) and China (Sun et al., 1987; Zhuge et al., 1990). Most studies revealed that the level of resistance in new cvs. increased but the spectrum resistance was not broad enough to be stable.

Systemic rice breeding work started as early as 1927 in Japan, and since then a number of B1-resistant cvs. have been released (Ito, 1965; Toriyama, 1972; Kiyosawa et al., 1975). The important ones are the Norin series, Norin 6 (Joshu × Senichi) and Norin 8 (Ginbozu× Asahi) developed in 1935 and 1936 respectively which were found resistant either to leaf or neck B1 but not to both. By crossing these two again, Norin 22 and Norin 23, which are resistant to both leaf and neck B1 were developed. Form the cv. Norin 22, a number of resistant genes were transferred to other B1 non-resistant cvs. This resulted in further improvement in getting nonlodging cvs. as well as other characters of both Norin 22 and 23. The following cvs. were identified : Koganenami (Norin 22 × Norin 23), Tacho (Norin 23 × Norin 22), Hatsunishiki and Honenwase (Norin 22 × Norin 1), Chokai (Norin 22× Rikku) and Chiyohikari and Yamabiko (Chukyoashi × Norin 22).

Ujihara (1951) developed B1 resistant cvs. Shinju and Futaba from an upland resistant cv. Sensho with the crosses made out of lowland cvs. Although cv. Sensho was used as a donor parent in many crosses in Japan, the progenies developed from it were never as resistant as Sensho. Further cv. Shuho (Futaba × Norin 6) was developed and again crossed with Norin 22 and five (Wakaba, Wasewakaba, Koganenishiki, Ulkonnishiki and Homarenishiki) excellent cvs. were developed. Again from the cross Sehuho × Rikku 132, Ginga was developed.

In China, breeding work began in early 1951 using two Chinese cvs. Toto and Reishiko as donors (Matsuo, 1952). Subsequently, Koyama (1952) released the Kanto series No.51

and Kanto No. 52 from Toto × Ginbozuchusei and Kanto No. 53, 54 and 55 from Norin 10 × Reishiko. Ito et al. (1961) selected Kusabue from the parents Kanto No. 53 × Norin 29 with improved quality and yield ability. Other cvs. developed are Yakara (Kanto No. 53 × Eiko); Teine (Kanto No. 52 × Eiko); Oyodo (Takara × Seusho 26) × Kanto No. 53; Mangetau-Mochi F_3 249 (Norin 25 × Toto × Norin 36) × Norin-mochi 45; Kagura-mochi $F_3$249 × Heinoku- mochi; Ugonishiki (Kanto No. 51 × Hatwanishiki). Kitamura (1961) developed several resistant cvs. using *indica* cv. Tadukan as a donor parent. Similarly, cv. Zenith from the US was used as a source of resistance for developing cvs. such as BC -68 and Fukanishi (Hirano, 1967). Again Nagai et al. (1973) developed Toride 1 and 2 from crosses made from cvs. TKM 9 and CO 25 developed at Tirur and Coimbatore respectively in the Indian state of Tamil Nadu. In South China Zhen-te series were obtained with multiple crossing with Toride -1, with Tetep as the resistant donor. Further, a local resistant source was used to develop Shuang Gui and Chin-hua-series as they have broad-spectrum B1 resistance (Sun et al., 1987).

The Yunnan province in China is considered as one of the rice germplasm centres in the world and rich in resistance source for rice B1 (CNRRI, 1988). The resistance source identified is from upland glutenous, *indica-japonica, indica* and *indica*-glutenous types such as Hanzinuo, Haobuka, Haolianggu, Haoyubangnan, Kaohenuo, Menghuizinuo and Xi-angnuo (Dai and Chen, 1983; Sun et al., 1987). The area around Tai Hu lake is yet another centre of rice genetic resource in China and the Jiangsu Academy of Agricultural Sciences could identify resistant rice cvs. namely Bodau, Erbailian, Heikezijing, Heuizuizi, Jian-nanwan, Queernuo, Yanzinuo, Zaoerhong and Zaomuxiquo (CNCRT, 1983).

In the US, Metcalf (1921) first attempted to control B1 by varietal resistance with the introduction of 12 Italian cvs. but found that their resistance did not last longer. Cralley and Adair (1942) found the cvs. Zenith and Arkansas Fortuna highly resistant in Arkansas, USA. Anderson et al. (1947) tested 19 cvs., 17 of which were Japanese and two American, and considered cv. Butte as resistant. Adair (1949) developed Nova through successive breeding programmes as crosses were made with Lacrosse along with other known resistant cvs. Zenith, Nira and Fortuna. This cv., however, was found susceptible to race 7 and 8 of *P.oryzae* in the US. Further, Atkins et al. (1965), through a sustained breeding programme in the US, involving Northrose, Zenith, MOR-500, Nato etc and several of the crosses, tested against the prevailing race 6 and determined the resistance of the various cvs. (Table 14). Subsequently, the resistant cv. Dawn was developed by Bollich et al. (1986).

In India, considerable work on breeding for resistance to B1 began in Tamil Nadu as early as in the 1920s and two promising cvs. CO 4 from Coimbatore and later TKM 9 from Tirur Rice Research station were released as resistant cvs. Later, ADT 10 developed from Aduthurai, also in Tamil Nadu, was crossed with CO 4 and the two resistant cvs. CO 25 and CO 26 rapidly replaced the then popular cv. ADT 10 during the 1950s. Subsequently, Sundararaman (1927) reported cvs. GEB 24, GEB 301 and CO 1 which were highly resistant. He also made attempts to develop resistant cvs. like susceptible Korangu Samba with the resistant GEB 24, CO 1 and CO 4. Further, Thomas (1938) found cvs. CO 4, CO 11, ADT 6 and 10998 highly resistant but later in 1939 he found CO 4 and CO 11 somewhat susceptible.

Table 14: Reaction of representative US rice cultivars to *Pyricularia oryzae* races 1-8, 10 and 16' (Atkins et al., 1965).

CI No.	Cultivar	Races of *Pyricularia oryzae*									
		1	2	3	4	5	6	7	8	10	16
Short grain											
1561	Caloro	R	S	S	S	R	S	S	S	R	S
1600	Colusa	R	S	S	S	R	S	S	S	R	
8988	Calrose	R	S	S	S	R	S	S	S	R	
Medium grain											
7787	Zenith	S	R	R	R	R	R	S	R	R	R
8310	Arkrose	R	S	S	S	R	S	S	S	R	S
8318	Magnolia	R	S	S	S	R	S	S	S	R	
8985	Lacrosse	R	S	R	R	R	S	S	S	R	S
8988	Nato	R	R	S	R	R	S	S	R	R	S
9155	Mo. R-500	R	R	R	R	R	R	S	R	R	
9407	Northrose	R	S	S	S	R	S	S	R	R	S
9416	Gulfrose	S	R	R	R	R	R	S	R	R	R
9559	Nova	R	MR	MR	R	R	MR	S	S	R	R
Long grain											
1344	Fortuna	S	R	S	S	R	S	S	R	R	
2702	Nira	S	R	R	R	R	MR	MR	MR	R	
8990	Blue bonnet 50	S	R	S	R	R	S	S	R	R	S
8993	Centruy Patna 231	S	R	S	R	R	R	MR	R	R	R
9013	Toro	S	R	S	R	R	S	S	S	R	R
9386	Vegold	MS	R	S	R	R	S	S	S		
9433	Belle Patna	S	R	S	R	R	S	S	S	R	R

'Some varietal reactions may be changed by subsequent tests R, resistant; MR, moderately resistant; MS, moderately susceptible; S, susceptible.

At the Central Rice Research Institute (CRRI), Cuttack, India, in 1954, the *japonica* × *indica* hybridization was initiated. The first cross was between CO 25 (CO 4 × ADT10) resistant to B1 and CO 13, susceptible to B1, resulting in two resistant cvs. CR 906 and CR 907. The second cross was between CO 25 resistant to B1 and susceptible to brown spot and BAM 10, resistant to brown spot but susceptible to B1. Three cvs. resistant to both B1 and brown spot were evolved (Padmanabhan and Ganguly, 1959; Padmanabhan, 1965b).

In the Indian states of Andhra Pradesh, Karnataka, Maharashtra, Uttar Pradesh and Jammu & Kashmir breeding for B1 resistance was undertaken. Either CO 4 or its progeny was used as a source of resistance and many popular cultures (crosses designated by numbers before testing and release) were released (Table 15) from time to time (Padmanabhan, 1965b). Several solutions including the development of multilineal lines and horizontal or field resistance based upon polygenic resistance have been suggested to overcome the problem of variability of the fungus. Apparently, the penetration phase in B1 is controlled by the polygenes. Work done at the CRRI by Padmanabhan et al. (1974) has shown that it is possible to combine a high degree of resistance to penetration (polygenic) with resistance to the spread of the disease inside the tissue which is controlled by major genes. There is evidence that in some of the moderately resistant cvs., resistance may be controlled by polygenes.

Table 15 · Resistant cvs identified by crossing CO 4 as one of the parents (Padmanabhan, 1965b)

Cross	Culture/AC No.
N 183 × CO 4	3552
BCP–2 × CO 4	5392
BCP–1 × CO 4	1849
CO 19 × CO 4	1807
P 1283 × CO 15	1816
MTU 19 × CO 4	10569
BAM 6 × CO 4	1762
BAM 6 × CO 4	1763
CO B × CO 4	6522
CO B × CO 4	6517
GEB-4 × CO 4	6578

Some of the important cvs. such as CO 4, CO 25, CO 26, TKM 1, ADT 25, Tetep, Tadukan and Zenith have been found to be highly resistant to Bl and they have been used as one of the parents in the breeding programmes in India. Several promising cvs. resistant to B1 have been evolved and many are being extensively tested in the field at various centres in the country. As a result, BJ 1, CR 905, CR 906, SM 8, CH 55, S 67 and AC 2489 are some of the promising cvs. evolved at different centres. Among the cvs. being improved are Supriya, IR 8, Jagannath and IR 20. Some of the recently released cvs. like Bharani and Vaigai exhibit high field resistance to B1.

The Directorate of Rice Research (DRR), Hyderabad, India (1986) identified a few resistant cultures for B1 from endemic areas as follows: RP 2243 - 11 - 1 (IR 8 × Surekha), RP 2161 - 15 - 41 (Rasi × WGL 22245), RP 2068-17-3-7 (Swarnadhan × Velluthacheera), CR 407 -1 (CR 94× Ratna), CR 571 (Zenith × Dular), RP 2071-18-1-1 (Swarnadhan × NLR 9676) and RP 2430-37-80-17 (Rasi × Chittraikar).

Mahadevappa (1990) reported that cultures such as CTH 1, CTH 3 and CTH 4 were found tolerant to both cold and B1 stresses in winter trials conducted at Bangalore, India. Among them CTH 1 was found to be promising with an attractive plant height (90 - 100 cm) and grain type (medium slender) besides being tolerant to B1.

In Taiwan most of the breeding lines released were the Taichung series 178, 179, 181, 182 and others (Taichung Station), Chianung 242, Chianung Yu 280 (Chiayi Station) and Kaohsiung 122, 135, 137, 164, 165 and 202 and Taipei 45 (Taipei Station) and Taitung 83 (Taitung Station). These progenies are crossed with a local N responsive cv. Ponlai.

In Thailand, Dasananda (1965) screened numerous cvs. against B1 and the resistant ones were used for breeding for B1 disease resistance. A number of crosses were made and the progenies were tested in 13 experimental stations. Due to successive screening, the susceptible cvs. were gradually eliminated and more resistant cvs. such as Dawk Mali, Leuang Tawng, Khaokod, SM 6, SM 8, etc., were identified. These are now in use.

In Colombia at CIAT, Roserro (1979) reviewed the work on breeding for B1 resistance. In all 1163 crosses were made from 1969 to 1976 and selection were made for superior genotypes with resistance to B1. The cvs. Colombia 1, Tetep, Dissi Hatif, Mamoriaka and

C 46-15 were used as sources of B1 resistance. Among them 14 lines were selected in early 1975 on the basis of yield, apparent B1 resistance and other desirable traits. However, their tolerance to B1 was not stable.

In Korea, a promising cv. Tongil, an *indica* × *japonica* hybrid, resistant to B1 at many centres for several years, broke down due to the shifting of the races. The virulent race was identified in the field around 1976 and most of the cvs. derived from Tongil became susceptible and succumbed to B1. Lee (1972) reported the breakdown of cv. Kwanok and analysed the annual changes of races, Lee et al. (1975) and Choi et al. (1989 a, b) reported the resistance of cvs. to rice B1 for both *japonica* and Tongil types.

In Egypt, cv. Giza 173, introduced from Japan, was popular for several years. However, it was severely attacked by leaf and panicle B1 in 1984 and subsequently planting of this cv. was abandoned (Tanaka, 1987). Reddy and Bastawisi (1990) reported that *indica* breeding lines (IR 28, IR 1626-203 [Giza 181]) retain resistance to the local B1 pathogen races.

In The Philippines, at the IRRI, the breeding work began in the 1960s as one of the important objectives. The 'miracle rice' - IR 8 introduced during 1966, was derived from a cross between Peta × Dee-geo-woo-gen. From 1962-1964, 8214 cvs. were screened at IRRI. Of these, 1457 were found highly resistant to Bl (IRRI, 1964) and were further tested seven times in the same nursery; finally, 450 were selected for tests at seven stations in different regions of the Philippines but only 75 proved resistant at all the stations (IRRI, 1966). These cvs. with a broad spectrum of resistance are used in breeding for B1 resistance. In 1969, two cvs. Carreon and Tetep, were used as they showed broad spectrum resistance (IRRI, 1970). However, Carreon proved to be a failure while several of the progenies of Tetep crosses showed resistance consistently at IRRI, Los Baños.

Numerous donor parents were used as source of resistance (Beachell and Khush, 1969; Khush and Beachell, 1972). They included Zenith, Sigadis, Tadukan, B 589, A 4-18 Dawn, H 105, T 172, Nahng, Mon-S4, Leuang Yai 34 and Gam Pai 15 from the world collection of germ plasm. Among them Dawn was used extensively. Through successive breeding programmes, the IRRI has been releasing from time to time a number of high yielding promising cvs. under the IR series namely, IR 28, IR 29, IR 32 and IR 34 which are resistant to B1.

Cultural practices

Cultural control includes the manipulation of various farming practices that favour the crop at the expense of the pathogen. Soil fertility (Doi et al., 1952), land improvement (Tanaka, 1961), deep ploughing (Nakagawa, 1962), plant spacing and split application of N-fertilizers (Nair, 1975; Nair and Chakrabarti, 1978; Venkatarao and Muralidaran, 1982b), crop production management (Kim and Lee, 1982), flooding (Kim et al., 1985), spacing and water management (Khin and Kim, 1990) reduce the Bl disease in specific situations. Kim and Kim (1990) reported that the benefit in yield from flooding comes from both reduced B1 infection and enhanced plant growth. Some of these practices will reduce the yield loss but cannot be considered the primary solution for long term rice B1 control.

Avoidance is considered as shifting of sowing time or date of planting (Padmanabhan and Ganguly, 1953) which enables the pathogen to prevent the susceptible stage of the plant and the favourable conditions for the fungus from coinciding. For instance, in Japan, farmers plant early to have less disease. The reduction in disease may be due to the air temperature which is too low at the tillering stage of early plantings and too high at the heading stage, thus preventing the vigorous disease development (Hashioka, 1950a, b; Kuribayashi and Ichikawa, 1952). Chandramohan and Palaniswamy (1963) correlated B1 incidence and the time of planting and noted that severe B1 occurred with low temperatures, high humidity and heavy dew. Although the high yielding IR 50 succumbed to B1 in recent years, farmers in South India still depend on this promising cv. for its excellent grain, keeping quality as well as productivity by resorting to early sowing or shortening the time of exposure to the pathogen.

Exclusion is to prevent the new races of the fungus from reaching an uninfected area and avoiding contact between a new race and the crop. For instance, in California, USA, there is no record of B1 although rice is grown commercially. Stringent regulation is in practice to restrict the movement of the B1 diseased material or fungus isolates to prevent their introduction.

Eradication involves elimination of the pathogen from infected areas by roguing, sanitation, crop rotation, removal of the collateral hosts and soil amendments. A knowledge of the life cycle of the pathogen, method of perennation, host range and habit of growth is essential for devising suitable measures to prevent the B1 disease.

Chemical control

The majority of the fungicides used in B1 control are protectants, prophylactic in their behaviour and are generally applied to seed, plant surface (foliage panicle, etc). They cannot penetrate the plant tissue in effective amounts. They act outside the plant prior to infection by the fungus. Eradicants are those that act to reduce the inoculum, the dormant or active pathogen in the host. These can also remain effective on or in the host for some time and function as protectants also. Therapeutants on the other hand, inhibit development of the disease syndrome in a plant when applied subsequent to invasion by a pathogen. Therapy can be by either physical such as solar energy and hot water treatment or by chemical means. Further, any successful fungicide must have some constant characteristics such as tenacity or adhesiveness to enable them to contact well and should last for a long period. They should not be phytotoxic.

Comprehensive literature has been provided on the control of B1 disease by chemicals (Fukunaga, 1965, 1966; Okamoto, 1965; Yoshino et al., 1990), fungistatic efficacy (Abeygunawardena and Pieris, 1958) and the equipment for application of fungicides (Hatai, 1965). In Japan, chemicals have been extensively used in controlling B1 and virtually all the rice fields have been saturated. The success is mainly due to a combination of special socio-economic and technical reasons. Further, an effective forecasting system is available with excellent agricultural extension services. Chemicals produced exceed their requirement and the cost of application is cheap, working out to 2-4% of the value of the produce. The semi-dwarf nature of the plant enables them to use a simple equipment and large scale application is done by aircraft through the well-organized communities. Besides these facilities, Japan has world renowned manufacturers of pesticides and is noted for its strong research and development work in fungicides.

Methods of controlling B1 have been developed and field tested in Japan (Okamoto, 1965), India (Padmanabhan et al., 1962, 1963), USA (Caffy and Lindberg, 1964), the Philippines (Nuque et al., 1971) and Malaysia (Chin, 1975).

In India, B1 has been controlled by dusting the plants with flower of sulphur (Narasimhan, 1936), ceresan-dust (Venkatakrishnaiah and Delvi, 1960a); treating seeds with agrosan GN (Venkatakrishnaiah and Delvi, 1960b) and seedlings with various chemicals (Srivastava, 1962).

Chemicals used in the control of B1 belong to diverse groups and act in different ways. The most important groups and chemicals used for B1 control are discussed here.

Copper compounds

Metcalf (1906) claimed that spraying with Bordeaux mixture (5-5-50) just before the emergence of earheads almost entirely controlled leaf B1 while Bokura in 1909 (Okamoto, 1965) tested other copper compounds. Subsequently, Waldee and Roberts (1947) recommended Bordeaux mixture and cuprous fungicides against B1 disease in Japan but no satisfactory results were obtained on account of their phytotoxicity to the rice plants. Hashioka and Kato (1955) found copper dust and Bordeaux mixture had no appreciable effect but reduced the elongation of the germ tube and appressorial formation of the pathogen considerably. Attempts towards systemic control through $CuSO_4$ and SiO_2 have been reported (Abe and Okmura, 1931; Akai, 1953). Padmanabhan (1965, 1966) gave an economic spray schedule (0.25% metallic copper in 370 1 of water at the rate of 251 1/ha) with a low volume nozzle (Padmanabhan et al., 1962). However, under environmental and physiological conditions highly favourable for B1, copper fungicides alone do not control B1 (Okamoto, 1965).

Mercuric compounds

Ogawa et al. (1953) discovered that phenyl mercuric acetate (PMA) is less phytotoxic to rice plants. This was used extensively in Japan as dust, wettable powder or emulsifiable concentrate for effective control of B1, while Okamoto and Hamaya (1962) found it to be very effective in seed treatment. Since PMA compounds are therapeutants, they have a strong preventive action against B1 fungus in inciting infection and sporulation on lesions. Infection is prevented not only on the treated part but also on the untreated parts as it is absorbed by the plant tissue. The mode of action of these compounds is not well established but their therapeutic effect which restricts the fungus growth is well known (Yamamoto, 1958 Okamoto et al., 1958, 1959). Although these compounds have slight or no toxic action on *japonica* rice cvs., they are toxic to *indica* cvs.

Organomercurial compounds

A blend of copper fungicides with organomercurial compounds was envisaged as it was less toxic than the copper fungicides used alone. Such blends are considered to be strong inhibitors of the respiratory enzymes of the B1 fungus but cause mammalian toxicity and are hazardous to the environment. Their use was banned in Japan in 1968 and subsequently in most other countries.

Organophosphorus compounds

Organophosphorus compounds and systemics were developed subsequently (Maeda et al., 1970; Nambu, 1972; Usugi and Tomizawa, 1992; Umeda, 1972; Tsumara, 1976; Watanabe, 1977; Froyd et al., 1978) and are very effective in B1 control. A few promising ones are Kitazin P (IBP), hinosan (edifenphos), oryzemate (probenazole), fujione (isoprothiolane), rabcide (pthalide), topsin M (thiophanatemethyl, Nf-44), EL-291 (tricyclazole), inezin (phenylphosphonothioate) and benlate (benomyl). Besides, several test compounds such as Hoe 22843, 22845 and 22985 are also effective. Mogi (1979) tried mixtures of benomyl and thiram and topsin-M thiram for seed treatment. Others tried amino acid derivatives (Homma et al., 1973). Although tricyclazole inhibits the mycelial growth (Tokousbatides and Sisler, 1978) it did not inhibit the sporulation and prevent spore release from the lesions (Kojama et al., 1988). Oxazoles (Pattanayak et al., 1978) were also used as fungicides against *P. oryzae*.

Kato (1974) reported that spraying of chemicals even after colonization of the pathogen gave satisfactory control of leaf Bl by reducing severe damage while Bandong et al. (1979) applied some of the systemic fungicides which controlled seedling Bl very effectively. Benomyl, thiophanate, thiophanate-methyl, PP 389 JF 5816 and Hoe 22843 appeared to be good as seedling protectants. Further, Otomo (1988) suggested a reduction in Bl epidemic with the newly induced systemics.

Khin et al. (1989) reported the fungicidal application on rice leaf Bl management. Loehken (1990) reported that systemic compounds like pyroquilon can be effectively used for seed treatment for upland rice as they protect the new growth for 45-50-d after sowing while granular water application into low land rice protects the whole plant for a similar duration. Further, excellent control of leaf Bl is possible with a low rate of pyroquilon applied during symptom appearance while delayed application results in poorer control even at high application rates (Chin and Bandhufalck, 1990).

Mode of action of fungicides

Uesugi and Fukunaga (1967) studied the mechanism of toxic action of alkylphenols and pentachlorophenol (PCP) on the mycelia of *P. oryzae*. The former did not inhibit mitochondrial respiration while the latter markedly increased indigenous respiration and glucose oxidation of intact cells. On the other hand, Sumina (1967) reported that β - acylphenylhydrazine derivatives are more difficult to permeate into Bl fungus spore than mycelia. The mode of action of PMA, Du-ter W-50 and Blasticidin-S (Ashrafuzzaman and Frederiksen, 1969), chlobenthiazone (Inoue et al., 1984 on *P. oryzae* was studied. Conidial germination was not affected and mycelium within the host tissue was not killed by the spray fungicides indicating that the effectiveness of the substances did not carry over into the second generation of conidia although the lesion size was restricted in growth.

Antibiotics

Work in Japan and India in recent years has shown that certain antibiotics can replace metallic fungicides. Yoshii (1949) was the first to report that a culture filtrate of *Cephalothecium* sp. could completely inhibit conidial gemination of the Bl pathogen and called it 'Cephalothecin'. The dried mycelial powder could also effect the development of leaf Bl in rice seedlings (Yoshii, 1950). Following this discovery a number of workers tested many antibiotics against *P. oryzae*. Notable among them are 'antiblastin' (Suzuki, 1954); 'antimycin-A' (Harada, 1955); 'blastimycin' (Watanabe et al., 1957); 'blasticidin-A' (Fukunaga et. al., 1955); 'blastimycin' and 'blasticidin' (Misato et al., 1959) and 'blasticidin-S' (Misato, 1961). Except blasticidin-S, none of these antibiotics can be put to use because of their instability or high toxicity to fish. Other antibiotics that have been reported to be inhibitory against *P. oryzae* include 'hygrostatin' and 'imocidin' (Furishiro et al., 1958; Inoye et al., 1958, 1959). Thirumalachar (1968) reported two antibiotics 'hamycin' and 'thiolutin' which are very effective against *P. oryzae*. Subsequently, Padmanabhan and Row (1966) reported that the culture broth of a species of *Streptomyces* had a powerful antibiotic action on the growth and conidial germination of *P. oryzae* even at 1 ppm level. Aureofungin, with high fungicidal activity, is very effective against seed-borne infection (Radhalkar and Neergard, 1969) as well as leaf Bl (Thirumalachar, 1974).

Fukunaga et al. (1968) reported that antimycin-A, blicidin-A and blimycin sprays of 100 µg/ml and above gave good control of leaf Bl. The residual effect were inferior to those of PMA. Further, Misato et al. (1968) showed that blicidin-S inhibited the oxygen uptake of the spores. It has better inhibitory activity on the conidial initiation as well as mycelial growth than PMA. Asakawa et al. (1968) examined the phytotoxicity of blicidin-S and its formulations. H-acid (8-amino-1-naphthol-3, 6-disulfonic acid) was found to reduce the phytotoxicity of the blicidin-S. Another successful discovery was 'kasugamycin', popularly known as 'kasumin', very effective on rice Bl at all the stages of infection (Okamoto, 1972). In China, 'quigfengmycin' was introduced (Shannghai Institute of Plant Pysiology, 1975) and was found to be effective against Bl control.

Fungicidal/antibiotic resistance

Development of fungicidal/antibiotic resistance in *P. oryzae* is well known. Resistance to PMA (Yoshii et al., 1958), chemicals like $CuSO_4$, HBO_4 and $HgCl_2$ (Yamazaki et al., 1964), organophosphorous compounds (Uesugi et al., 1969) as well as Kasugamycin (Katagiri and Uesugi, 1974, 1978) has also been reported. Taga et al. (1978) studied the analysis of kasugamycin resistance in the perfect state (*P. oryzae*) while Taga et al. (1979) identified the three different loci controlling kasugamycin resistance in *P. oryzae.*

The appearance of resistant strains my be due to spontaneous mutations, application of fungicides (Miura et al., 1976; Ito and Yamaguchi, 1977) and competition between sensitive and resistance strains (Ito and Yamaguchi, 1979). Further, the inihibitory effect of isoprothiolane on the metabolism of phosphorimidate by the Bl fungus isolates in relation to fungicide sensitivity was studied (Miyagi et al., 1985).

The problem of fungicide resistance has been less encountered with the conventional fungicides which generally are multisite inhibitors while increased frequency occurred with the introduction of systemic fungicides that are single site inhibitors (Dekker, 1977). Blast fungus isolates resistant to kasugamycin (Miura et al., 1975; Miura, 1984); organophosphorous compounds (Katagiri et al., 1980) and edifenphos (Lalithakumari and Kumari, 1987) have been reported. Further, Chien (1987) suggested a simple model for solution to fungicide resistance.

Cross resistance against fungicides

Lalithakumari et al. (1987) reported the analysis of various cellular components showed a significant increase in all the macromolecular contents, namely DNA, RNA, proteins, lipids and phospholipids of edifenphos (EDDP)-resistant mutants of *P. oryzae.* The efflux of electrolytes was reduced in the resistant mutants as compared to the control. However, phospholipids, the target sites of inhibition for EDDP, seem to be unaffected in the mutant as evidenced from the higher values of phosphatidylcholine, as compared to the sensitive strain. Here, the mechanism of resistance to this fungicide is not related to the target sites. It is likely that the resistance in *P. oryzae* to EDDP influences cell membrane permeability as evidenced from the reduction efflux of electrolytes from the resistant mutants. Although plasmid DNA was isolated in both sensitive strain and resistant mutant of *P. oryzae*, the intensity was more in the latter and this was interpreted as the rapid amplification of plasmids in the presence of edifenphos.

Further, Lalithakumari and Annamalai (1990) reported cross resistant studies to select a suitable chemical for breaking the EDDP-resistance. The results on the sensitivity of selected EDDP- resistant mutants of *P. oryzae* against untreated fungicides were presumably due to pleiotropic mutation. All the mutations of *P. oryzae* were sensitive to ziram through a number of chemicals to which each mutant showed varied sensitivity. Thus, to over come failure of disease control due to EDDP-resistance, ziram was recommended as an alternative chemical.

Several fungicides/antibiotics have been released from time to time for rice Bl control and their formulation, group, mode of action have been summarized in Table 16.

Table 16: Various fungicides/antibiotics used on rice blast control (Manibhushanrao, 1990)

Fungicide/ Antibiotic	Trade name	Formulation	Group	Mode of action	Reference
Actidione	_	85-100%	Cycloheximide	Systemic	Ashrafuzzaman and Frederiksen (1969)
Benomyl	Benlate	50 WP	Benzamidazole	Systemic	Ashrafuzzaman and Frederiksen (19769); Bandong et al. (1978)
Bordeaux mixture	–	Liquid 1% (5:5:52) $CuSO_4$ & lime	Copper	Contact	Padmanabhan et al. (1962)
Carbendazim	Bavistin	50 WP	2 (Methoxy-carbomoyl) Benazamidazole	Broad spectrum systemic	
Carboxin	Vitavax	75 W	Oxathilin derivative	Systemic	Ashrafuzzaman and Frederiksen (1969); Venkata-Krishnaiah and Delvi (1960a)
Copper oxy-chloride	Biltox, Fitolan	50 W	Copper derivative	Contact, curative	Padmanabhan et al. (1962)
Cuprous oxide	Perenox	50 W	-do-	Contact	-do-
Edifenphos	Hinosan	50 EC	Organo-phosphate	-do-	Mogi (1979)
EL-291	–	–	Benzathiazole	Systemic	Froyd et al. (1976, 1978)
Fuji-one	Isoprothiolane	–	Non-mercuric	Submerged application; inhibits penetration	Tsumura (1976); Araki and Miyagi (1976)
Inezin	–	–	Organo-phosphorus	Contact	Uesugi and Tomizawa (1972)
Kitazin P	13 P	48 E	-do-	Systemic & contact	Chiba et al. (1973) Maeda et al. (1970)

Contd...

Fungicide/ Antibiotic	Trade name	Formulation	Group	Mode of action	Reference
Mancozeb	Dithane M-45	80 W+zinc	Dithio carbamate	Contact	Ashrafuzzaman and Frederiksen (1969)
Meneb	Dithane M-22	80%	-do-	-do-	-do-
Methoxy ethyl mercuric chloride	Ceresan	WP	Methoxy-ethyl mercuric chloride	-do-	Padmanabhan et al. (1962)
	Cersesan-lime	1:5		-do-	Venkatakrishnaiah and Delvi (1960a)
Dryzemate 8G	Probenazole	Granule	Saccharine derivative	Submerged application; inhibits mycelial growth and penetration	Watanabe et al. (1977); Reddy (1984)
Phenylmercuric acetate (PMA)	Agrosan	300% WP	Organo-mercurial	Contact	Ogawa et al. (1953); Okamoto and Yamamoto (1955); Okamoto et al. (1958, 1959); Ashrafuzzaman and Frederiksen (1969)
	Agrosan GN	Dust	Phenylmercuric acetate + ethyl mercuric Chloride	-do-	Venkatakrishnaiah and Delvi (1960b); Padmanabhan et al. (1962)
PP 389 JF 5816	–	50% WP	Methyl-tetrazole	Systemic	Bandong et al. (1978)
Rabicide	Fthalide	50 WP	Non-mercuric	Systemic	Numbu (1972); Vehara and Tsuzaki (1972)
Rhodane compounds No. 176, 393	–	Liquid	Non-mercuric	Preventive	Okamoto and Sekiguchi (1962)
Sulphanilamide	–	–	Sulphanilamide	Prophylactic	Srivastava (1962)
Thiophanate-methyl	Fungitox	75% WP	–	Systemic	Bandong et al. (1978)
Thiourea	–	–	Sulphur associated	-do-	-do-
Tricyclazole	E2-291	WP	Benzathezole	Systemic	Tsuzuki et al. (1975); Froyd et al. (1976)

Contd...

Fungicide/ Antibiotic	Trade name	Formulation	Group	Mode of action	Reference
Antibiotics/Combination					
Antiblastin	–	–	–	Inhibited to growth	Suzuki (1954)
Antimycin-A	–	–	–	-do-	Harada (1955)
Aureofungin	–	Liquid	–	Growth inhibition	Thirumalachar (1974)
Blasticidin-A	–	WP	–	Chemical instability	Fukunaga et al. (1955); Takeuchi et al. (1958)
Blasticidin-S	–	WP	2-4% Benzyl amino-benzene sulfonate	Inhibits spore germination/ protein synthesis	Fukunaga (1963); Miato (1961)
Blastimycin	–	–	–	Inhibited the growth	Ashrafuzzaman and Frederiksen (1969); Watanabe (1967)
Cephalothecin	–	–	–	-do-	Yoshii (1949)
Kasugamycin	Kasumin	2% liq. or WP granules	–	Preventive & Curative	Padmanabhan and Row (1966); Okamoto (1972)
Kasugamycin+ Copper oxychloride	Karusan	(75.6 & 45%)	–	-do-	Mogi (1979)
Kasugamycin + Fehalide	Kasurabicide	WP (1.2 & 20%)	–	-do-	-do-
Gingfengmycin	–	–	–	Curative	Shangai Inst. Pl. Physiology (1975)

Botanical pesticides

Pesticides of plant origin are less toxic to plants, systemic in nature, easily biodegradable but exhibit some phytotoxic effect. The tropical flora which had to defend themselves against various types of microorganisms during the process of evolutions, are rich in antimicrobial substances. Ohno et al. (1983) tested new antifungal natural products, podoblastins, and found them effective against the rice B1 disease. Tewari and Dath (1984) reported that leaf extracts of henna or tree mignonette *Lawsonia inermis* contain 2-hydroxy-1, 4-naphthoquinone (lawsone) as well as extracts of betel leaf *Piper betel* are toxic to Bl pathogen. Neither lawsone nor its structural analogue are toxic to rice plants. Admittedly, there is scope to identify many plants possessing antimicrobial substances toxic to potential rice pathogens. More critical evaluation of botanical pesticides with field grown plants is required.

Homma et al. (1980) reported that lecithin soybean seeds effectively controls Bl fungus, while Iwata et al. (1988) studied the inhibitory effect of components of lecithin on *P. oryzae*. Although lecithin has no fungicidal action it inhibits effectively penetration of hyphal cells.

Phosphatidylethanolamine (PE) is one of the four components which has activity similar to lecithin. It has no inhibitory effect on conidial germination at any concentration, but inhibits appressorial formation at 1 ppm by about 29%. (70% at 1000 ppm) and hyphal penetration by about 90% while 1-100 ppm can protect the rice Bl by about 50% indicating a role in PE in the inhibitory effect of lecithin.

Biocontrol

Intensified use of agrochemicals, of late, has resulted in the accumulation of toxic compounds hazardous to humans and environment and also build up of resistance by the pathogens. This necessitated the need for innovative disease control strategies such as the use of biocontrol agents as a viable alternative in modern disease management programmes. Biocontrol of rice B1 through hyperparasites and seed bacterization has been initiated in recent years.

Hashioka and Fukita (1969) studied the hyperparasitic potential of *Gliocladium deliquescence* on *P. oryzae* and described the mode of parasitism. The pseudoappressoria produced by the former never penetrated the host fungus cells, acted mostly as holdfasts and were killed by direct hyphal contact. *Trichoderma* spp. invaded and coiled around the host cells but showed less mycoparasitism.

Chung (1988) found that seed bacterization and seed treatment with *Trichoderma* spp. reduced the incidence of seedling Bl. Gnanamanickam et al. (1989) reported that leaf and neck B1 infections were reduced by spraying with fluorescent pseudomonad bacterial cell suspensions (10^8 CEU/ml) when plants were 20, 30, 40-d old.

Evaluation of long term control

The world rice production was about 470 million tonnes (t) in 1988 and the challenge before scientists is to increase it to 560 million by 2000 AD and to 780 million by 2200 AD—a more than 68% increase—just to keep in step with the population growth. This can be done largely with the introduction of new modern rice cvs., effective cultural practices, advancement of appropriate technology for increasing productivity and effective prevention of damage due to major pests and diseases like Bl that limit the yields.

Another possibility for increased production is by retarding the disease outbreak and developing potent fungicides for Bl control. As an estimate, about nine Bl-specific fungicides have been identified and marketed by at least seven reputed companies from the US, UK and Japan. There will be added costs of production in areas where fugicides are used to control Bl. For instance, in Korea, the amount of fungicide used for Bl control is much higher than that for other major diseases (RDA, 1989).

Earlier, most research on Bl concentrated on selection and screening of Bl-resistant lines and cvs. rather than on analyzing the Bl resistance genes, except in a few cases. In the last three decades, some examples of breakdown of resistance to the rice Bl fungus due to pathogenic variability have come to notice. For instance, the total breakdown of the resistance on cv. Tongil and its sister lines in Korea with the spread of a new super virulent Bl fungus race after the 1977 typhoons was totally unexpected and the intensive crop production programme was closed. Against this background, Crill et al. (1982) suggested a long term control programme for rice B1 in Korea i.e., use of monogenic/durable resistance after exploring all known alternatives like cultural practices, chemical control and use of field resistance. The concept of race prediction and gene rotation proposed by Crill and Khush (1979) was also successfully adopted in Korea. This provides the stabilizing selection and guides the evolution of the pathogen to rice to cause minimal loss from Bl in rice.

Besides, integrated disease management (IDM)-blending the triad namely cultural, biological and the use of resistant cvs. in right manner is the latest trend in tackling B1, a disease of high magnitude. By adopting IDM, farmers should cut down their plant protection cost, enjoy a more balanced and safe environment and still have a healthy crop. Again, combining host resistance and judicious use of fungicides with effective forecasting, increases durability of B1 control measures by slowing down the pathogen's genetic adaptation.

Research Highlights

Despite impressive publications from 1961 to 1990 on the rice B1, (Fig. 28) it still remains the worlds's most important disease of rice. Although intensive efforts are made for its control, B1 makes the cultivation of rice impossible in many areas. Much progress has been made recently on B1 management which includes both fundamental and applied research. New discoveries have been made on cell biology and molecular genetics of the B1 fungus in recent years. Hamer et al. (1988) found that the conidia of *M. grisea* contain a mucilaginous substance released from the spore tip which probably has a role in attaching the conidium to the hydrophobic surface of the rice leaf. This appears to be the first report of such a mechanism in plant pathogenic fungi. Further, Howard and Ferrari (1989) showed that melanin has no role in the adhesion of appressoria, but allows it to maintain high turgor pressure, the primary mechanism by which fungus penetrates the host surface.

Leung et al. (1988) identified genes in the B1 fungus that control host species and cv. specificity. Valent (1990) characterized the populations of the B1 fungus using RFLPs and a library of mapped clones is now available. Successful transformation system for *M. grisea* was achieved using the plasmid vector pAN7-1. Further, with MGR-DNA fingerprinting, reliable genetic and pathogenic diversity in B1 fungus was achieved (Hamer et al. 1989). Cutinase gene of *M. grisea* was cloned to rice cv. (Sweigard et al. 1992a) and found that it is not required for pathogenicity, suggesting the fungus will be adopting, different strategies of penetration. Such highly precise molecular techniques will help in understanding the molecular basis of variation of the pathogen.

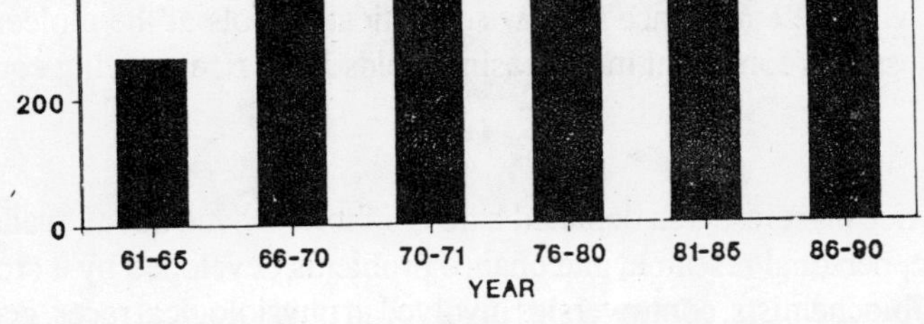

Fig. 28: Number of rice B1 publications during 1961-90 (modified from Bonman and Ahn, 1990 & IRRI library).

Further, near isogenic lines (NILs) with Bl resistance genes have been identified for fundamental studies of the genetics and pathology of Bl resistance (Mackill et al. 1988). The NILs with an *indica* background may also prove useful as improved Bl differentials as well as in understanding Bl resistance in the tropics.

One of the applied aspects of Bl study will be the recognition of the importance of field resistance (Toriyama, 1975; Ahn and Ou, 1982) and developing methods to measure quantitative resistance in the greenhouse (Toriyama, 1975) and in the field (Notteghem, 1985; Marchetti, 1983) and studying its components and inheritance in both irrigated and upland rice cvs. (Notheghem, 1985). Again, the application of durable resistance to Bl disease (Johnson, 1979), recognizing the cvs. with durable Bl resistance (Lee el al., 1987) and durability specific to the rice-growing environment (Bonman and Mackill, 1988) are well known. Incidently, cv. IR 36 has been durably resistant to B1 to low land tropical areas (Yah and Bonman, 1986).

The failure of horizontal resistance to control rice B1 was documented and the possible strategy of rotating monogenes for resistence based on prediction studies was fully explored.

Gene rotation is a new and effective method of B1 control and becomes important in view of the recent development of first generation hybrid rice cvs. which are the excellent material for rapid incorporation of new resistant genes.

Another significant milestone in Bl research is the use of simulation models in forecasting epidemics. Although much quantitative knowledge is available on the epidemiology of temperate rice Bl (Hashimoto et al., 1982, 1984; Ohata et al., 1982, 1987; Choi, 1987; Choi et al., 1988; Koshimizu, 1988; Hayashi and Koshimizu, 1988; Ishiguro and Hashimoto, 1988, 1990; Koizumi and Kato, 1990; Teng et al., 1990) not much is known of tropical rice (Manibhushanrao and Krishnan, 1991; Manibhushanrao et al., 1990). These developed models have the potential to guide within-season decisions and may eventually be useful in predicting the 'Bl potential' of different environments and finding out how the potential would alter with changes in agronomic practices.

Further, IRRI and the US Environmental Protection Agency are involved in a collaborative research on modelling of rice B1 and climate changes, using the historical weather data of 61 locations in seven Asian countries. Utilizing the geographic information systems technology to display the data on risk maps, they are planning altogether a newer approach to pin point the potential losses due to rice B1 rather studying the control methods.

In recent years, with the tremendous increase in our knowledge of Bl disease, the frequency of epidemics by breakdown of resistance in promising high yielding rice cvs. as well as changes in the virulence pattern of races can be minimised through effective forecasting and with the emergence of new sophisticated tools at the molecular level, Bl should become less of a constraint in increasing yields in all rice growing countries.

In Retrospect

The path of rice blast research depicted here is a fabric woven out of available past and present literature, personal research, and chance problems developed by a cross section of rice pathologists/biochemists, controversies involved in physiological races, generic/species names and even the spelling of rice B1 pathogen have now been mostly resolved. Earlier views on genetic variability also changed considerably.

The break-down of host-resistance, development of super-races and fungicidal resistance etc. are well established. Of late, advent of molecular techniques with greater precision, established many facts for which precise means of examining variation and population dynamics of the pathogen in the field are available. Much new knowledge on some of the exciting areas such as cell and molecular genetics, cloning genes, RFLP mapping etc. was made available. In view of these rapid advances made in the past few years, the rice B1 disease has become an ideal model system for studies in plant pathology.

Bibliography

Abe, T. 1931. On the effects of sun light on infection of rice plants by *Piricularia oryzae*. *Forsch. Geb. Pflkrankh. Kyoto*, **1** : 46-53.

Abe, T.1933. On the relationship of atmospheric humidity to infection of rice plant by *Pyricularia oryzae* Br. et Cav. *Forsch. Geb. Pflkrankh. Kyoto*, **2** : 98-124.

Abe, T. 1935. On the resistance of conidia of *Pyricularia oryzae* to low temperatures. *Ann. Phytopathol. Soc. Japan*, **5** : 206-215.

Abe, T. 1936. Comparison of pathogenicity in different culture strains of *Piricularia oryzae* and varietal susceptibility of the rice plant to blast disease. *Ann. Phytopathol. Soc. Japan*, **6** : 15-26.

Abe, T. and E. Okamura. 1931. On the effect of copper sulphate upon the susceptibility of the rice plant to the blast disease. *Forsch. Geb. Pflkrankh. Kyoto*, **1** : 54-70.

Abe, T.and Y. Yasuda. 1962. On the appressorium formation of conidia of rice blast fungus and its pathological significance. *Rep. Kyoto Univ. Agric.*, **13** : 36-44.

Abe, S., S. Kiyosawa and S. Ono. 1974. Inheritance of blast resistance of upland rice varieties. I. Inheritance of an upland rice variety Norin 4. *Bull. Ibarki Agric. Exp. Stn.*, 15 : 47-64.

Abeygunawardena, D.V.W. and J.W.L. Peiris. 1958. Studies on the fungicidal control of rice blast. I. The fungistatic efficacy of certain organic and inorganic fungicides on *Piricularia oryzae*. *Indian Phytopathol.*, **11** : 116-121.

Abumiya, H. 1955. On the infection types of leaf blast disease of the rice pant. *Jubilee Publ. Commem.* 60th Birthday of Profs. Y. Tochinai and T. Fukushi. Sapporo, Japan, 197-201.

Abumiya, H. 1958. Phytopathological studies on the breeding of rice varieties resistant to blast disease. II. On the process of brown necrosis formation at leaf lesion. *Bull, Tohuku Natl. Agric. Exp. Stn.* **14**: 15-21.

Abumiya, H. 1959. Phytopathological studies on the breeding of rice varieties resistant to blast disease. III. On the resistance for the lesion development in leaf blast with their classifying method and on the inheritance of resistant of the Japanese and foreign rice varieties. *Bull. Tohoku Natl. Agric. Exp. Stn.*, **17** : 1-101.

Abumiya, H. and T. Kobayashi, 1955. Phenolic compounds in rice leaves in relation to resistance to blast. *Ann. Phytopathol. Soc. Japan*, **20** : 35-36.

Adair, C.R. 1949 . Report of the Arkansas Agricultural Experirmental Station (unpublished) (cited by Atkins et al., 1965).

Adyanthaya, N. R. and G. Rangaswami. 1952. The distribution of silica in relation to resistance and 'blast' disease in rice. *Madras Agric. J.*, **39** : 198-204.

Ahn, C. J. and H. C. Chung. 1962. Studies on the physiological races of rice blast fungus *Piricularia oryzae* in Korea. *Seoul Univ. J. Biol. & Agric., Ser.* D, **11** : 77-83.

Ahn, S.W. 1982. The slow blasting resistance to rice blast. In : Proc. symp. on Rice resistance to blast. IRAT/GERDAT. Montepellier, France, pp. 343-370.

Ahn, S.W. and D.V. Sehu 1990. International evalution for durable blast resistance. *Int. Rice Res. Conf.* Seoul, South Korea, p. 23.

Ahn, S.W. and H.S. Chung. 1974. Effect of near-ultraviolet irradiation on sporulation by *Pyricularia oryzae* Cavara on cultural media. *Ann. Phytopathol. Soc. Japan*, **40** : 337-343.

Ahn, S.W. and S.H. Ou. 1982. Quantitative resistance of rice blast disease. *Phytopathology,* **72** : 279-282.

Ahn, S.W. and D.V.Seshu. 1987. Geographical variation of virulence patterns of the rice blast fungus. In: 11th *Int. Cong. on Pl. Prot. Manila, Philippines.* p. 152. (Abstr.).

Ahn, S.W., J.D. Yu and C. Q. Torres. 1988a. Classification of leaf blast lesions. *Int. Rice. Res. Newsl.,* **13** : 15-16.

Ahn, S.W., D.V. Seshu, N.K. Park and C.H. Kim. 1988b. Analysis of multilocation evaluation for rice blast resistance of Korean varieties. In : 5th *Int. Cong. Pl. Path. Kyoto, Japan,* p. 265.

Akai, S. 1939. On the ash figures of leaves of rice plants transplanted from the different kinds of nursery beds and their susceptibilities to blast disease. *Ann. Phytopathol. Soc. Japan,* **9** : 223-235.

Akai, S. 1953. Studies on Helminthosporium blight of rice plants. VII. On the relation of silicic acid and supply to the outbreak of Helminthosporium blight and blast disease in rice plants. *Ann. Phytopathol. Soc. Japan,* **17** : 109-112.

Akatsuka, T., O. Kodama, H. Kato , Y. Kono and S. Takeuchi, 1983. 3-Hydroxy-7-oxo-sandaracopimaradiene (Oryzalexin A), a new phytoalexin isolated from rice. *Agric.Biol. Chem.,* **47** : 445-447.

Akatsuka, T.,O.Kodama, H.Sekido, Y.Kono and S. Takeuchi, 1985. Novel phytoalexins (Oryzalexins A, B and C) isolated from rice blast leaves infected with *Pyricularia oryzae.* Part I : Isolation, characterization of biological activities of oryzalexins. *Agric. Bio. Chem.,* **49** : 1689-1694.

Akimoto, S. 1939. Varietal differnce in absorption of nitrogen and silica in relation to rice blast. *Agric. & Hort. Tokyo,* **14** : 2279-2290.

All China cooperation of Research on Physiological Races of *Pyricularia oryzae* (ACCR). 1980. Research on physiological races of rice blast fungus in China. *Acta Phytopathol. Sin.,* **10** : 72-82.

Ambalanathan, B. 1981. Physiology of host parasite interaction in the blast disease of rice. Studies on permeabiliity changes. Ph. D. Thesis, Univ. of Madras, Madras. p. 90.

Amin, K. S. and G. Venkatarao. 1979. Rice blast control by nitrogen management. *Phytopath. Z.,* **96** : 140-145.

Andersen, A. L. and B. W. Henry. 1946. The use of wetting and adhesive agents to increase the effectiveness of conidial suspensions for plant inoculations. *Phytopathology,* **36** : 1056-1057.

Andersen, A. L., B. W. Henry and E. C. Tullis. 1947. Factors affecting infectivity, spread and persistence of *Pyricularia oryzae* Cav. *Phytopathology,* **37** : 94 : 110.

Anikeev, V. G., Ya. P. Bunko, N. A. Tikhonova, O. A. Platnova and V. A. Shkalikov. 1985. Conditions of wintering over the pathogen of rice blast in Primorski Krai. *Izv. Timiryazev. S-KH Akad.,* **3** : 139-142.

Anstead, R. D. 1924. Rep. Madras Presidency for 1922-23. p. 30.

Aoki, Y. 1935. On physiologic specialization in the rice blast fungus, *Piricularia oryzae* Br. et Cav. *Ann. Phytopathol. Soc. Japan,* **2** : 107-120.

Apparao, A, 1956a. *Studies on the blast disease of paddy.* Ph. D. Thesis, Univ. *of Madras,* Madras, India. p. 117.

Apparao, A. 1956b. The role of pH in nitrogen utilization by *Piricularia oryzae. Experientia,* **12** : 215-218.

Apparao, A. 1959. Iron-copper antagonism and growth of *Piricularia oryzae. Curr. Sci.,* **28** : 336-337.

Apparao, A.1962. Nutritional factors governing sporulation of *Piricularia oryzae* in culture. *J. Indian Bot. Soc.,* **41** : 604-622.

Apparao, A. 1964. Nitrogenous manuring in relation to blast disease of rice. *Proc. Indian Acad. Sci.,* **59** : 173-184.

Apparao, A., L. Saraswathi-Devi and S. Suryanarayanan. 1955. Growth requirements of *Piricularia oryzae* Br. et Cav. *J. Indian Bot. Soc.,* **34** : 37-42.

Arai, N. and R. Yoshino, 1987. Studies on the sporulation of rice blast fungus. I. Relation between sporulation and temperature. *Ann. Phytopathol. Soc. Japan,* **53** : 371-372.

Arase, S., M. Nozu, S. Itoi and F. Kagayama. 1984. Pathogenicity of *Pyricularia* sp. from bamboo and *P. oryzae* to bamboo grown in the fields. *Bull. Fac. Agric. Shimane Univ. Japan,* **18** : 183-187.

Arase, S., S. Kinoshita, M. Kano, M. Nozu, E. Taneka and S. Nishimura. 1990a. Studies on host-selective infection mechanism of *Pyricularia oryzae* Cav. (2). Production of susceptibility inducing factor(s) from germinating spores and their phytotoxicity. *Ann. Phytopathol. Soc. Japan,* **56**. 322-330.

Arase, S., K. Kondo, Y. Honda, M. Nozu and S. Nishimura. 1990b. Studies on host-selective infection mechanism of *Pyricularia oryzae* Cav. (3). Light-dependency of leaf necrosis formation by toxins (3) from germinating spores. *Ann. Phytopathol. Soc. Japan,* **56** : 346-350.

Arase, S., Y. Yamaji, M. Iedome, Y. Honda and M. Nozu. 1993. Effect of thermal and cycloheximide pre-treatments on light-dependent necrosis formation to rice leaves by *Pyricularia oryzae* toxin (s). *J. Phytopathology,* **138** : 108-110.

Arimoto, Y., Y. Homma, N. Ohtsu and T. Misato. 1976. Studies on chemically induced resistance of plants to diseases. I. The effect of soaking of rice seed in dodecyl DL-alanine hydrochloride seedling infection by *Pyricularia oryzae. Ann. Phytopathol. Soc. Japan,* **42** : 397-400.

Asaga, K., T. Iutura, T. Kobayashi, K. Shindo, K. Iwata, K. Shibata and H. Koide. 1975. Occurrence of rice blast and causal races on breeding lines with resistance *Pi-z'* derived from Toride No. **1.**, Kanto-Tosan. *Pl. Prot. Soc. Proc.,* **22** : 14-15.

Asai, G. N. 1960. Intra- and inter-regional movement of uredospores of black stem rust in the upper Mississippi river valley. *Phytopathology,* **50** : 535-541.

Asai, G. N., M. W. Jones and F. G. Rorie. 1967. Influence of certain environment factors in the field on infection of rice by *Pyricularia oryzae. Phytopatholgy,* **57** : 237-241.

Asakawa, M., T. Misato and K. Fukunaga. 1968. Research and development of antibiotic for the control of plant diseases. III. Formation of antibiotics for the rice blast control. *Bull. Natl. Inst. Agric. Sci. Japan,* **22** : 153-241.

Ashrafuzzaman, M. H. and R. A. Frederiksen. 1969. Chemical Control of rice blast. Bull. MP. 923. A & M Univ. Texas, USA, pp. 1-12.

Astolfi. G. 1828. Congettura sulla malattia del brusone che infesta del riso.... *Ann. Univ. Tech.* Milano 6, 198. (*Sp. Pub. Sec. Mycol.,* Disease Survey USDA 6, **13,** 1954.)

Asuyama, H. 1965. Morphology, taxonomy, host range, and life cycle of *Piricularia oryzae.* In : *The rice blast disease,* Johns Hopkins Press, Baltimore & Maryland. pp. 9-22.

Atkins, J. G. 1962. Prevalence and distribution of pathogenic races of *Piricularia oryzae* in the US. *Phytopathology,* **52** : 2. (Abstr.).

Atkins, J. G. and T. H. Johnston. 1965. Inheritance in rice of reaction to races 1 and 6 of *Piricularia oryzae. Phytopathology,* **44** : 993-995.

Atkins, J. G., C. N. Bollich, T. H. Johnston, N. E. Jodon, H. M. Beachell and G. E. Templeton. 1965. Breeding for blast resistance in the United States. In : *The rice blast disease,* Johns Hopkins Press, Baltimore & Maryland. pp. 333-341.

Atkins, J. G., A. L. Robert., C. R. Adair, K. Goto, T. Kozaka, R. Yanagida, M. Yamada and S. Matsumoto. 1967. An International set of rice varieties for differentiating races of *Pyricularia oryzae. Phytopathology,* **57** : 297-301.

Awoderu, V. A. 1970. Identification of races of *Pyricularia oryzae* in Nigeria. *Pl. Dis. Reptr.* **54** : 520-523.

Awoderu, V. A. and O. F. Eusruoso. 1974. Reduction of grain yield of two rice varieties infected by the rice blast disease in Nigeria. *Nigerian Agri. J.,* **11** : 170-173.

Aylor, 1982. Modelling spore dispersal in a barley crop. *Agric. Meterol.* **26** : 215-219.

Ayyamperumal, P., V. Veerraju and A. Mahadevan. 1968. Occurrence of inhibitors in rice leaves. *Curr. Sci.,* **37** : 502-503.

Azizi, A.F.M. 1972. Genetic behaviour of resistance to blast disease *Piricularia oryzae* Cav. and some other agronomic characters. Ph. D. Thesis. Fac. Agric. Ain-Shams Univ. Cairo, Egypt. p. 135.

Bailey, A. G. and A. Van Eijnatten. 1961. Corn gray spot caused by *Piricularia grisea. Phytopathology,* **51** : 197-198.

Baldacci, E. 1947. Malattic del riso e brusone-V La malattia da *Piricularia oryzae. Risicultura,* **35** : 95-136.

Bandong, J. M. and S. H. Ou. 1966. The physiological races of *Pyricularia oryzae* Cav. in the Philippines. *Philipp. Agric.,* **49** : 655-667.

Bandong, J. M., F. L. Nuque, B. A. Estrada, F. A. Elazegui, S. H. Ou and J. P. Crill. 1979. Chemical disease control evaluation—1978. International Rice Research Institute (IRRI). P. O. Box. 933, Manila, Philippines. p. 51. (Mimeographed).

Bapreddy, D. and S. Vaheeduddin. 1956. Dissemination of blast spores. *Sci. Cult.* **21** : 677.

Barksdale, T. H. and G. N. Asai. 1961. Diurnal spore release of *Pyricularia oryzae* from rice leaves. *Phytopathology,* **51** : 313-317.

Barksdale, T. H. and M. W. Jones. 1965. Minimum values of dew period and temperature required for infection of *Pyricularia oryzae. Phytopathology,* **55** : 503.

Barr, M. E. 1977. *Magnaporthe, Telimenella* and *Hyponectria* (Physosporellaceae). *Mycologia,* **69** : 952-966.

Bastiaans, L. 1991. Quantifing the effects of rice leaf blast on leaf photosynthesis, crop growth, and final yield using an ecophysiological model of crop growth. In : *Rice blast modelling and forecasting*. Int. Rice Res. Inst. (IRRI), Box 933, Manila, Philippines, pp. 89-90.

Beachell, H. M., and G. S. Khush. 1969. Objectives of rice breeding programme at IRRI. *Sabrao Newsl.,* **1** : 69-80.

Berger, R. D. and J. W. Jones, 1985. A general model for disease progress with functions for variable latency and lesion expansion on growing hosts. *Phytopathology,* **75** : 792-797.

Bhardwaj, C. L. and B. M. Singh, 1982. Pathogenic variation in *Pyricularia oryzae* in Himachal Pradesh. *Indian Phytopathol.,* **35** : 165-170.

Bhardwaj, C. L. and B. M. Singh, 1983. The stability of resistance to *Pyricularia oryzae* Cav. in rice *Indian Phytopathol.,* **36** : 422-426.

Bhat, R. V. 1970. Studies on the genus *Drechslera*. Ph. D. Thesis. Univ. of Madras, Madras, India. p. 185.

Bhattacharya, D. and S. K. Bose. 1981. Studies on standardising the condition of sporulation in *Pyricularia oryzae. Indian Phytopathol.* **34** : 382-383.

Bhattacharya, M. and K. R. Samaddar. 1977. Fugistasis, lysis and survival of *Pyricularia oryzae* and *Helminthosporium oryzae* in soil. *Phytopath. Z.,* **89** : 60-67.

Bidaux, J. M. 1978. Screening for horizontal resistance to blast (*Pyricularia oryzae*) in Africa. In : *Rice in Africa* (eds. I.W. Buddenhagen and G. J. Persley) Academic Press, London. pp. 159-174.

Biffen, R. H. 1905. Mendels law of inheritance and wheat breeding. *J. Agric. Sci.,* **1** : 4-48.

Biossonnet-Mennes, M. and H. LeCoq. 1976. Transmission du virus par fusion de protoplastes chez *Pyricularia oryzae* Briosi et. Cav. *Physopl. Vegetable,* **14** : 251-257.

Bird, P.M. and I. A. Callow. 1982. Phytoalexin production in the incompatible reaction of rice to *Pyricularia oryzae. Int. Rice Res. Newsl.,* **7** : 5-6.

Bokura. U. 1930. On the relation of the balance of three principal fertilizers to rice blast. *Pl. Prot. Tokyo,* **17** : 332-340.

Bollich, C. N., J. G. Atkins, J. E. Scott and B. D. Webb. 1966. Dawn—a blast resistant, early mature long grain rice variety. *Rice J.,* **69** : 14-18.

Bonman, J. M. and S. W. Ahn. 1990. Managing rice blast disease—a continuing challenge. Int. Rice Res. Inst. (IRRI) Box 933, Manila, Philippines (Mimeographed).

Bonman, J. M. and D.J. Mackill. 1988. Durable resistance to rice blast disease. *Oryza,* **25** : 103-110.

Bonman, J. M., B. A. Estrada and J.M. Bondong. 1987a. Correlation between leaf and neck blast resistance in rice. *Phytopathology,* **77** : 1723,

Bonman, J. M., B. A. Estrada and J. M. Bandong. 1989. Leaf and neck blast resistance in tropical lowland rice cultivars. *Pl. Dis.,* **73** : 388-390.

Bonman, J. M., B. A. Estrada and I. R. Denton. 1985. Blast management with upland rice cultivar mixtures. In : 2nd International upland rice conference. Indonesia. p. 7.

Bonman, J. M., T. I. Vergel de Dios and M. M. Khin. 1986. Physiologic specialization of *Pyricularia oryzae* in the Philippines. *Pl. Dis.,* **70** : 767-769.

Bonman, J. M., T. I. Vergel de Dios, J. M. Bandong and E. J. Lee. 1987b. Pathogenic variability of monoconidial isolates of *P. oryzae* in Korea and Philippines. *Pl. Dis.,* **71** : 127-130.

Bonman, J. M., J. M. Bandong, Y. H. Lee, E. J. Lee and B. Valent. 1989. Race-specific partial resistance to blast in temperate japonica rice cultivars. *Pl. Dis.,* **73** : 496-499.

Bourett, T. M. and R. J. Howard. 1990. *In vitro* development of penetration structures in the rice blast fungus, *Magnaporthe grisea. Can. J. Bot.*, **68** : 329-342.

Bourett, T.M., and R.J. Howard, 1992. Action in penetration pegs of the fungus rice blast pathogen *Magnaporthe grisea. Protoplasma.* **168** : 20-26.

Bourett, T.M., M.A. Picollelli and R.J. Howard. 1993. Postembedement labelling of interacelluar concanavalin A-binding sites in freeze-substituted fungal cells *Expt. Mycol,* **17** : 223-235.

Bourke, P. M. A. 1970. Use of weather information in the prediction of plant disease epiphytotics. *A. Rev. Phytopath.,* **8** : 345-370.

Bremer, H., H. Ismen, G. Karel, K. Ozakan and M. Ozakan. 1948. Beitrage zur kennt nisder parasitischen pilze der Turkei Teil III. *Rev. Fac. Sci. Univ. Istanbul,* Ser. B, **13** : 1-53.

Briosi, G., and F. Cavara. 1892. *I. Funghi Parassiti* No. 188. Pavia, Italy.

Buck, K. W. 1987. Viruses of plant pathogenic fungi. In : *Genetics and Plant Pathogens.* (P. R. Day, and G. L. Jellis, eds.), Blackwell Scientific Publications, Oxford. pp. 111-126.

Buckley, T. A. and E. F. Allen. 1951. Notes on current investigations, April to June, 1951. *Malaysia Agri. J.,* **34** : 133-141.

Burleigh, J. R., M. G. Eversmayer and A. P. Roelfs. 1972. Development of linear equations for predicting wheat leaf rust. *Phytopathology,* **62** : 947-953.

Butt, D. J. and D. J. Royle. 1974. Multiple regression analysis in the epidemiology of plant diseases. In : *Epidemics of plant diseases : Mathematical analysis and modelling.* (J. Kranz, ed.), Springer-Verlag, New York. pp. 78-114.

Caffey, H. R. and G. D. Lindberg. 1964. The effect of chemical control of blast on yield, milling percent and mercury residue. 56th *Ann. Prog. Rept. Rice Expt. Stn.* Crowby, Louisiana. pp. 278-285.

Campacci, C. A. 1950. Rice blast. *Biologico,* **16** : 128-130.

Cartwright, D. W., P. Langcake and J. P. Ride. 1980. Phytoalexin production in rice (*Oryza sativa*) and its enhancement by a dichlorocyclopropane fungicide. *Physiol. Pl. Pathol.,* **17** : 259-267.

Cartwright, D. W., P. Langcake, R. J. Pryce., D. P. Leworthy and J. P. Ride. 1977. Chemical activation of host defence mechanism as a basis for crop protection. *Nature,* **267** : 511-513.

Cartwright, D. W., P. Langcake, R. J. Pryce., D. P. Leworthy and J. P. Ride. 1981. Isolation and characterization of two phytoalexins from rice as momilactones A and B. *Phytochem.,* **20** : 535-537.

Castano, J., D. R. McKenzie and R. R. Nelson. 1989. Components analysis of race no-specific resistance to blast disease of rice caused by *Pyricularia oryzae. J. Phytopathology,* **127** : 89-99.

Caten, C. E. 1981. Parasexual processes in fungi. In : *The fungal nucleus* (K. Gull and S. G. Oliver, eds.), Cambridge Univ. Press, Cambridge. pp. 87-100.

Cavara, F. 1891. Fungi Longbardiae exsiccati No. 49. (cited by Padwick, 1950).

Chai, R., M. Jin and Q. Zhang. 1989. Disease control of rice blast by mixing different resistant varieties. 10th Symposium on rice blast control. RDA, S.Korea. pp. 83-84.

Chakrabarti, N. K. 1971. Relation between meteorological factors and development of foliar blast of rice. *Oryza,* **8** : 393-397.

Chakrabarti, N. K. 1976. Diseases of site and their control. In : *Proc. Natl. seminar on control of pests and diseases of rice and jowar in India. Pesticides,* **2** : 11-15.

Chakrabarti, N. K. and R. D. Wilcoxson. 1970. Effect of light on sporulation by *Pyricularia oryzae. Phytopathology,* **60** : 171-172.

Chandramohan, J. and S. Palaniswamy. 1963. Incidence of blast disease of rice in relation to time of planting. *Rice News Teller,* **11** : 86-91.

Chang, T. T., M. K. Wang, K. M. Lin and C. P. Cheng. 1965. Breeding for blast resistance in Taiwan. In : *The rice blast disease,* Johns Hopkins Press, Baltimore & Maryland. pp. 371-377.

Chen, H., J. Fan and Y. Duan. 1987. Studies on the classification of rice varieties for blast resistance. *J. Yunnan Agric. Univ.* **2** : 15-19.

Chen, Y. S., H. C. Ren and C. T. Fang. 1964. Studies on the carbon and nitrogen nutrition of *Piricularia oryzae* and *Helminthosporium oryzae. Acta Phytopath. Sin.,* **7** : 165-174.

Chen, Z. J. 1964. Cytological studies of rice blast fungus *Piricularia oryzae* Cav. I. *J. Agric. & For.,* **13** : 1-6.

Chen, Z. J. 1965. Cytological studies of rice blast fungus *Piricularia oryzae* Cav. II. *J. Agric. & For.*, **14** : 1-14.

Chen, Z. Q. 1989. Investigations on the mechanism of resistance to rice blast disease. *J. South China Agric. Univ.*, **10** : 82-91.

Chetia, M. 1964. Air spora and their role in sterility in plants. I. Air spora over paddy fields. II. Sources of air spora over paddy fields. *J. Univ. Gauhati.*, **15** : 121-133; 134-147.

Chiba, S. and K. Yamashita. 1957. Influence of the rate of nitrogen and potassium on the growth of rice plant in potassium-deficient soil. *Rep. Soc. Pl. Prot. North Japan*, **8** : 25-27.

Chiba, S., J. Chiba, K. Shimada and H. Kagawa. 1972. Epidemiological studies on rice blast disease. 1-2. *Ann. Phytopathol. Soc. Japan*, **38** :15-21; 299-305.

Chiba, S. J. Chiba, K. Shimada, H. Kagawa and H. Fukujyu. 1973. Effect of O, O-disopropyl-S-benzyl thiophosphate (Kitazin P) on the control of rice disease. *Bull. Amori. Agric. Expt. Stn.*, **18** : 18-34.

Chida, T. and H. D. Sisler. 1987. Restoration of appressorial penetration ability by melanin precursors in *Pyricularia oryzae* treated with antipenetrants and in melanin-deficient mutants. *J. Pestic. Sci.*, **12** : 49-55.

Chien, C. C. 1967. Studies on the physiological races of the rice blast fungus, *Pyricularia oryzae* Cav. *Bull. Taiwan Agric. Res. Inst.*, **26** : 1-63.

Chien. C. C., W. H. Tsai, Y. Z. Yang and C. Liu. 1984. Studies on the epidemiology of rice blast disease in central areas of Taiwan. *J. Agric. Res. China*, **33** : 169-180.

Chin, K. M. 1975. Fungicidal control of the rice blast disease. *Malaysian Agric. J.*, **50** : 221-228.

Chin, K. M. 1985. Response of *Pyricularia oryzae* populations to integrated disease management. In: *Integrated Pest Management*, (B. S. Lee et al. eds.) pp. 183-190.

Chin, K. M. 1987. A simple model of selection for fungicide resistance in plant pathogen populations. *Phytopathology*, **77** : 666-669.

Chin, K. M. and A. Bhandhufalck. 1990. Application timing of disease control agents on rice. (B. T. Grayson et al. eds.) *Soc. Chem. Ind. Elsevier Applied Sci., London*. pp. 131-142.

Chin, S. M. and Y. C. Teng. 1976. *J. Agri. Res. Taiwan*, **25** : 249-258.

CNCRT (China National Coordinating Research Team on rice blast). 1983. Evaluation of rice varietal resistance to blast (*Pyricularia oryzae Cav.*) *6th Symp. on rice blast control*. pp. 1-10.

China National Rice Research Institute (CNRRI). 1988. Realization of rice cropping in China. *Zhejiang Sci. & Tech. Publ. House*. p. 167.

Chinte, P. T., 1965. The nutritional requirements and physiological responses of *Pyricularia oryzae in vitro*. M. Sc. Thesis, College of Agric. Univ. of the Philippines. p. 76.

Choi, J. E., J. S. Park and N. K. Park. 1989a. Resistance of varieties to rice blast in Korea. I. Japonica type of varieties. *J. Agric. Sci. Chungnam Natl. Univ., Korea*, **16** : 1-18.

Choi, J. E., J. S. Park and N. K. Park. 1989b. Resistance of varieties to rice blast in Korea. II. Tongil type of rice varieties. *J. Agric. Sci. Chungnam Natl. Univ. Korea*, **16** : 129-137.

Choi, W. J. 1987. A computer simulation model for rice leaf blast. M. S. thesis, Seoul Natl. Univ., Suweon, Korea. p. 78.

Choi, W. J., E. W. Park and E. J. Lee. 1987. A computer simulation model for rice leaf blast. *Phytopathology*. **77** : 1777 (Abstr.).

Choi, W. J., E. W. Park and E. J. Lee. 1988. LEAFBLAST—a computer simulation model for leaf blast on rice. *Korean J. Pl. Path.*, **4** : 25-32.

Chu, O. M. Y. and H. W. Li. 1965. Cytological studies of *Pyricularia oryzae* Cav. *Bot. Bull. Acad. Sin. Taipei*, **6** : 116-130.

Chumley, F. G. and B. Valent. 1990. Genetic analysis of melanin deficient, nonpathogenic mutants of *Magnaporthe grisea*. *Mol. Pl. Microbe Interact.*, **3** : 135-143.

Chung, H. S. 1974. New races of *Pyricularia oryzae* in Korea. *Korean J. Pl. Prot.* **13** : 19-23.

Chung, H. S. 1988. Recent advances of studies on rice fungal disease. World situation of rice diseases and recent advances in their studies. 5th *Int. Phytopathol. Cong.*, Kyoto, p. 6 (Abstr.).

Corbetta, G. 1961. II brusone del riso. *Riso*, **10** : 28-31.

Cralley, E. M. and C. R. Adair. 1942. Rice blast in Arkansas. *Pl. Dis. Reptr.*, 26 : 149-150.

Crawford, M. S., F. G. Chumley, C. G. Weaver and B. Valent. 1986. Characterization of the heterokaryotic and vegetative diploid phases *of Magnaporthe grisea. Genetics,* **114** : 1111-1129.

Crill, J. P. and G. S. Khush. 1979. Effective and stable control of rice blast with monogenic resistance. *ASPAC Food & Fertilizer Tech. Centre Ext. Bull. No.* **128** *Taiwan.* p. 13.

Crill, J. P. Y. S. Ham and H. M. Beachell. 1982. The rice blast disease in Korea and its control with race prediction and gene rotation. In : *Evaluation of the gene rotation concept for rice blast control.* IRRI, Box 933, Manila, Philippines. pp. 123-130.

Dai, F. and J. Chen. 1983. Resistance of rice varieties from Yunan Provience to rice blast and varietal leaf blight. 6th Symp. on *Rice blast control.* pp. 29-30.

Dasananda, S. 1965. Breeding for blast resistance in Thailand. In : *The rice blast disease,* Johns Hopkins Press, Baltimore & Maryland. pp. 379-396.

Day, P. R. 1973. *Genetics of host parasite interaction.* W. H. Freeman and Company, San Francisco. p. 238.

DeFaria, J. C., A. S. Prabhu and F. J. P. Zimmerman, 1982. Effect of nitrogen fertilization and fungicidal sprays on blast and yield of upland rice (*O. sativa* cv. IAC-47). *Pesqu. Agropec. Brasl.,* **17** : 847-852.

Deighton, F. C. 1965. *Pyricularia* Vs *Piricularia. Common. Phytopathol. News.,* **6** : 2.

Dekker, J. 1977. Resistence. In : *Systemic fungicides* (R. W. Marsh, ed.), Longman, London & New York. pp. 176-197.

Dhua, U. 1989. Relation between size, colour and spore liberation from rice blast lesions. *Phytopath. Z.,* **124** : 27-30.

DRR (Directorate of Rice Research), Hyderabad. 1986. Research highlights for 1986. p. 13.

Dodoff, D. N., and I. C. Kovacevski,. 1930. Preliminary study of the blast disease of rice in Bulgaria. *Bulgarian Agri. Soc. Sci. Trans.,* **25** : 61.

Doi, Y. and N. Suzuki. 1952. Histochemical studies on rice blast lesions. *Preli. Rep.* 1951 *work Sec. Pl. Inst. Agric. Sci.,* pp. 268-286.

Doi, Y., T. Sado and T. Kodama. 1952. Relationship between methods of planting and disease of rice. *Chugoku-Shikoku Agr. Expt. Stn. Bull.,* **1** : 74-82.

Duan, Y., Y. Zhu, H. Hen and N. Xinzhen. 1987. Inheritance and mechanism of resistance to rice blast. *J. Yunnan Agric. Univ.* **2** : 9-16.

Dube, H. C. 1971. Electrolytic changes in blasted rice. *Curr. Sci.,* **42** : 249-250.

Dugger, B. M. 1909. *Fungus diseases of plants.* Boston, Ginn & Co. p. 508.

Ellingtore, A. H., B. C. Wu and W. Robertson, 1990. Inheritance of avirulence/virulence in a cross of two isolates of *Magnaporthe grisea* pathogenic to rice. *Phytopathology,* **80** : 108-111.

El Refaei, M. I. 1977. Epidemiology of rice blast disease in tropics with special reference to the leaf wetness in relation to disease development. Ph. D. Thesis. *Indian Agri. Res. Inst.* (IARI), *New Delhi.* p. 180.

Eroutor, P. G., E. C. Jr. Gilmore, C. N. Bollich and M. A. Marchetti. 1978. Inheritance of resistance to race IB-54 and IG-1 of *Pyricularia oryzae* in rice. *Crop Sci.,***18** : 505-506.

Exconde, O. R. and A. D. Raymundo. 1970. Further studies on the assessment of yield losses due to rice blast. *Philippine Phytopathology.,* **16** : 66-74.

Ezuka, A. 1972. Field resistance of rice varieties to blast disease. *Rev. Pl. Proc. Res.,* **5** : 1-12.

Ezuka, A. 1973. Resistance of rice varieties. Akishino-mochi and Takasaago-mochi to blast disease. *Bull. Tokai-Kinki Natl. Agric. Exp. Stn.,* **25** : 57-64.

Ezuka, A. 1979. Breeding for and genetics in blast resistance in Japan. In : *Rice Blast Workshop, Int. Rice. Res. Inst.* (IRRI), Box 933, Manila Philippines. pp. 27-48.

Ezuka, A., T. Yunoki., Y. Sakurai., H. Shrinada and K. Toriyama. 1969a. Studies on the varietal resistance to rice blast. 1. Tests for genotype of true resistance. *Bull. Chugoku Agric. Exp. Stn.* No. 4, 1-32.

Ezuka, A., T. Yunoki., Y. Sakurai., H. Shindo and K. Toriyama. 1969b. Studies on the varietal resistance in paddy fields and upland nursery beds. *Bull. Chugoku Agric. Expt. Stn.,* **4** : 35-54.

Farkas, G. L. and Z. Király. 1958. Enzymological aspects of plant disease. I. Oxidative enzymes. *Phytopath. Z.*, **31** : 251-272.

Farkas, G. L. and Z. Király. 1962. Role of phenolic compounds in the physiology of plant disease and disease resistance. *Phytopath. Z.*, **44** : 105-150.

Fatemi, J. and R. R. Nelson. 1977. Intra-isolate heterokaryosis in *Pyricularia oryzae*. *Phytopathology*, **67** : 1523-1525.

Feng, D., G. Peng, Q. Luo and Q. Yang. 1989. Changes of physiological races of blast fungus and loss in resistance of rice varieties. 1984-88. In : *Abstracts of rice blast control*, Korea, pp. 62-63.

Ferault, A. C., D. Spire, F. Rapilly, J. Bertrandy, M. Skajennikoff and P. Bernaux. 1971. Observation de particules virales dans des souches de *Piricularia oryzae* Briosi et Cav. *Ann. Phytopathol.*, **3** : 267-269.

Ferraris, T. 1913. 'Brusone' del Riso c la *'Piricularia oryzae'* Br. et Cav. Nota preliminare. *Flora italica*, p. 779.

Friesland, H. and H. Schrodter. 1988. The analysis of weather factors in epidemiology. In : *Experimental techniques in plant disease epidemiology* (J. Kranz and J. Rotem, eds.), Springer-Verlag, New York. pp. 115-132.

Froyd, J. D., L. R. Guse, Y. Kushiro. 1978. Methods of applying tricyclazole for control of *Pyricularia oryzae* on rice. *Phytopathology*, **68** : 818-822.

Fukunaga, K. 1965. Fungicide development for host control. In : *Rice blast disease*, Johns Hopkins Press, Baltimore & Maryland. pp. 409-414.

Fukunaga, K. 1966. Antibiotics and new fungicides for control of rice diseases. In : *Plant. Disease*. 11th. *Pacific. Sci. Congr. Tokyo*. pp. 157-180.

Fukunaga, K., T. Misato, I. Ishii, M. Asakawa and M. Katagiri. 1968. Research and development of antibiotics for the rice blast control. *Bull. Natl. Inst. Agric. Tokyo*, **22** : 1-94.

Fukunaga, S., T. Misato and I. Ishii. 1955. Blasticidin, a new antiphytopathogenic fungal substance. *Bull. Agric. Chem. Japan*, **19** : 181-188.

Fukutomi, M., P. Park, M. Tsuda and A. Veyama. 1976. Ultrastructure of the ascus and ascospore walls in the perfect state of *Pyricularia* sp. *Trans. Mycol. Soc. Japan*, **17** : 74-76.

Fulton, H. R. 1908. Diseases affecting rice in Louisiana. *Bull. Louisiana Agri. Exp. Stn.* No. **105**. p. 28.

Furishiro, K. S., K. Shimizu, H. Sasaki, M. Minneyota and T. Fujisawa, 1958. Hygrostatin a new antibiotic substance. *Chem. Abstr.*, **58** : 24-39.

Gálvez, E., E. Uillerma and J. C. Lozano T. 1968. Identification of races of *Pyricularia oryzae* in Colombia. *Phytopathology*, **58** : 294-296.

Gangopadhyay, S. 1983. *Current concepts on fungal diseases of rice*, Today and Tommorrow's Publishers, New Delhi. p. 349.

Gangopadhyay, S and S. Y. Padmanabhan. 1987. *Breeding for disease resistance in rice*. Oxford and IBH Publishing Co., New Delhi. p. 340.

Ganguly, D., M. S. Balakrishana and S. Y. Padmanabhan. 1954. Effect of manuring on the incidence of blast. *Rice News Teller*, **2** : 130-138.

Genovesi, A. D. and C. W. Magill. 1976. Heterokaryosis and parasexuality in *Pyricularia oryzae* Cav. *Can. J. Microbiol.*, **24** : 531-536.

Gera, F. A. 1846. Sulla Mallattie Chiamata indistintamente col nome brusone e sopra i rimediīdā ūsarsi a previnirle e distruggrle. Lombardo-Venete Giorn. Agr., Milan, 1848 Fasc. 35. (*U.S. Dept. Agr. Mycol. Dis. Ser. Spl. Publ.*, **6** : 47. 1954).

Giatgong, P. 1961. Studies on blast disease of rice and grasses. *Madras Agric. J.*, **48** : 25.

Giatgong, P. and R. A. Frederiksen. 1967. Variation in pathogenicity of *Piricularia oryzae*. *Phytopathology*, **57** : 460 (Abstr.).

Giatgong, P. and R. A. Frederiksen. 1968. Chromosome number and mitotic division in *Pyricularia oryzae*. *Phytopathology*, **58** : 728 (Abstr.).

Giatgong, P. and R. A. Frederiksen. 1969. Pathogenic variability and cytology of monoconidial subcultures of *Pyricularia oryzae*. *Phytopathology*, **59** : 1152-1157.

Gill, M. and J. M. Bonman. 1988. Effect of water deficit on rice blast. I. Influence of water deficit on components of resistance. *J. Pl. Protect. Trop.* **5** : 61-66.

Girbardt, M. 1965. Eine Zielschnitt methode fur Pilzzellen. *Mikroskopie,* **20** : 254-264.

Gnanamanickam, S. S., R. C. Reys and T. W. Mew. 1989. Biological control of rice blast (Bl) with antagonistic bacteria. *Int. Rice Res. Newsl.,* **14** : 34-35.

Gnanamanickam, S. S., R. Valasubramanian, K. V. Thara, Savithri Singh, R. Sivaraj, A. Chatterjee, M. Levy, S. Muthukrishan and T. W. Mew. 1993. Molecular approach for biological control of rice blast and sheath blight. 6th Int. Phytopath. Congress. Montreal, Canada. (Abstr.).

Goto, I. 1959, 1960. On the inheritance of resistance to the blast disease. Speeches given at annual conference of the Phytopathol. Soc. Japan. (cited by Takahashi, 1965).

Goto, I. 1970. Genetic studies on resistance of rice plant to blast fungus. I. Inheritance of resistance in crosses. Sensho X H-79 and Imochi-Shirazu X H-79. *Ann. Phytopathol. Soc. Japan.* **36** : 304-312.

Goto, K. 1955. History of the blast disease and changes in methods of control. *Agric. Impro. Bureau, Min. Agri. Fores. Japan,* **5** : 1-2.

Goto, K. 1960. Progress report of the joint research on the rice blast fungus, *Pyricularia oryzae* (Fascicle 1). *Pl. Inst. Dis. Insect. Forecast. Ser., No.,* **5,** pp. 1-89.

Goto, K. 1963. On the races of rice blast fungus, *Piricularia oryzae,* in Japan. *Ann. Phytopathol. Soc. Japan,* **28** : 61.

Goto, K. 1965a. Physiological races of *Piricularia oryzae* in Japan. In : *The rice blast disease,* Johns Hopkins Press, Baltimore & Maryland. pp. 237-244.

Goto, K. 1965b. Estimating losses from rice blast in Japan. In : *The rice blast disease,* Johns Hopkins Press, Baltimore & Maryland. pp. 195-202.

Goto, K. and T. Yamanaka, 1957. Studies on the rice blast fungus. *Nat. Inst. Agric. Sci. Bull.,* **11** : 72-81.

Goto, K. and S. Yamanaka. 1960a. *Pyricularia* on sweet vernal grass. *Interim Rep. Natl. Inst. Agric. Sci. Div. Pl. Path. Tokyo,* **13** : 88-90.

Goto, K., S. Yamanaka and S. Kobayashi, 1954. Blast disease of *Zizania latifolia. Ann. Phytopathol. Soc. Japan,* **18** : 160.

Goto, K. et al. 1964. Joint work on the races of rice blast fungus, *Piricularia oryzae* (Fascicle 2) *Pl. Inst. Dis. Sur. Spl. Rep.,* **18,** p. 132.

Goto, K. et. al. 1967. U. S. Japan Cooperative research on the international pathogenic races of the rice blast fungus, *Pyricularia oryzae* Cav., and their international differentials. *Ann. Phytopathol. Soc. Japan,* **33** : 1-14.

Govindarao, P. and D. Koteswararao, 1958. A new record of *Piricularia* from India. *Indian Phytopathol.,* **10** : 107-109.

Govindaswamy, V. V. 1964. Temperature and blast incidence. *Madras Agric. J.,* **51** : 255-257.

Gregory, P. H. 1968. Interpreting plant disease dispersal gradients. *Ann. Rev. Phytopath.,* **6** : 189-212.

Griffiths, D. A. 1969. Rice blast disease in Malaya : Spore germination and host penetration. *Pl. Dis. Reptr.,* **54** : 490-499.

Gunther, H. 1986. Simulation of the epidemiology of *Pyricularia oryzae* in rice : a first attempt. *Agric. Univ. Wageningen, Dept. Theor. Prod. Ecol.,* p. 80.

Guochang, S. and S. Shuyuan. 1989. A new inoculation technique for rice blast. *Int. Rice Newsl.,* **14** : 15.

Hackel, H. 1984. Zur Mussung der Benetzungsdauev von Pflanzen Vertahren und Ergefnissu. *Meterol. Resch.,* **37** : 97-104.

Hamer, J. E., B. Valent and F. G. Chumley. 1989. Mutations at the *SMO* genetic locus affect the shape of diverse cell types in the rice blast fungus. *Genetics,* **122** : 351-361.

Hamer, J. E., B. Valent. and F. G. Chumley. 1990. Mutations that alter spore morphology and appressorium development in *Magnaporthe grisea* (cited by Hamer et al., 1989).

Hamer, J. E., R. J. Howard, F. G. Chumley and B. Valent. 1988. A mechanism for surface attachment in spores of a plant pathogenic fungus. *Science*, **239** : 288-290.

Han, S. H., Y. H. Lee and S. H. Yu. 1989. Variation in the pathogenicity of protoplast regenerated isolates of *Pyricularia oryzae*. *Korean J. Pl. Pathol.*, **4** : 156-160.

Hansford, C. Q. 1943. Contributions towards the fungus flora of Vganola. V. Fungi imperfecti. *Proc. Linn. Soc. London* 1942-43. **1** : 34-67.

Hara, K. 1916. *Piricularia oryzae* on barley. *Byochu-gai zasshi*, **3** : 693-694.

Hara, K. 1918. *Diseases of the rice plant*. p. 218.

Harada, Y. 1955. Studies on a new antibiotic for rice blast control. *Ann. Meet. Agric. Chem. Soc. Japan.* (cited by Okamoto, 1965).

Hashiguch, S. and H. Kato. 1983. Quantificational method applied to the forecasting of the rice blast occurrence. *Pl. Protect.*, 37 : 425-429.

Hashiguch, S., H. Kato and S. Kiyosawa. 1973. Effect of climatic factors on progress of rice blast. Research on the counter measures by breeding to 'breakdown' of true resistance against rice blast disease. *Min. Agric. Forest. Japan*, pp. 366-370.

Hashimoto, A., K. Hirano and K. Matsumoto. 1984. Studies on the forecasting of rice blast development by application of the computer simulation. *Spl. Bull. Fukushima Pref. Agric. Exp. Stn.*, **2** : 1-104.

Hashimoto, A., K. Matsumoto and T. Adachi. 1982. Some trials for forecasting rice leaf blast epidemics by application of the computer simulation. *A. Rep. Soc. Pl. Protec. North Japan*, **33** : 12-14.

Hashioka, Y. 1943a, b, c. Studies on the rice blast disease in the tropics. (a). Influence of temperature of air and soil upon the resistance of the rice plants to the blast disease.
(b). Relation of temperature and humidity to prevalence of the disease.
(c). Relation of the age of the rice plants to blast resistance and its significance as a controlling factor for the prevalence of the disease. *J. Soc. Trop. Agric.*, **15** : 53-65; 99-111; 161-177.

Hashioka, Y. 1944a. b. Studies on the rice blast disease in the tropics. (a). Influence of the temperature on blast resistance of leaves of the rice plants grown under different nutritional conditions.
(b). Relation of temperature of leaf blast resistance and different varieties of rice plants collected from the districts in various latitudes. *J. Soc. Trop. Agric.* **16** : 163-176; 196-204.

Hashioka, Y. 1948. The microclimate of the paddy field in connection with prevalence of rice blast disease. *Agri. Meteorol. Japan*, **6** : 25-29.

Hashioka, Y. 1950a. Studies on the mechanism of prevalence of rice blast disease in the tropics. *Tech. Bull. Taiwan Agric. Res. Inst.* No. **8**, p. 237.

Hashioka, Y. 1950b. On the resistance of rice plants to rice blast in relation to air temperature. *Nagaku*, **4** : 6.

Hashioka, Y. 1965. Effects of environmental factors on development of causal fungus disease. In : *The rice blast disease*, Johns Hopkins Press, Baltimore & Maryland. pp. 153-161.

Hashioka, Y. 1971. Notes on *Pyricularia*. I. Three species parasitic to Musaceae, Cannaceae, and Zingiberaceae. *Trans. Mycol. Soc. Japan*, **12** : 126-135.

Hashioka, Y. 1972. Fine structure of the rice blast. IX. Scanning electron-microscopical observations on appressoria of the rice blast fungus and other species of *Pyricularia*. *Res. Bull. Fac. Agric. Gifu. Univ.*, **33** : 65-73.

Hashioka, Y. 1973. Notes on *Pyricularia*. II. Four species and one variety parasitic to Cyperaceae, Gramineae, and Commelinaceae. *Trans. Mycol. Soc. Japan*, **14** : 256-265.

Hashioka, Y. and T. Fukita. 1969. Ultrastructural observation on mycoparasitism of *Trichoderma, Gliocladium* and *Acremonium* to phytopathogenic fungi. *Rep. Tottori Mycol. Inst.*, **7** : 8-18.

Hashioka, Y. and H. Ikegami. 1961. Amino acid contents in rice leaves in relation to blast susceptibility of rice seedlings predisposed by growth temperature and nitrogen nutrition. *Ann. Phytopathol. Soc. Japan*, **26** : 50.

Hashioka, Y. and H. Ikegami. 1969. Fine structure of the rice blast. V. Surface structure of the rice blast fungus and the diseased leaves. *Res. Bull. Fac. Agric. Gifu Univ.*, **28** : 40-50.

Hashioka, Y. and K. Inagaki. 1968. Karyological characters of the different species and an interspecific heterokaryon of *Pyricularia*. *Ann. Phytopathol. Soc. Japan,* **34** : 391.

Hashioka, Y. and K. Kato. 1955. Hofchen Briefe. **8** : 177-191.

Hashioka, Y. and N. Kitano. 1971. Fine structure of the rice blast. VIII. Structure and ultrastructure of the initial infection of neck-blast. *Res. Bull. Fac. Agric. Gifu Univ. Japan,* **31** : 83-97.

Hashioka, Y., H. Ikegami and T. Murase. 1968. Fine structure of the rice blast. III. The mode of invasion of *Pyricularia oryzae* into rice epidermal cells. *Res. Bull. Fac. Agric. Gifu. Univ.,* **26** : 23-30.

Hashioka, Y., H. Ikegami, O. Horino and T. Kamei. 1967a, b. Fine structure of the rice blast. I. Electron micrographs of *Pyricularia oryzae* and the epidermal cell wall of rice plants. *Res. Bull. Fac. Agric. Gifu Univ. Japan,* **24** : 66-77; 78-90.

Hastings de Gutierrez, L. 1955. Presence of *Pyricularia oryzae* on sugar cane in Costa Rica. *Pl. Dis. Rep.,* **38** : 695.

Hatai, N. 1965 Equipment for application of fungicides for blast control. In : *The rice blast disease.* Johns Hopkins Press, Baltimore, Maryland. pp. 415-420.

Hatai, N. et. al. 1955. Deposit and adherence of chemicals applied. Studies on spraying and dusting. *Agric. Impr. Bur. Min. Agric. Fores. Japan,* pp. 12-16.

Hayashi, T. and Y. Koshimizu. 1988. Computer programme BLASTEM for forecasting occurrence of rice leaf blast. *Tohoku Natl. Agric. Exp. Stn.* **78** : 123-138.

He, Y., Y. Wang, J. Li, C. Li, S. Kiyosawa, T. Higashin and N. Horishe. 1989. Gene analysis of blast resistnace of a Yunnan upland rice variety Haunaihuan. *Oryza,* **26** : 173-182.

He, Z., C. Shi., Z. Shen, D. Auang and J. Wang. 1989. Analysis of peroxidase isoenzymes in rice varieties susceptible to rice blast. *Chinese J. Rice Sci,* **3** : 95-96.

He, Z. H. and Z. T. Shen. 1990. Near-isogenic pairs of indica rices with blast (Bl) resistant genes. *Int. Rice Res. Newsl.* **15** : 7.

Heath, M. C., F. J. Howard, B. Valent and F. G. Chumley. 1992. Ultrastructural interactions of one strain of *Magnaporthe grisea* with goosegrass and weeping lovegrasses. *Can. J. Bot.,* **70** : 779-787.

Heath M. C., B. Valent, R. J. Howard and E. G. Chumley. 1990. Correlations between cytologically detected plant fungal interactions and pathogenicity of *Magnaporthe grisea* towards weeping lovegrass. *Phytopathology,* **80** : 1382-1386.

Hebert, T. T. 1971. The perfect stage of *Pyricularia grisea. Phytopathology,* **61** : 83-87.

Helentjaris, T., M. Slocum, S. Wright, S. Schaefer and J. Nienhuis. 1986. Construction of genetic linkage maps in maize and tomato using restriction fragment length polymorphisms. *Theor. Appl. Genet,* **72** : 761-769.

Hemmi, T. 1930. On the relation of soil moisture to the development of the rice blast disease. *Japan. J. Bot.,* **5** : 4 (Abstr.).

Hemmi, T. 1933. Experimental studies on the relation of environmental factors to the occurrence and severity of blast disease in rice plants. *Phytopath. Z.,* **6** : 305-324.

Hemmi, T. 1949. Studies on the rice blast disease. Asakura Co., Tokyo, pp. 269-293.

Hemmi, T. 1949. Studies on rice blast fungus of Indian corn. *Ann. Phytopathol. Soc. Japan,* **13** : 23-25.

Hemmi, T. et al. and T. Abe. 1931. On the relation of temperature and continuous wetting to the infection of rice plants by *Pyricularia oryzae. Forsch. Geb. Pflkrankh.,* **1** : 33-45.

Hemmi, T. and T. Abe. 1932. Studies on the rice blast disease. II. Relation of the environment to the development of blast disease. *Bull. Dept. Agri. For. Japan,* No. **47**, p. 207.

Hemmi, T. and J. Imura. 1939. On the relation of air humidity to conidial formation in the rice blast fungus, *Pyricularia oryzae* and the characteristics in the germination of conidia produced by the strains showing different pathogenicity. *Ann. Phytopathol. Soc. Japan,* **9** : 147-156.

Hemmi, T., T. Abe, J. Ikeya and Y. Inoue. 1936. Studies on the rice blast disease. IV. Relation of the environment to the development of blast disease and physiologic specialization in the rice blast fungus. *Mater. Rural Improv. Dep. Agric. For., Japan,* No. 105, p. 145 (cited by Hashioka, 1965).

Henry, B. W. and A. L. Andersen. 1948. Sporulation by *Pyricularia oryaze*. *Phytopathology*, **88** : 265-278.

Higashi, T. and K. Kushibuchi. 1978. Genetic analysis of field resistance to leaf blast (*Pyricularia oryzae*) in rice. *Japan J. Breed*, **28** : 277-286.

Higashi, T. and T. Saito. 1985. Linkage group of field resistance genes of upland rice variety "Sensho" to leaf blast caused by *Pyricularia oryzae* Cav. *Japan J. Breed.*, **35** 438-448.

Hilda, A. and S. Suryanarayanan, 1976. Cross-protection in the blast disease of *Panicum repens* L. *Proc. Indian Acad. Sci.*, **84** : 215-225.

Hilda, A. and S. Suryanarayanan, 1977. Relationship between prefound/post-infectional antifungal substances in leaves of *Panicum repens* L. and compatibility with *Pyricularia* spp. *Proc. Indian Acad. Sci.*, **85** : 257-268.

Hirano, K. and K. Goto. 1963. Pathogenesis and ecology of panicle blast of rice plant. *Bull. Natl. Inst. Agric. Sci. Ser*. C. **16** : 1-58.

Hirano, K. 1967. Recent problems in rice breeding for blast resistance in Japan. In : *Rice diseases and their control by growing resistant varieties and other measures. Agric. For & Fish. Res. Coun.*, pp. 103-111.

Hirayama, T., S. Horie, H. Nagayama and K. Matsuda. 1978. Studies on cellulases of a phytopathogenic fungus. *Pyricularia oryzae* Cav. II. Purification and properties of a β-glucosidase. *J. Biochem.*, **84:** 27-37.

Hirayama, T., S. Horie, H. Nagayama, K. Matsuda and K. Tamari. 1976. Number and interrelation of components of C_x enzyme from *Pyricularia oryzae. Agric. Biol. Chem.*, **40** : 2137-2142.

Hirooka, T. Y. Miyagi, F. Araki, H. Kunoh and H. Ishizaki. 1982. The effect of isoprothiolane on the emergence of infection pegs from appressoria of *Pyricularia oryzae. Pestic. Sci.* **13** : 379-380.

Hirst, J. M. 1952. An automatic volumetric spore trap. *Ann. Appl. Biol.*, **39** : 257-265.

Hoette, S. 1936. Pitting diseases of bananas in Australia. *Proc. Roy. Soc. Vict.*, N. S., **48** : 90-95.

Homma, Y., Y. Arimoto and H. T. Mijato. 1980. Effect of soyabean lecithin on each growth state of rice blast fungus. *Pyricularia oryzae* Cav. in its life cycle. *J. Pestic. Sci.*, **5** : 611-613.

Homma, Y., T. Shinda and T. Misato. 1973. Studies on the control of plant diseases by amino derivatives I. Effect of n-lauroyl-L-valine on rice blast. *Ann. Phytopathol. Soc. Japan*, **39** : 90-98.

Honma, N., K. Ishihara, N. Ogaswara and K. Tamari. 1968. The effects of nitrogen compounds on the sporulation of a slant culture of *Pyricularia oryzae. Nigata Norin* Kenkyu. **20** : 89-100.

Hori, M. 1963. Studies on the forecasting of rice blast with special reference to the experimental forecasting. *Spec. Res. Rep. Dis. and Ins. Forec. Min. Agr. Forest. Japan*, **14** : 1-76.

Hori, M. 1965. Forecasting techniques on the basis of morphological and chemical characters of the host plant. *Ann. Phytopathol. Soc. Japan*, **31** : 301-306.

Hori, M., T. Arata and Y. Inoue. 1960. Studies on the forecasting method of blast disease. VI. Forecasting by the degree of accumulated starch in the sheath of rice plant. *Ann. Phytopathol. Soc. Japan*. **25** : 2.

Hori, M., K. Uchino and Y. Kijima 1957. Correlation between the number of silicified cells in upper most leaves of rice plants and neck blast. *Pl. Prot.*, **11** : 333-336.

Hori, S. 1898. Blast disease of rice plants. *Sp. Rep. Imp. Agri. Exp. Stn. Tokyo*, **1** : 1-36.

Horie, T. 1981. System ecological studies on crop-weather relationships in photosynthesis, transpiration and growth. *Bull. Natl. Inst. Agric. Sci. Ser.*, **28** : 1-181.

Horino, O. and S. Akai. 1965. Comparison of the electron micrographs of conidia of *Helminthosporium oryzae* Ito & Kurib and *Pyricularia oryzae* Cav. *Trans. Mycol. Soc. Japan*, **6** : 41-46.

Howard, R. J. and M. A. Ferrari. 1989. The role of melanin in appressorium function. *Exp. Mycol.*, **13** : 403-418.

Howard, R. J., T. M. Bourett and M. A Ferrai. 1991a. Infection by *Magnaporthe :* an *in vitro* analysis. In : *Electron microscopy of plant pathogens* (eds. K. Mendgen and D. E. Lesemann) Springer-Verlag, New York . pp. 251-264.

Howard R. J., M. A. Ferrai, D. H. Roch and N. P. Money. 1991b. Penetration of hard substrates by a fungus employing enormous turgor pressures. *Proc. Natl. Acad. Sci.* USA. **88.** 11281-11284.

Hsieh, S. C. and C. C. Chen. 1967. Recent status of rice breeding for blast resistance in Taiwan with special regard to races of blast fungus. In : *Rice diseases and their control growing resistant varieties and other measures. Trop. Agric. Res. Ser.,* **1** : 65-81.

Hsieh, S. C., M. H. Lin and H. L. Liang. 1967. Genetic analysis in rice. VIII. Inheritance of resistance to races 4, 22, and 25 of *Pyricularia oryzae. Bot. Bull. Acad. Taipei,* Spl. No. **8**, pp. 255-260.

Huang, C. H. and C. C. Chien. 1961. Investigation of physiological races in the blast fungus (*Pyricularia oryzae*) *Spl. Publ. Tawain Agric. Res. Inst.,* **3** : 31-37.

Hughes, S. J. 1958. Revisiones hyphomycetum aliquot cum appendice de a nominibus rejiciendis. *Can. J. Bot.,* **36** : 727-836.

Hunst, P. L., F. M. Latterell and A. E. Rossi. 1986. Variation in double stranded RNA from isolates of *Pyricularia oryzae. Phytopathology,* **76** : 674-678.

Hwang B. K. and E. J. Lee. 1980. A note on quantitative variation in pathogenicity of *Pyricularia oryzae* in the greenhouse. *Korean J. Pl. Prot.,* **19** : 1-4.

ICAR (Indian Council of Agricultural Research, New Delhi). 1954. *Agr. Res. Rep.* 1952-53. pp 2-45.

IITA (International Institute of Tropical Agriculture). 1979. *A. Rep.* 1978.

Ikata, S., S. Taguchi, M. Yoshida and H. Iguchi. 1938. Results of experiments on the control of rice blast disease Rept. 5. *Farm Bull. Dep. Agri. For. Japan,* No. 130.

Ikeda, M. 1933. Influence of application of various nitrogenous fertilizers on silica content of rice plants. *Totori Soc. Agri. Sci.,* **4** : 265-270.

Ikegame, H. and I. Takagi. 1972. Pathogenicity of *Pyricularia* and *Cochliobolus* conidia compressed by hydrostatic pressure. *Ann. Phytopathol. Soc. Japan,* **38** : 95-99.

Ikehashi, H. and S. Kiyosawa. 1981. Strain-specific reaction of field resistance of Japanese rice varieties revealed with Philippine strains of rice blast fungus *Pyricularia oryzae* Cav. *Japan J. Breed.,* **31** : 293-301.

Imura, J. 1938. On the effect of sunlight on the enlargement of lesions of the blast disease. *Ann. Phytopathol. Soc. Japan,* **8** : 23-33.

Imura, J. 1940. On the effect of sunlight on the incubation periods and development of blast disease and Helminthosporium disease of rice plants. *Ann. Phytopathol. Soc. Japan,* **10** : 16-26.

Inbe, T. and S. Matsumoto. 1985. Inheritance of resistance of rice varieties to the blast fungus strains virulent to the variety Reiho. *Japan J. Breed.,* **35** : 332-339.

Ingold, C. T. 1964. Possible spore discharge mechanism in *Pyricularia. Trans. Br. Mycol. Soc.,* **47** : 573-575.

Ingold, C. T. 1965. *Spore liberation.* Clarendron Press, Oxford. p. 210.

Inoue, E. 1963. On the turbulent structure airflow within crop canopies. *J. Meterol. Soc. Japan,* **41** : 317-326.

Inoue, S., J. Kato, V. W. I. Jordan and K. J. Brent. 1987. Inhibition of appressorial adhesion of *Pyricularia oryzae* to barley leaves by fungicides. *Pestic. Sci.,* **19** : 145-152.

Inoue, S., T. Uematsu and T. Kato. 1984. Mechanism of rice blast control of chlobenthiazone. II. Effects of chlobenthiazone on the infection process by *Pyricularia oryzae. J. Pestic. Sci.,* **9** : 689-695.

Inoue, T., Y. Okamoto and Y. Nishikado. 1958. *Rep. Agric. Res. Coun.,* 1958-59, Japan.

Inoue, Y. 1939. Comparison of the cellulose decomposition by culture strains of the rice blast fungus *Piricularia oryzae* Br. et Cav. *Ann. Phytopathol. Soc. Japan,* **9** : 33-40.

Inoue, Y. 1943. On the epidemic outbreak of rice blast in 1941 (Oita Prefecture). *Ann. Phytopathol. Soc. Japan,* **12** : 181-190.

Inukai, T. et al. 1992a, b. (i) Blast resistant genes *Pi-2 (t)* and *Pi-z* may be allelic. (ii) Blast resistant gene *Pi-2(t)* in near-isogenic line C104 PKT. *Rice Genet. Newsl.* **9** : 90-92; 94-94.

IRAT (Institut de Recherches Agronomiques Tropicales et des Cultures Vivieres). 1977. A. *Rep.* 1976.

IRRI. 1964. *A. Rep.* 1963.

IRRI. 1965. *A. Rep*. 1964.

IRRI. 1966. *A. Rep*. 1965.

IRRI. 1967. *A. Rep*. 1966.

IRRI. 1968. *A. Rep*. 1967.

IRRI. 1975. *A. Rep*. 1974.

IRRI. 1976. *A. Rep*. 1975.

IRRI. 1977. *A. Rep*. 1976.

IRRI. 1979. *A. Rep*. 1978.

IRRI. 1988. Standard evaluation system for rice. (3rd ed.) P. O. Box 933, Manila, Philippines. p. 54.

Ishi, S., C. Hirano, Y. Iwata, M. Nakazatra and H. Miyagawa 1962. Isolation of benzoic and salicylic acids from rice plants as growth inhibiting factors for the rice stem borer (*Chilo suppressalis*) and some rice plant fungus pathogens. *J. Appl. Entom. Zool. Japan*, **6** : 281-288.

Ishiguro, K. 1986. Composing a simulation model for epidemics of rice panicle blast. *Nogyo Gijutu*, **41** : 491-495.

Ishiguro, K. and A. Hashimoto. 1988. Development of a simulation model for rice panicle blast epidemics. (1) Composition of basic model. *Bull. Fukishima Pref. Agric. Res. Stn.*, **27** : 1-19.

Ishiguro, K. and A. Hashimoto. 1989. Recent advances of forecasting of rice blast epidemics using computers in Japan. *Trop. Agric. Res.* Ser. **22** : 153-162.

Ishiguro, K. and A. Hashimoto. 1990. Development of a simulation model for rice panicle blast epidemics. (2) Its verification and sensitive analysis. *Bull. Fukushima Pref. Agric. Res. Stn.*, **29** : 15-70.

Ishiguro, K., S. Yoshitake, S. Watanabe, F. Nemoto and A. Hashimoto. 1988. Temporal dynamics of thalide on rice leaves in the field and its efficacy for preventing leaf blast. *A. Rep. Pl. Prot. North Japan*, **39** : 63-65.

Ishizuka, Y. and Y. Hayakawa. 1951. Resistance of rice plant to the imochi disease (rice blast disease) in relation to its silica and magnesia contents. *J. Sci. Soil & Manure*, **21** : 253-260.

Ito, I. and T. Yamaguchi, 1977. Occurrence of kasugamycin resistant rice blast fungus influenced by the application of fungicides. *Ann. Phytopathol. Soc. Japan*, **43** : 101-103.

Ito, I. and T. Yamaguchi. 1979. Competition between sensitive and resistance strains of *Pyricularia oryzae* Cav. against kasugamycin. *Ann. Phytopathol. Soc. Japan*, **45** : 40-46.

Ito, R. 1965. Breeding for blast resistance in Japan. In : *The rice blast disease*. Johns Hopkins Press, Baltimore & Maryland. pp. 361-370.

Ito, R. *et al.* 1961. On the new rice variety 'Kusabue'. *J. Kanto-Tosan Agri. Exp. Stn.*, **18** : 21-33.

Ito, S. 1918. On the binomials of the fungus in Japan. *Bot. Nag. Tokyo*, **32** : 303-308.

Ito, S. and H. Hayashi, 1931. On the relation of silica supply to rice blast. *J. Sapporo Soc. Agri. Sci.*, **22** : 460-461.

Ito, S. and K. Kuribayashi, 1931. Studies on the rice blast disease. *Farm Bull. Dept. Agri. For. Japan.* No. 30. p. 81.

Ito, S. and M. Sakamoto, 1939, 1940, 1941. Studies on rice blast. *Res. Hokkaido Univ. Bot. Lab. Fac. Agric. Rep.*, 1939-43.

Itoi, S., F. Sato, J. Yamamoto, T. Uchita and C. Noda. 1979. Overwintering of *Pyricularia oryzae* on the living bamboo grass leaves and pathogenicity of the rice blast fungus, *P. oryzae* Cav. to bamboo and bamboo grasses. *Ann. Phytopathol. Soc. Japan*, **45** : 375-385.

Itoi, S., M. Nozu, F. Sato, J. Yamamoto, C. Noda and T. Uchita. 1978. *Pyricularia* sp. parasitic to bamboo and bamboo grasses. *Ann. Phytopathol. Soc. Japan*, **44** : 209-213.

Iwano, M. 1987. Suppression of rice blast infection by incompatible strain of *Pyricularia oryzae* Cav. *Bull. Tohuku Natl. Agric. Exp. Stn.*, **75** : 27-39.

Iwano, M. 1990. Distribution of pathogenic races of rice blast fungus and varietal resistance to blast in Yunnan Province. *Pl. Prot.*, **44** : 57-61.

Iwasaki, S., H. Muro, S. Nozoe, S. Okuda and Z. Sato. 1972. Isolation of 3, 4-dihydro-3, 4, 8-trihydroxy-1 (2H)-naphthalenone and tenuazonic acid from *Pyricularia oryzae* Cav. *Tetrahedron Lett.*, **1** : 13-16.

Iwasaki, S., S. Nozoe, S. Okuda, S. Sato and I. Kozaka. 1969. Isolation and structural elucidation of a phytotoxic substance produced by *Pyricularia oryzae* Cav. (pyriculol). *Tetrahedron Lett.*, **45** : 3977-3980.

Iwasaki, S., H. Muro, K. Sasaki, S. Nozoe, S. Okuda and Z. Sato. 1973. Isolation of phytotoxic substance produced by *Pyricularia oryzae* Cav. *Tetrahedron Lett.*, **37** : 297-306.

Iwata, M., Y. Suzuki, T. Watanabe, S. Mase and Y. Sekizawa. 1980. Effect of probenazole on the activities of enzymes related to the resistant reaction in rice plants. *Ann. Phytopathol. Soc. Japan*, **46** : 297-306.

Iwata, T., S. Ymanaki and T. Narita. 1955. Pathogenicity of rice blast fungus to gramineous plants. *Ann. Phytopathol. Soc. Japan*, **19** : 175. (Abstr.).

Iwata, T., M. Shimoyama, Y. Homma and Y. Arimoto. 1988. Inhibitory effect of components of lecithin on *Pyricularia oryzae*. 5th *Int. Phytopath. Cong.*, Kyoto, (Abstr.).

Izawa, K., M. Uchiyama and Y. Takahashi. 1970. Production of phytoalexin-like substance at the phenomenon of the blast resistance. *Ann. Phytopathol. Soc. Japan*, **36** : 159.

Jennings, P. R. 1963. Problems of rice blast requiring further attention : A concluding survey to the symposium on rice blast disease. *Int. Rice Comm. Newsl.*, **12** : 16-21.

Jin, M., X. Tao and Q. Zhang. 1986. Distribution of physiological races of blast fungus *Pyricularia oryzae* and changes of rice composition. *Zhejiang Agric. Sci.*, **5** : 222-226.

Jodon, N. E. and J. G. Atkins. 1966. Duplicate blast resistant genes tested for linkage relationships. *Int. Rice Comm. Newsl.* **15** : 28-32.

John, V. T. 1977. Rice diseases of major economic importance in India. *Seeds & Farms*, **3** : 5-8.

Johnson, R. 1979. The concept of durable resistance. *Phytopathology*, **69** : 198-199.

Johnson, T. W. 1954. *Piricularia oryzae* on *Eremochloa ophiuriodes* in Florida. *Pl. Dis. Rep.*, **38** : 796.

Jo. .ianandan, D. and E. R. B. Shanmugasundaram. 1968. Studies on cellulase of *Pyricularia oryzae*. *Enzymologia*, **35** : 11-18.

Jothianandan, D. and E. R. B. Shanmugasundaram. 1969. Studies on proteinase of *Pyricularia oryzae*. *Phytopath. Z.* **65** : 318-324.

Kahn, R. P. and J. L. Libby, 1958. The effect of environmental factors and plant age on the infection of rice by the blast fungus *Pyricularia oryzae*. *Phytopathology*, **48** : 25-30.

Kapoor, A. S. and B. M. Singh. 1983. Estimation of yield losses in rice cultivation Norin 18 due to neck blast. *Indian Phytopathol.* **36** : 136-137.

Katagiri, M. and Y. Uesugi. 1974. Evaluation of susceptibility of rice blast fungus to kasugamycin. *Ann. Phytopathol. Soc. Japan*, **40** : 106-107.

Katagiri, M. and Y. Uesugi. 1978. *In vitro* selection of mutants of *Pyricularia oryzae* resistant to fungicides. *Ann. Phytopathol. Soc. Japan*. **44** : 218-219.

Katagiri, M., Y. Uesugi and Y. Umehara. 1980. Development of resistance to Organophosphorous fungicides in *P. oryzae* in the field *J. Pest Sci.*, **5** : 417-421.

Kato, H. 1974. Epidemiology of rice blast disease. *Rev. Pl. Prot. Res.*, **7** : 1-20.

Kato, H. 1976. Some topics in a disease cycle of rice blast and climatic factors. In : *Proc. Symp. Climate and rice*. Int. Rice Res. Inst. (IRRI), Box 933, Manila, Philippines, pp. 417-421.

Kato, H. 1979. The perfect state of *Pyricularia oryzae* Cav. 10th *Int. Cong. Pl. Prot. and 71st Ann. Meet. Am. Phytopathol. Soc.* p. 143 (Abstr.).

Kato, H. and A. E. Daimond. 1966. Factors affecting sporulation of the rice blast fungus, *Piricularia oryzae*. *Phytopathology*, **56** : 864-865.

Kato, H. and T. Kozaka. 1974. Effect of temperature on lesion enlargement and sporulation of *Pyricularia oryzae* in rice leaves. *Phytopathology*, **64** : 828-830.

Kato, H. and T. Yamaguchi. 1980. Host range and interactions of *Pyricularia* spp. from various cereals and grasses. *Proc. Kanto-Tosan Pl. Prot. Soc. Japan*, **27** : 14-15.

Kato, H., T. Sasaki and Y. Koshimizu. 1970. Potential for conidium formation of *Pyricularia oryzae* in lesions on leaves and panicle of rice. *Phytopathology*, **60** : 608-612.

Kato, H., T. Yamaguchi, and N. Nishihara. 1976. The perfect stage of *Pyricularia oryzae* Cav. in culture. *Ann. Phytopathol. Soc. Japan*, **42** : 507-510.

Kato, H., T. Yamaguchi and N. Nishihara. 1977. Seed transmission, pathogenicity and control of blast fungus and susceptibility of ragi to *Pyricularia oryzae* from grasses, cereals and mioga. *Ann. Phytopathol. Soc. Japan,* **43** : 392-401.

Kato, K. 1977. The perfect state of *Pyricularia* spp. *Rev. Pl. Prot. Res.,* **10** : 20-28.

Kato, T., T. Hirukawa, Y. Yamaguchi, T. Namai. 1988. Blast resistance in rice and oxygenated unsaturated fatty acid. II. *Kagaku to Seibutsu.* **26** : 850-854.

Kato, T., M. Tsunakawa, N. Sasaki. H. Aizawa, K. Fujita, Y. Kitahara and N. Takahashi. 1977. Growth and germination inhibitors in rice husk. *Phytochem.,* **16** : 45-48.

Kato, T., C. Kabuto, N. Sasaki, H. Tsunakawa, H. Aizawa, K. Fujita, Y. Kato, Y. Kitahara and N. Takahashi, 1973. Momilactones, growth inhibitors from rice, *Oryza sativa* L. *Tetrahedron Lett.,* **39** : 3861-3864.

Katsube, T. and Y. Koshimizu. 1970. Influence of blast disease on harvests of rice plants. 1. Effect of panicle infection on yield components and quality. *Bull. Tohoku Agric. Expt. Stn.,* **39** : 55-96.

Katsuya, K. 1959. Susceptibility of wild and foreign cultivated rice to the blast fungus, *Piricularia oryzae. Ann. Rep.* 1958, *Natl. Inst. Genet., Japan,* **9** : 48-49.

Katsuya, K. 1960. Susceptibility of wild rice to the blast fungus, *Piricularia oryzae. Ann. Rep.* 1959, *Natl. Inst. Genet. Japan,* **10** : 76-77.

Katsuya, K. 1961. Studies on the susceptibility of *Oryza* to the blast fungus, *Piricularia oryzae. Ann. Phytopathol. Soc. Japan,* **26** : 153-159.

Katsuya, K. and S. Kiyosawa. 1969. Studies on mixture inoculation of *Pyricularia oryzae* on rice. II. Inter-strain differences of mixture inoculation effect. *Ann. Phytopathol. Soc. Japan,* **35** : 299-307.

Kaur, S. 1972. Inheritance of resistance to blast disease of rice. Ph. D. Thesis. Utkal university, Bhubaneswar, p. 209.

Kaur, S., S. Y. Padmanabhan and M. Rao. 1976. Induction of resistance to blast disease (*Pyricularia oryzae*) in high yielding rice variety, Ratna (IR 8/TKM 9). *Mut. Breed. Newsl.,* **7** : 7-8.

Kawakami, T. 1901. On the rice blast disease. *J. Sapporo Soc. Agric. Fores.,* **2** : 1-47.

Kawakami T. 1902. On the blast disease of rice. *J. Sapporo Soc. Agric. Fores.,* **3** : 1-3.

Kawamura, E. 1940. Reaction of certain species of the genus *Oryza* to the infection of *Piricularia oryzae. Sci. Fac. Terkult. Bull. Kyushu Univ.* No. **9** : pp. 157-166.

Kawamura, E. and K. Ono. 1948. Studies on the relation between the preinfection behaviour of rice blast fungus, *Piricularia oryzae* and water droplets on rice plant leaves *Bull. Natl. Agri. Expt. Stn.,* No. **4** : pp. 1-12.

Khin, C. H. and C. K. Kim. 1990. Rice blast development under different water-management practices in fields in Korea. *Korean J. Pl. Pathol.,* **6** : 51-57.

Khin, C. H., C. K. Kim and E. J. Lee. 1989. Effect of host resistance and fungicide application on rice leaf blast management. *Korean J. Pl. Pathol.,* **5** : 377-382.

Khush, G. S. and H. M. Beachell. 1972. Breeding for disease and insect resistance at IRRI. In : *Rice Breeding,* Int. Rice Res. Ins. Box 933, Mainla, Philippines. pp. 309-322.

Kim, B. S. and E. W. Park. 1987. Effect of leaf blast and panicle blast on grain yield of rice. *Phytopathology,* **77** : 1777. (Abstr.).

Kim, C. H. and C. K. Kim. 1990. Rice blast development under different water-management practices in fields in Korea. *Korean J. Pl. Pathol.,* **6** : 51-57.

Kim, C. K. and C. H. Kim. 1991. Predicting rice blast outbreaks in Korea. In : *Rice blast modelling and forecasting.* Int. Rice Res. Inst. (IRRI) Philippines pp. 53-67.

Kim, C. H., D. R. MacKanzie and M. C. Rush. 1988. Field testing computerized forecasting system for rice blast disease. *Phytopathology,* **78** : 931-934.

Kim, C. H., M. C. Rush and D. R. MacKenzie. 1985. Effect of water management on the epidemic development of rice blast. *Phytopathology,* **75** : 501.

Kim, C. K. 1982. Improved methods for rice blast forecasting. *Korean J. Pl. Prot.,* **21** : 19-22.

Kim, C. K. 1983. Investigation of incubation period of neck blast in rice plants. *Korean J. Pl. Prot.,* **22**: 287-292.

Kim, C. K. 1987a. Effect of inoculum density on penetration of *Pyricularia oryzae*. *Res. Rep.* RDA, Korea. **29** : 214-216.

Kim, C. K. 1987b. Disease forecasting : Past and future in special reference to rice leaf blast. *Korean J. Pl. Pathol.*, **3** : 304-312.

Kim, C. K. and S. C. Lee. 1982. Reduction of the incidence of rice neck blast by integrated soil improvement practice. *Korean J. Pl. Prot.*, **21** : 15-18.

Kim, C. K. and H. S. Min. 1989. Conidia release and dispersal pattern of *Pyricularia oryzae* under cloudy or rainy conditions., *Int. Rice. Res. Newsl.*, **14** : 34.

Kim, C. K. and S. Mogi. 1985. Effect of temperature treatments on the penetration and disease development in the leaf epidermis by the rice blast fungus, *Pyricularia oryzae* Cavara. II. Difference in percent penetration, hyphal growth and lesion formation by pre and post disposing temperature *Korean J. Pl. Pathol.*, **1** : 122-127.

Kim, C. K. and S. Mogi. 1986. Effect of temperature treatment on the penetration and disease development in the leaf epidermis by the rice blast fungus *Pyricularia oryzae* Cavara. 1. Effect of pre and post-disposing temperature on the disease development of rice blast. *Ann. Phytopathol. Soc. Japan*, **52** : 184-192.

Kim, C. K. and S. Mogi. 1987. Effect of temperature treatments on the penetration and disease development in the leaf epidermis by the rice blast fungus, *Pyricularia oryzae* Cavara. III. Effect of percent penetration and hyphal growth on lesion formation under different temperature regimes. *Korean J. Pl. Prot.*, **26** : 43-47.

Kim, C. K. and R. Yoshino. 1987. Epidemiological studies of rice blast disease caused by *Pyricularia oryzae* Cavara. I. Measurement of the amount of spores released from a single lesion. *Korean J. Pl. Pathol.*, **3** : 120-123.

Kim, C. K., H. S. Min and R. Yoshino. 1989. Epidemiological studies on rice blast disease caused by *Pyricularia oryzae* Cavara. IV Conidia release under rainy conditions. *Korean J. Pl. Path.*, **5** : 60-64.

Kim. C. K., H. S. Min and R. Yoshino. 1990. Epidemiological studies of rice blast disease caused by *Pyricularia oryzae* Cavara. V. Successive conidia release from different aged lesions in the nature. *Korean J. Pl. Path.*, **6**: 214-218.

Kim. C. K., R. Yoshino and E. J. Lee. 1987. Epidemiological studies of rice blast disease caused by *Pyricularia oryzae* Cavara. II. Sporulation and conidia release from naturally infected lesions. *Korean J. Pl. Prot.*, **26** : 37-41.

Kim, C. K., R. Yoshino and S. Mogi. 1975. A trial of estimating a number of leaf blast lesions on rice plants on the basis of the number of trapped spores and wetting period of leaves. *Ann. Phytopathol. Soc. Japan*, **41** : 492-499.

Kimura, S. and T. Kobayashi. 1980. Fluctuation of blast resistance in rice varieties under low air temperature conditions. *Res. Bull. Hokkaido Natl. Agric. Expt. Stn.* **128** : 13-21.

Kingslover, C. H., T. H. Barksdale and M. A. Marchetti. 1984. Rice blast epidemiology. *Pa. Agric. Exp. Stn. Bull.*, No : 853 p. 33.

Kitamura, A. 1961. Genetic study on sterility caused by hybridization between *japonica* and *indica*. *Recent Dev. Threm. Bull.*, **3** : 23-25.

Kitani, K. 1941. On the bacterium associated with the rice blast disease. *Ann. Phytopathol. Soc. Japan*, **11** : 136-146.

Kiyosawa, S. 1966. Studies on inheritance of resistance of rice varieties to blast. III. Inheritance of resistance of a rice variety Pi No. 1. *Japan J. Breed.*, **16** : 243-250.

Kiyosawa, S. 1967a. The inheritance of resistance of the Zenith type varieties of rice to blast fungus. *Japan J. Breed.*, **17** : 99-107.

Kiyosawa, S. 1967b. Inheritance of resistance to rice variety Pi No. 4 to blast. *Japan J. Breed.*, **17** : 165-172.

Kiyosawa, S. 1967c. Genetic studies on host-pathogen relationship in the rice blast disease. In : proceedings of a symposium on rice diseases and their control by growing resistant varieties and other measures. *Agric. For. & Fish Res. Coun. Tokyo*, pp. 137-153.

Kiyosawa, S. 1968a. IV. Genetic analysis for the host parasite relationship in the blast disease of rice. In Genetics, evolution and breeding in rice. In : 12th *Int. Cong. Genet. Tokyo, Japan,* **2** : 3. (Abstr.).

Kiyosawa, S. 1968b. Inheritance of blast resistance in some Chinese rice varieties and their derivatives. *Japan J. Breed.,* **18** : 193-205.

Kiyosawa, S. 1969a. Inheritance of resistance of rice varieties to a Philippine fungus strain of *Pyricularia oryzae. Japan J. Breed.,* **19** : 61-73.

Kiyosawa, S. 1969b. Inheritance of blast resistance in West Pakistani rice variety, Pursur. *Japan J. Breed.,* **19** : 121-128.

Kiyosawa. S. 1970. Inheritance of blast resistance of the rice varieties Homare Nishiki and Ginga. I. Resistance of Homare Nishiki and Ginga to the fungus strain ken 54-4. *Bull. Natl. Inst. Agric. Sci.,* **21** : 73-105.

Kiyosawa, S. 1972a. The inheritance of blast resistance transferred from some *indica* varieties of rice. *Bull. Natl. Inst. Agric. Sci.,* **22** : 1-21.

Kiyosawa, S. 1972b. Genetics of blast resistance in : *Rice Breeding*. Int. Rice Res. Inst., (IRRI) Box 933, Manila, Philippines. pp. 203-225.

Kiyosawa, S. 1972c. Mathematical studies on the curve of disease increase—a technique for developing epidemic development. *Ann. Phytopathol. Soc. Japan,* **38** : 30-40.

Kiyosawa, S. 1974a. Studies on genetics and breeding of blast resistance in rice. *Misc. Publ. Natl. Inst. Agric. Sci.,* **1** : 1-58.

Kiyosawa, S. 1974b. Inheritance of blast resistance of a U. S. rice variety Dawn. *Japan J. Breed.,* **24** : 117-124.

Kiyosawa, S. 1976a. Pathogenic variation of *Pyricularia oryzae* and their use in genetic and breeding studies. *Sabrao J.,* **81** : 53-67.

Kiyosawa, S. 1976b. Methods of testing and gene analysis of blast resistance of rice varieties. *Oryza,* **13** : 1-32.

Kiyosawa, S. 1978. Identification of blast-resistance genes in some rice varieties. *Japan J. Breed.,* **28** : 287-296.

Kiyosawa, S. 1979. Breeding for disease resistance in crop plants and its fundamental studies. *Agric. Hort.,* **54** : 1427-1432.

Kiyosawa, S. 1981. Gene analysis for blast resistance. *Oryza,* **18** : 196-203.

Kiyosawa, S. 1982. Genetics and epidemiological modelling of breakdown of plant disease resistance. *Ann. Rev. Phytopathol.,* **20** : 93-117.

Kiyosawa, S. 1985. Tests for blast resistance of rice varieties in China, Brazil, The Philippines and others. *Japan J. Breed.,* **35** : 162-163.

Kiyosawa, S. and C.I. Cho 1980. Indentification of blast-resistant genes in Korean rice variety, Tonigil, *Japan. J. Breed* **30** : 73-82.

Kiyosawa, S. and H. Fujimaki. 1967. Studies on mixture inoculation of *Pyricularia oryzae* on rice I. Effects of mixture inoculation and concentration on the formation of susceptible lesions in the injection inoculation. *Bull. Natl. Inst. Agric. Sci.,* Tokyo, No : **17** : pp. 1-20.

Kiyosawa, S. and Z.Z. Ling. 1984. Identification of genes for rice blast resistance in Chinese differential varieties with Japanese fungus strains. *Scientia Sincia* **27** : 273-283.

Kiyosawa, S. and V.V.S. Murty. 1969. The Inheritance of resistance in Indian rice variety, HR 22. *Japan J. Breed.,* **19** : 269-276.

Kiyosawa, S. and M. Shiyomi. 1972. A theoretical evaluation of the effect of mixing resistant variety with susceptible variety of controlling plant diseases. *Ann. Phytopathol. Soc. Japan,* **38** : 41-51.

Kiyosawa, S. and M. Yakoo. 1970a. b. I. Inheritance of blast resistance of the rice variety Toride-1 selected from the cross Norin 8X TKM-1. II. Inheritance of blast resistance of the rice variety Toride-2 bred by transfering resistance of the Indian variety CO. 25. *Japan J. Breed.,* **20** : 129-132; 181-186.

Kiyosawa, S., S. Matsumoto and S. C. Lee. 1967. Inheritance of resistance of rice variety Norin 22 to two blast fungus strains. *Japan J. Breed.,* **17** : 1-6.

Kiyosawa, S., D.T. Mackill, J.M. Bonman, Y. Tanaka and Z.Z. Ling 1986. An attempt of classification of worlds rice varieties based on reaction pattern to blast fungus strains *Bull. Natl. Agrobiol Resour.* **2** : 13-39.

Kiyosawa, S., H. Yasuda and I. Shintani. 1977. Blast nursery tests for field resistance of rice varieties with different genotypes for true resistance to blast. *Ann. Phytopathol. Soc. Japan,* **43** : 517-523.

Kiyosawa, S., H. Ikehashi, H. Kato and Z.Z. Ling. 1981. Pathogenicity tests of Philippine isolates of blast fungus using two sets of rice varieties. *Japan. J. Breed.,* **31** : 367-376.

Kiyosawa, S., Y. Orimoto, Y. K. He and Z. Ling. 1983a. Estimation of genotypes for blast resistance of rice varieties by testing with differential varieties. *Oryza,* **20** : 216-222.

Kiyosawa, S., Y. Orimoto, Y. K. He and Z. Ling 1983 b. Inheritance of blast resistance of an IRRI's rice variety IR 1905-81-3-1. *Japan J. Breed.,* **33** : 31-39.

Kiyosawa, S., D. J. Mackill, J. M. Bonman, Y. Tanaka and Z. Z. Ling. 1986. An attempt of classification of world's rice varieties based on reaction pattern to blast fungus strains. *Bull. Nat. Inst. Agrobiol. Res. Japan,* **2** : 13-39.

Ko, J. H., B. K. Hwang and B. G. Hwang. 1993. Restriction fragment length polymorphism of mitochondrial and nuclear DNA's among Korean races of *Magnaporthe grisea. J. Phytopathology,* **138** : 41-54.

Kobayashi, J. 1968. Observation of the onset of leaf blast in the field. *A. Rep. Pl. Prot. North Japan,* **19** : 7 (Abstr.).

Kobayashi, J. 1984. Studies on epidemic of rice leaf blast, *Pyricularia oryzae* Cav., in its early stage. *Bull. Akita Agric. Expt. Stn.,* **26** : 1-84.

Kobayashi, Y. 1963. Studies on the forecasting method of the rice blast by means of chemical analysis of leaves. Special report on disease and insect forecasting. *Min. Agric. For. Japan,* **15** : 1-100.

Koga, H. 1983. Comparison of spore germination and appressorial formation of *Pyricularia oryzae* Cav. on rice leaves in compatible and incompatible combinations. *Proc. Assoc. Pl. Prot. Hokuriku* No. **31**. pp. 7-12.

Koga, H. 1988. Ultrastructure of compatible and incompatible reactions of rice to *Pyricularia oryzae* Cav. In : 5th *Int. Cong. Pl. Path., Kyoto, Japan,* P. 239. (Abstr.).

Koga, H. 1989a. Histopathology of resistant and susceptible rice plant inoculated with *Pyricularia oryzae* Cavara. JARQ **23** : 8-15.

Koga, H. 1989b. Hypersensitive cell death and autofluorescence in rice leaf-sheath epidermal cells under attack by *Pyricularia oryzae* Cav. *Ann. Phytopathol. Soc. Japan,* **55** : 471. (Abstr.).

Koga, H. and T. Kobayashi. 1980. A whole-leaf clearing and staining technique to observe the invaded hyphae of blast fungus and host responses in rice leaves. *Ann. Phytopathol. Soc. Japan,* **46** : 679-681.

Koga, H., T. Kobayashi and O. Horino. 1982. Electron-microscopical observation of rice leaves infected with *Pyricularia oryzae* Cav. in compatible and incompatible combinations. I. Fine structure of invaded hyphae in host cells. *Ann. Phytopathol. Soc. Japan,* **48** : 281-289.

Koga, H. O. Horino and T. Kobayashi. 1986. Ultrastructural studies on the resistance mechanisms of rice plants infected with *Pyricularia oryzae* Cav. *Bull. Hokiriku Natl. Expt. Stn.* **28** : 1-56.

Koga, H. and R. Yoshino. 1988 Observations of rice panicle blast by fluorescent labelling. *Ann. Phytopathol Soc. Japan,* **54** : 229-232.

Koga, H. and R. Yoshino 1990. A scanning electron-microscropic study of conidiophores of *Pyricularia oryzae* Cav., on rice panicles. *Proc. Assoc. Pl. Prot. Hokuriku* No. **38** : pp. 3-8.

Koga, H. and R. Yoshino 1991a The ultrastructure of rice panicles infected with *Pyricularia oryzae* Cav. *Proc. Assoc. Pl. Protect.* Hokuriku. No. **39** pp. 17-22.

Koga, H. and R. Yoshino 1991 b. The ultrastructure of rice spikelets infected with *Pyricularia oryzae* Cav. *Proc. Assoc. Pl. Prot. Hokuriku* No **39** pp. 23-29.

Koh, Y. J., B. K. Hwang and H. S. Chung. 1987. Adult plant resistance of rice to leaf blast. *Phytopathology,* **77** : 232-236.

Koizumi, S. 1988. unpublished data. (cited by Koizumi and Kato, 1990).

Koizumi, S. and H. Kato. 1985. Canopy structure and distribution of leaf blast lesions within rice canopy in relation to cultivation method and growth stage of rice. *Ann. Phytopathol. Soc. Japan,* **51** : 319.

Koizumi, S. and H. Kato. 1986. Variation of canopy structure and distribution of blast lesions after heading of rice. *Ann. Phytopathol. Soc. Japan,* **53** : 518.

Koizumi, S and H. Kato. 1987. Effect of mixed plantings of susceptible and resistant rice cultivars on leaf blast development. *Ann. Phytopathol. Soc. Japan,* **52** : 28-37.

Koizumi, S. and H. Kato. 1991. Dynamic simulation of blast epidemics using a multiple canopy, spore dispersal model. In : *Rice blast modelling and forecasting* Int, Rice Res. Inst. (IRRI) Box 933, Manila, Philippines pp. 75-88.

Koizumi, S. and H. Kato. 1990. Distribution of rice blast lesions and susceptible tissues in rice canopy. *Bull. Natl. Agric. Res. Cent.* Tsukuba (in press).

Kojima, Y., S. Maeno, S. Takanashi, K. Takama, H. Yamamoto, and S. Kavada 1988. Effect of tricyclazole on the spore release of rice blast fungus, *Pyricularia oryzae* and the secondary infection. In: 5th *Int. Cong. Pl. Path., Kyoto, Japan,* p. 332. (Abstr.).

Konishi, S. 1933. On the physiologic specialization in the rice blast fungus *Pyricularia oryzae* Br. et Cav. *Forch. aus. dem. Gcb. Pflkrankh. Kyoto,* **2** : 55-77.

Kono, T. 1977. Studies on the utilization method of electronic computer in the forecast work on disease and insect outbreak. IV. The system calculating forecast values using multiple regression analysis. *Bull. Hiroshima Agric. Exp. Stn.,* **39** : 1-20.

Kono, Y., S. Takeuchi, O. Kodama, H. Sekido and T. Akatsuka. 1985. Novel phytoalexins (Oryzalexins A, B and C) isolated from rice blast leaves infected with *Pyricularia oryzae.* Part II. Structural studies of oryzalexins. *Agric. Biol. Chem.,* **49** : 1695-1701.

Koshimizu, Y. 1982. A forecasting method for leaf blast outbreak by the use of AMeDAS data. *Kongetsu No. Noyaku,* **26** (1-4) : 46-53, 64-67, 74-83, 68-78.

Koshimizu, Y. 1983. A forecasting method for leaf blast outbreak by the use of AMeDAS data. *Pl. Prot.,* **37** : 454-457.

Koshimizu, Y. 1988. A forecasting method for occurrence of rice leaf blast with AMeDAS data *Bull. Tohoku Natl. Agric. Exp. Stn.,* **78** : 67-121.

Kowon, S. H., J. H. Oh, Y. S. Ham and J. H. Kim. 1975. Genetic study on the resistance of blast disease in variety Tongil. *Korean J. Breed.* **7** : 153-157.

Koyama, T. 1952. On the breeding of highly resistant varieties to rice blast by the hybridization between Japanese varieties and foreign varieties in Japanese type of rice. *Japan J. Breed.,* **2** : 25-30.

Kozaka. T. 1965. Control of rice blast by cultivation practices in Japan. In : *The rice blast disease.* Johns Hopkins Press, Baltimore, Maryland. pp. 421-438.

Kozaka, T. 1975. Reaction of rice varieties to major races of *Pyricularia oryzae* in Asian countries. In: *Rice in Asia. Univ. Tokyo, Press, Japan,* pp. 318-326.

Kozaka, T. and Y. Sonuku. 1963. Out break of the rice blast disease and contents of free amino acids in the plant body in relation to different environments. *Ann. Phytopathol. Soc. Japan,* **28** : 90.

Kozaka, T. and T. Teraoka. 1977. Ethylene evaluation by rice plants infected with *Pyricularia oryzae* Cav. in relation to the stunting of diseased plant. *Ann. Phytopathol. Soc. Japan,* **43** : 549-556.

Kozaka, T., S. Matsumoto, and M. Yamada, 1970. Reactions of foreign rice varieties to major races of *Pyricularia oryzae* in Asia. *Bull. Natl. Inst. Agric. Sci.,* No. **24** : 113-152.

Kramer C.L., M.G. Eversmeyer and T.I. Collins. 1976. A new 7-day spore sampler. *Phytopathology,* **66** : 60-61.

Krishnan, P., T.T. Muralidharan, K. Manibhushanrao and B.C. Reddy. 1992. Epidemiological studies on rice blast disease-I. Spore releasing capacity of various types of B1 lesions and spore release pattern against weather parameters. *Int. J. Trop. P1. Dis.* **10** : 153-165.

Krishnaswami, C. S. 1952. Influence of nitrogen, phosphorus and potash on incidence of blast disease of rice. *Madras Agric. J.,* **39** : 205-214.

Krishnaswami, C. S., C. V. Govindaswamy and N. R. Adyanthaya. 1959. Blast and foot-rot diseases of rice in Madras state. *Rep.* ICAR, July 1959.

Krishnaswamy, V., S. Edition and K. Sivaprakasam. 1972. A new grass host for *Pyricularia oryzae* Cav. *Curr. Sci.,* **41** : 262.

Krishnaswamy, V., C. V. Govindaswamy and P. Vidhyasekaran. 1969. Studies on the breakdown of resistance in the blast resistant variety CO 29 in Tamil Nadu. *Madras Agri. J.* **56** : 347-354.

Kuilman, L. W. 1940. Uitstoeling en bloei van de Rijstplant. III. Vatbaarheid van de (opvoeding-soplossing gekweekte) varieteit Oentoeng voor *Piricularia oryzae* Cav. *Mededelinen Algemeen Proefstation voor den Landbouw, Batavia,* **45** : 27-32.

Kulkarni, N. B. and M. K. Patel, 1956. Study of the effect of nutrition and temperature on the size of spores in *Piricularia setariae* Nisikada. *Indian Phytopathol.,* **9** : 31-38.

Kumamoto, Y. 1948. Effects of cultural solutions of rice blast and jute anthracnose organisms on the occurrence of the diseases. *Nongpau Taiwan Agric. Res. Inst. Formosa,* **4** : 192-195.

Kumar, S. 1988. Variation in rice blast fungus and shift in pathogenicity due to periodical sub-culturing. *Oryza.* **25** : 52-56.

Kumar, S. and K. Sridhar. 1984. Phytoalexin production in rice—*Pyricularia oryzae* interaction. Induction by wounding. *Acta. Phytopath. Acad. Sci. Hung.,* **19** : 245-249.

Kumar, S. and R. Sridhar. 1985. Significance of cell wall phenols in the resistance of rice against blast. *Curr. Sci.,* **54** : 874-876.

Kumar, S. and R. Sridhar, 1987. Significance of epicuticular wax in the specificity of blast fungus to rice varieties. *Int. J. Trop. Pl. Dis.,* **5** : 131-139.

Kumar, S. and R. Sridhar. 1990. Phytoalexins in rice-*Pyricularia oryzae* interaction. Factors affecting phytoalexin production. *Acta Phytopathol. Entamol.* Hung., **28** : 56-69.

Kuo, T. T., Y. S. Lee, H. F. Yunan and H. W. Li. 1968. Nutritional aspects of host-parasite relationship in the rice blast fungus. *Bot. Bull. Acad. Sin.* **9** : 36-46.

Kuo, T. T., L. H. Yang, P. Z. Yeh, and H. W. Li. 1967. Studies on induced mutants of *Piricularia oryzae* and on their pathogenicity. *Bot. Bull. Acad. Sin.,* **8** : 20-29.

Kuribayashi, K. 1928. Studies on overwintering, primary infection, and control of rice blast fungus. *Piricularia oryzae. Ann. Phytopathol. Soc. Japan,* **2** : 99-117.

Kuribayashi, K. and H. Ichikawa. 1941. Relation of the floating conidia of rice blast fungus to prediction of the disease. *J. Pl. Pro.* **28** : 309-314; 404-416.

Kuribayashi, K. and H. Ichikawa. 1952. Studies on the forecasting of the rice blast disease. *Bull. Nagano Pref. Agric. Exp. Stn.* **13** : 1-229.

Kuribayashi, K. and H. Terazawa. 1953. Injection as an artificial inoculation method in rice blast disease. *Pro. Asso. Pl. Pro., Hokuriku,* **3** : 9-10.

Kuribayashi, K. and H. Terazawa. 1956. Relation between the rice blast fungus and barley. *Proc. Assoc. Pl. Prot., Hokuriku,* **4** : 16-17.

Kuribayashi, K., H. Ichikawa, and M. Terazawa. 1953. Study of Physiological specialization in rice blast fungus. *Rep. Nagano Agri. Expt. Stn.,* 1951 and 1952, p. 31.

Kurosaki, Y. 1978. Role of brown discoloured tissues around blast lesion in the defence reaction of rice plant against *Pyricularia oryzae* Cav. *Ann. Phytopathol. Soc. Japan,* **44** : 670-673.

Kuwatsuka, S. and Y. Oshima. 1961; 1962. Studies on polyphenols of rice plants part I and II. *J. Agric. Chem. Soc. Japan,* **35** : 67-71; 71-74.

Lalithakumari, D and P. Annamalai. 1990. Edifenphos resistance in *Pyricularia oryzae* and *Drechslera oryzae*. In : *Proc. Int. symp. on Managing resistance to agrochemicals. Am. Chem. Soc.* pp. 249-263.

Lalithakumari, D and D.S.S.V.R. Kumari. 1987. Screening for field resistance in *Pyricularia oryzae* to edifenphos. *Indian Phytopathol.,* **40** : 342-347.

Lalithakumari, D., K. Mathivanan and D.S.S.V.R. Kumari. 1987. Cross resistance of edifenphos resistant mutants of *Pyricularia oryzae* against some pesticides. In : 11th *Int. Cong. Pl. Protec.,* Manila. p. 61. (Abstr.).

Langcake, P., and S.G.A. Wickins. 1975. Studies of the action of the dichlorocyclopropanes on the host - parasite relationship in the rice blast disease. *Physiol. Pl. Pathol.* **7** : 113-126.

Latterell, F. M. 1972. Two views on pathogenic stability in *Pyricularia oryzae*. *Phytopathology,* **62** : 771. (Abstr.).

Latterell, F. M. 1975. Phenotypic stability of pathogenic races of *P. oryzae* and implication for breeding for blast resistant varieties. In : Proc. Seminar on horizontal resistance to the blast disease of rice. CIAT, Cali, Columbia, pp. 199-234.

Latterell, F.M. and A.E. Rossi. 1986. Longevity and pathogenic stability of *Pyricularia oryzae*. *Phytopathology.,* **76** : 231-235.

Latterell, F. M., E.C. Tullis, and J.W. Collier. 1960. Physiologic races of *Piricularia oryzae* Cav. *Pl. Dis. Reptr* . **44** : 679-683.

Latterell, F.M., E.C. Tullis, R.J. Otten, and A. Gubernick. 1954. Physiologic races of *Piricularia oryzae*. *Phytopathology,* **44** : 495. (Abstr.).

Leach, C. M. 1962. Sporulatin of diverse species of fungi under near ultraviolet radiation *Can. J. Bot.* **40** : 151-161.

Leach, C. M. 1980. Influence of humidity, red-infrared radiation, and vibration on spore discharge by *Pyricularia oryzae*. *Phytopathology,* 70 : 201-205.

Leaver, F.W., J. Leal and C. R. Brewer. 1947. Nutritional studies of *Piricularia oryzae*. *J. Bacteriology,* **54** : 401-408. (Abstr.).

Lebrun,M. H., J. Orcival and C. Duchartre. 1984. Resistance of rice of tenrazonic acid, a toxin from *Pyricularia oryzae*. *Rev. Cytol. Biol. Veget. Bot.,* **7** : 249-259.

Lee, E. J. 1972. Epidemic outbreak of blast disease in the resistant variety of rice; Kwanok. *Korean J. Pl. Prot.,* **11** : 1-4.

Lee, E. J. and S. Y. Cho. 1990. Variation in races of rice blast disease and varietal resistance in Korea. Int. Rice Res. Inst. (IRRI), Box 933, Manila, Philippines (Mimeographed).

Lee, E. J., W. J. Joo and B. J. Chung. 1975. Indentification and annual changes of races of *Pyricularia oryzae* in Korea. *Korean J. Pl. Prot.* **14** : 199-204.

Lee, E. J., Y. H. Lee and J. D. Ryu. 1988. Blast race analysis in Korea. Korea-IRRI Collaborative Project on Rice Research. RDA Suweon, Korean. 22-31.

Lee, E. J., J. D. Ryu, W. H. Yeh, S. S. Han and Y. H. Lee. 1987. Proposal of a new method for differentiating pathogenic races of *Pyricularia oryzae* Cavara in Korea. *Res. Rep. RDA* (P. M. & U). **29** : 206-213.

Lee, S.C., and S. Matsumoto. 1966. Studies on the physiologic races of rice blast fungus in Korea during the period of 1962-1963. *Ann. Phytopathol. Soc. Japan,* **32** : 40-45.

Lee, S. C., J. S. Shim, and E. J. Lee. 1968. Studies on the nuclear cytology of *Pyricularia oryzae* Cav *Korean J. Pl. Prot.,* **5** : 19-26.

Lee, U. C., N. Matsuyama and T. Yamaguchi. 1975. Polysaccharide-degradation ability of rice blast fungus of a different pathogenicity and virulence. *Ann. Phytopathol. Soc. Japan,* **41** : 215-218.

Lee, Y. S. and L. C. Wu. 1967. The nitrogen nutrition of *Piricularia oryzae*. *Bot. Bull. Acad. Sin., Taipei,* **8** : 66-72.

Legg, B. J., and F. A. Powell. 1979. Spore dispersal in a barley crop : A mathematical model. *Agric. Meteorol.,* **20** : 47-67.

Leong, S., U. Lehtinen, R. Karjalainen, D. Skinnen, P. Tooly, S. Leong and A. Illingloe. 1990. Transformation of the rice blast fungus *Magnaporthe grisea* to hygromycin B resistance. *Curr. Genetics.* **17** : 409-441.

Leung, H., E. S. Borromeo, M. A. Bernardo and J. L. Notteghem. 1988. Genetic analysis of virulence in the rice blast fungus *Magnaporthe grisea*. *Phytopathology,* 78 : 1227-1233.

Levy, M., J. Romao, M. A. Marchetti and J. E. Hamer. 1991. DNA Fingerpringting with a dispersed repeated sequence resolves pathotype diversity in the rice blast fungus. *The Plant Cell* **3** : 95-102.

Liang, W. J. 1979. Effects of meteorological factors on spore germination, appressorium formation and invasion of the rice blast fungus. *Pyricularia oryzae*. *Natl. Sci. Counc. Mon.,* **7** : 810-819.

Lin, K. M., and P.C. Lin 1961. Survey of intervarietal varition in resistance to blast disease. *Spl. Pub. Taiwan Agric. Res. Ins.*, **3** : 38-44.

Lin, K. R. and G. J. Liu 1957. On the over wintering of rice blast (*Pyricularia oryzae* Cav.) in Whei-li district, Szechuan. *Acta Phytopath. Sini.*, **3** : 201-207.

Lin, S., R. Feng, X. Kong, W. Xu, J. Sheng, Z. Chang and W. Wan. 1980. Studies on disease resistance in rice breeding. Blast resistance of Ken rice and the release of Zhong-dan No. 1, No. 2 and No. 3. *Scientia Agric. Sinica*, **1** : 1-14.

Ling, K. C., and S. H. Ou, 1969. Standardization of the international rice numbers of *Pyricularia oryzae. Phytopathology*, **59** : 339-342.

Liu, X. Q., H. S. Li, Y. X. Hu and G. Z. Zhang. 1989. Physiological and biochemical studies on rice varieties used for differentiating resistance to rice blast. *Acta Phytopath. Sin.*, **16** : 31-36.

Locci, R. 1970. Fungal diseases of rice. 1. Investigation on the rice blast fungus by scanning electron-microscopy. *Riso*, **19** : 99-110.

Loganathan, M., and V. Ramaswamy 1984. Effect of blast on IR 50 in late samba. *Int. Rice Res. Newsl.*, **9** : 6.

Loehken, A. 1990. The benefit of seed/early season fungicide application for the management of rice blast. In : *Pest management in rice*, (B. T. Grayson et al. eds.) *Soc. Chem. Ind. Elsevier Applied Science*, London. pp 77-86.

Lokeswari, T. S. 1987. Studies of factors affecting permeability changes in the blast disease of rice. Ph.D. Thesis. Univ. of Madras, Madras, India. p. 170.

Loprieno, N. 1964. I mutanti nutrizionali nello studio dei repporti ospitepatogene nelle fitopatie da microorganismi. *Agricultura Ital. Pisa*, p. 15.

Luo, K., S. Y. Huang, Q. M. Xiao and B. X. Wang. 1979. Preliminary assay on virulence of rice blast fungus in Human Provience. *Acta Phytopath. Sin.*, **9** : 83-92.

Lutterll, E. S. 1954. An undescribed species of *Piricularia* on sedges. *Mycologia*, 46 : 810-814.

MacCartney, H. A. and B.D.L. Fitt. 1985. Construction of dispersal models. In : Mathematical modelling of crop disease (C.A. Gilligan, ed.). *Adv. pl. Pathol.*, **3** : 107-143.

MacKenzie, D. R. 1979. The multiline approach to the control of some cereal diseases. In : *Proc. Rice Blast Workshop*. IRRI, Los Baños, Philippines, pp. 199-216.

MacKenzie, D. R. 1981. Breeding for disease resistance in cereals. 28th *Ann. Meet. Conf. Pest. Manage. Soc.*, St. Catharines, Ontario, pp. 63-69.

MacKenzie. D. R., and R. L. Villareal, 1977. The spread of rice blast disease. *Proc. Am. Phytopathol. Soc.* **3** : 309 (Abstr.).

Mackill, D. J. and J. M. Bonman. 1986. New hosts of *Pyricularia oryzae. Pl. Disease* **70** : 125-127.

Mackill, D. J. and J. M. Bonman, 1990. Use of near-isogenic lines in blast resistance research. In : *Int. Rice Res. Conf.* Seoul, S. Korea. p. 16.

Mackill, D. J., J. M. Bonman, H. S. Suh and R. Srilingam. 1985a. Genes for resistance to Philippine isolates of the rice blast pathogen. *Int. Rice Genet. Newsl.*, **2** : 80-81.

Mackill, D. J., J. M. Bonman, H. S. Suh and R. Srilingam. 1985b. Number of blast resistance genes in some indica rice cultivars. *Agron. Abstr.*, **85** : 62.

Mackill, D. J., J. M. Bonman, P. D. Tenorio and T. I. Vergel De Dios. 1988. Near-isogenic *indica* rice lines with blast resistance genes. *Rice Genet. Newsl.*, **5** : 98-101.

Madden, L. V. and M. A. Ellis, 1988. How to develop plant disease forecasters? In : *Experimental techniques in plant disease epidemiology* (J. Kranz and J. Rotem, eds.). Springer-Verlag, New York, pp. 191-208.

Maeda, T., A. Abe, K. Kaliki and T. Misato. 1970. Studies on the mode of action of organophosphorus fungicide. Kitazin Part II. Accumulation of an amino sugar derivatives of Kitazin treated mycelia of *Pyricularia oryzae. Agric Biol. Chem.* **34** : 700-710.

Mahadevappa. M. 1990. Cold and blast tolerant rice for winter. The Hindu : **113** : (257) 24.

Maheswariamma, S. and J. Samraj. 1973. Observations on the changes in the atmospheric spore content of *Pyricularia oryzae* Cav. and on the intensity of blast incidence on rice. *Agric. Res. J. Kerala.*, **11**: 155-158.

Malca M. I. and J. H. Owen. 1957. The grey leaf spot disease of St. Augustine grass. *Pl. Dis. Reptr.*, **41** : 871-875.

Malik, S. A., and M. A. Khan. 1943. Parasitic fungi of the North-West Frontier Province. *Ind. J. Agri Sci.*, **13** : 522-527.

Manibhushanrao, K. 1971a. Studies on the cultures of *Pyricularia oryzae* Cav. I. Morophology and pathogenicity. *Riso,* **20** : 283-297.

Manibhushanrao, K. 1971b. Studies on the cultures of *Pyricularia oryzae* Cav. II. Utilization of nitrogen and carbon compounds. *Phytopath. Z.*, **72** : 172-182.

Manibhushanrao, K. 1971c. Studies on the cultures of *Pyricularia oryzae* Cav. III. *In vitro* production of pectolytic and cellulolytic enzymes. *Phytopath. Z.*, **72** : 203 -212.

Manibhushanrao, K. 1979. Low night temperature on the susceptibility of International rice differentials to race of *Pyricularia oryzae.* In : *Proc. Symp. Host-pathogen-interaction.* Univ. of Madras, Madras, pp. 309-312.

Manibhushanrao, K. 1990. Contribution to blast and sheath blight diseases of rice in India. D. Sc. Thesis Univ. of Madras, Madras, p. 126.

Manibhushanrao, K. and P. R. Day. 1972. Low night temperature and blast disease development on rice. *Phytopathology,* **62** : 1005-1007.

Manibhushanrao, K. and L. Jesudhas. 1984. Conidium ontogeny in *Pyricularia. Phytopath. Z.*, **110** : 235-244.

Manibhushanrao, K. and P. Krishnan. 1991. Epidemiology of blast (EPIBLA) : a simulation model and forecasting system for tropical rice in India. In : *Rice blast modelling and forecasting*. Int. Rice Res. Inst. Box 933, Manila, Philippines. pp. 31-38.

Manibhushanrao, K. and S. H. Ou. 1979. Genetic recombination in the blast pathogen *Pyricularia oryzae* Cav. *Philippine Phytopathology,* **15** : 122-126.

Manibhushanrao. K. and S. H. Ou. 1980. Karyological studies in *Pyricularia oryzae* Cav. *Riso* **29** : 305-308.

Manibhushanrao, K. and S. Suryanarayanan. 1971. Isolation of an inositol-less mutant of *Pyricularia oryzae* Cav. by ultraviolet irradiation. *Phytopath. Z.*, **72** : 67-75.

Manibhushanrao. K. and M. Zuber. 1978. Disease resistance in cereals. *Acta. Phytopathol. Acad. Sci. Hung.* **13** : 313-335.

Manibhushanrao, K., Y. Joe, and P. Krishnan. 1989. Rice blast : Epidemiology and application of its techniques in disease management. *Rev. Trop. Pl. Path.*, **6** : 119-149.

Manibhushanrao, K., R. Kalyanasundaram and S. Suryanarayanan. 1971. Antigenic comparison of cultures of *Pyricularia oryzae* Cav. *Phytopath. Z.*, **70** : 121-126.

Manibhushanrao, K., M. Zuber and N. Matsuyama. 1988. Phenol metabolism and plant disease resistance. *Acta. Phytopathol. Entamol. Hung.*, **23** : 103-114.

Manibhushanrao, K., P. Krishan, K. N. Ponnuswamy, S. Natarajan and B. Reddy. 1990. Development of computerized simulation model (EPIBLA) for rice leaf blast epidemic. In : *Proc. Int. Symp. Rice Res. New Frontiers.*, DRR, Hyderabad. pp. 10.

Mao, J., M. He, Z. He, S. He and Y. Liang. 1989. Effect of cultivar mixture on blast disease and yield of rice. *10th sym. on Rice blast control.* p. 85.

Marchetti, M. A. 1983. Dilatory resistance on rice blast in USA rice. *Phytopathology,* **73** : 645-649.

Marchetti. M.A., M.C. Rush and W. E. Hunter. 1976. Current status of rice blast in the Southern United States. *Pl. Dis. Rep.* **60** : 721-725.

Matsumoto, K. 1979. Analog simulation analysis on the efficiency of chemical control for rice leaf blast epidemics. *Bull. Natl. Agric. Expt. Stn.* Ser. E. 15.

Matsumoto, S., T. Kozaka, and M. Yamada. 1969. Pathogenic races of *Piricularia oryzae* Cav. in Asia and some other countries. *Bull. Natl. Inst. Agric. Sci. Tokyo,* **23**, 1-36.

Matsuo, T. 1952. Genecological studies on cultivated rice. *Bull. Natl. Inst. Agric, Sci. Tokyo,* **3** : 1-111.

Matsuura, K. 1967. Inhibitory action of fungicides on the sporulation of *Piricularia oryzae* on the host lesions and on the sponge matrix. *Ann. Phytopathol. Soc. Japan,* **33** : 5-11.

Matsyuama. N. 1975. The effect of ample nitrogen fertilizer on cell-wall materials and its significance to rice blast disease. *Ann. Phytopathol. Soc. Japan,* **41** : 56-61.

Matsuyama. N. 1983. On the detection of momilactones A and B in healthy and blast infected leaves by GLC. *Ann. Phytopathol. Soc. Japan,* **49** : 200-205.

Matsuyama, N. and A.E. Diamond. 1973. Effect of nitrogen fertilizer on biochemical processes that could effect the lesion size of rice blast. *Phytopathology,* **63** : 1202-1203.

Matsyuama, N. and T. Kozaka. 1971. Comparative gel electrophoresis soluble proteins and enzymes of rice blast fungus *Pyricularia oryzae* Cav. *Ann. Phytopathol. Soc. Japan,* **37** : 259-265.

Matsuyama, N. and S. Wakimoto. 1985. Purification and characterization of anti blast substance, S-1, formed mainly in blast-resistant lower rice leaves. *Ann. Phytopathol. Soc. Japan,* **51** : 498-500.

Matsuyama, N. and S. Wakimoto. 1987. The production of anti-blast substances in blast-infected and UV-irradiated rice leaves *Ann. Phytopathol, Soc. Japan,* **53** : 499-453.

Matsuyama, N. and S. Wakimoto. 1988. Isolation and identification of diterpenoid anti-blast substances produced in the blast infected rice leaves. *Ann. Pytopathol. Soc. Japan,* **54** : 183-188.

Matsuyama, N. and T. Yamaguchi. 1981. Alteration of virulence and agressiveness during serial transfer and simplified storage of rice blast fungus *Pyricularia oryzae* Cav. *Ann. Phytopathol. Soc. Japan,* **47** : 457-463.

Matsuyama, N. H. Kato and T. Yamaguchi. 1977. Comparison of the isoenzyme patterns of extracellular enzymes in *Pyricularia* strains from Gramineous and Zingiberaceous plants. *Ann. Phytopathol. Soc. Japan,* **43** : 419-425.

Maximos, M. A. 1969. Inheritance of reaction to blast and some other characters in certain cross in rice., *Oryza sativa* L. M.Sc. Thesis *Fac. Agric.* Ain-Oshams Univ. Cairo, Egypt. p. 121.

Mayee, C.D. 1974. Perpetuation of *Pyricularia oryzae* through infected seeds in Vidarban region. *Indian Phytopathol.,* **27** : 604-605.

Mc Couch, S. R., G. Kochert, Z. H. Yu, Z. Y. Wang, G. S. Kush, W. R. Coffman and S. D. Tanksley. 1988. Molecular mapping of rice chromosomes. *Thoer. Appl. Genet.,* **76** : 815-829.

McIntosh, A.E.S. 1951. *Ann. Rep. Dept. Agric. Malaya 1949.* p. 87. *(Rev. Appl. Mycol.,* **30** : 508-509).

McRae, W. 1922. *Imp. Mycol. Sci. Rep.* Pusa Agric. Res. Inst., India. 1921-22, pp. 44-50.

Mehta, P. R., B. Singh and S. C. Mathur. 1952. A new leaf spot disease of bajra (*Pennisetum typhoides* Stapf and Hubbard) caused by a species of *Piricularia. Indian Phytopath.ol,* **5** : 140-143.

Mello-Sampayo, T., E. Vianna and M. Silva. 1954. Ensaios preliminares sôbre a determinacao de resistência de algumas formas cultivadas de arroz *Piricularia oryzae* Br. et Cav. *Ministerio de Economia Lisbon,* p. 38 *(Rev. Appl. Mycol.,* **34** : 481-482).

Mendgen, K. and H. Desining 1993. Transely Review No. 48. Infection structures of fungal plant pathogens-a cytological and physiological evaluation. *New Phytol.* **124** : 193-213.

Meredith, D. S. 1962. *Piricularia musae* Hughes in Jamaica. *Trans. Brit. Mycol. Soc.,* **45** ; 137-142.

Metcalf, H. 1906. A preliminary report on the blast of rice, with notes on other rice diseases. *Bull. South Carolina Agri. Expt. Stn.,* No. 121-p. 43.

Metcalf, H. 1907. The pathology of rice plant. *Science,* **25** : 264-265.

Metcalf, H. 1913. Diseases of rice. In : *Diseases of tropical plants,* (M.T. Cook, ed.) pp. 99-166.

Metcalf, H. 1921. The story of a plant introduction *J. Washington Acad. Sci.,* **11** : 474.

Misato, T. et al., 1959. Antibiotics as protectant fungicides against rice blast. 2. The therapeutic action of blasticidin-S. *Ann. Phytopathol. Soc. Japan.* **24** : 302-306.

Misato, T., I. Ishi, M. Asakawa, M. Katagiri and K. Fukunaga. 1968. II. Mode of action of blasticidin-S, a new antibiotic for rice blast control. *Bull. Natl. Inst. Agric. Sci. Japan,* **22** : 95-152.

Misawa, T. and K. Takahashi. 1968. Phytoalexin production in various rice - cultivar combinations. *Ann. Phytopathol. Soc. Japan,* **34** : 179.

Misawa, T. and N. Matsuyama. 1960. Studies on conidial dispersal of rice blast fungus. *Ann. Phytopathol. Soc. Japan,* **25** : 3 (Abstr.)

Miura, H. 1984. Ecological study on the kasugamycin resistant rice blast fungus. *Spec. Bull. Yamagata Pref. Exp. Stn.,* **14** : 1-44.

Miura, H., H. Ito and S. Takahashi. 1975. Occurrence of resistant strains of *Pyricularia oryzae* to kasugamycin as a cause of diminishing fungicidal activity to rice blast. *Ann. Phytopathol. Soc. Japan,* **41** : 415-417.

Miura, H., M. Katagiri, Y. Yamaguchi and H. Ito. 1976. Mode of occurrence of kasugamycin resistant rice blast fungus. *Ann. Phytopathol. Soc. Japan,* **42** : 117-123.

Miyake, Y., T. Hirooka and F. Araki. 1985. Inhibitory effect of isoprothiolane on metabolism of a phosphoramidate by isolates of *Pyricularia oryzae* Cav. in relation to fungicide sensitivity. *Pest Biochem., Physiol.,* **23** : 102-107.

Miyake, Y., and M. Adachi. 1922. Chemische Untersuchungen über die Widerstandsfähigkeit der Reisarten gegen die 'Inokrankheit' 1-2 bericht. *J. Biochem., Japan,* **1** : 223-247.

Miyake, Y. and M. Ikeda. 1932. Influence of silica application on rice blast. *J. Sci. Soil & Manure, Tokyo,* **6** : 53-76.

Miyazaki, K. 1928, 1930. Morphological and physiological studies on rice blast occurıng on plants treated with excessive nitrogen. Parts 1 and 2. *Agri. & Hort. Japan,* **3** : 754-764; **5** : 439-445.

Mizusawa, Y. 1959. Electron microscopy of ultrathin sections of conidia of *Pyricularia oryzae. J. Pl. Prot.,* **13** : 159-160.

Moens, P. B. and F. D. Perkins. 1969. Chromosome number of a small protist : accurate determination. *Science,* **166** : 1289-1291.

Mogi, S. 1979.Chemical control of rice blast in Japan. In: *Proc. Lec. Meet. on rice blast.* RDA, Seweon, Korea. pp. 317-348.

Mogi, S. and K. Yanagita. 1970. Aggressiveness of strains in rice blast fungus. *Ann. Phytopathol . Soc. Japan,* **36** : 157 (Abstr.).

Mohanty, C.R. and S. Gangopadhyay. 1982. Variation of leaf angle and incidence of rice blast dısease. *J. Agric Sci.,* **98** : 225-227.

Montoya, M., A. Carlos and S.W. Ahn. 1984. Yield loss caused by *Pyricularia* in cica-B rice under simulated enhanced drought conditions and epidemiological parameters for evaluating the levels of infection. *Acta. Aron.,* (Palmira) **34** : 23-47.

Mori, K. and M. Waku. 1985. Synthesis of oryzalexins A, B and C, the diterpenoidal phytoalexins isolated from rice blast leaves infected with *Pyricularia oryzae. Cav. J. Turrialba,* **41** : 5653-5660.

Moscebriofrio, L. and A. Moralesabanto. 1975. Influence of potassium on the resistance of rice tissues to the attack of *Pyricularia oryzae.* Cav. *J. Turrialba,* **25** : 316-322.

Muralidharan, K. 1974. Studies on the blast disease with reference to polyphenoloxidase, Ph.D. Thesis University of Madras, Madras. p. 84.

Muralidharan, K., and G.V. Venkatarao. 1980. A simple method of forecasting outbreak of rice blast. *Indian Phytopathol.* **33** : 560-564.

Muramatu, Y. and T. Koyanagi. 1977. Attempts to utilize computers for forecasting rice diseases and pests. *Pl. Protect.* **31** : 59-63.

Murata, J., K. Kuribayashi, and I. Kawai. 1933. Studies on the control of rice blast. III. Influence of home made manure on blast disease. *Nogi Kairyo Shiryo,* **64** : 1-138.

Nagai, I and A. Imamura. 1931. Morphology of the 'neck' of the panicle as related to the resistance against blast disease in rice varieties. *Ann. Agric. Expt. Stn. Chosen.,* **5** : 284-304.

Nagai, I., H. Fujimaki, and M. Yokoo. 1973. Breeding and some genetic findings for multiracial resistance to blast by using japonicaX indica hybrid of rice. *J. Agric. Res. Quart.,* **7** : 63-70.

Nagarajan, S. 1988. Epidemiology and loss of rice, wheat and pearl millet crops due to diseases. In : *Int. Symp. on Crop losses and diseases outbreaks in tropics and counter, measures. Trop. Agric. Res. Centre, Japan,* p. 209 (Abstr.).

Nair, J. K. 1975. Epidemiology of blast disease of rice. Ph.D. Thesis. Orissa Univ. Agric. & Tech., Bhubaneshwer, India. p. 240.

Nair, J. K., and N. K. Chakrabarti. 1978. Spread of blast infection in nature under different plant population through adjustment of spacing in rice. 3rd. Int. Cong. Pl. Path. Munchen, (Abstr.).

Nakagawa, K. 1962. Ecological studies on rice blast with special reference to the effect of deep ploughing. *Pl. Dis. & Insect Pests*, **2** : 1-71.

Nakajima, T., K. Tamari, K. Matsuda, and H. Tanaka and N. Ogasawara. 1970. Studies on the cell wall of *Pyiricularia oryzae*. Part II. The chemical constituents of the cell wall. *Agric Biol. Chem* . **34** : 553-574.

Nakatomi, S. 1926. On the variability and inheritance of the resistance of rice plants to rice blast disease. *Japan J. Gen.*, **4** : 31-38.

Namai, T., S. Yamanaka, and T. Misawa. 1977. Studies on the sporulation of rice blast fungus. *Pyricularia oryzae* Cavara, II. Effect of irradiation at different stages of sporulation. *Ann. Phytopathol. Soc. Japan*, **43** : 175-182.

Nambhu, K. 1972. Rabcide, a newly developed fungicide for controlling rice blast. *J. Pesticide Inform.*, **10** : 73-76.

Narasimhan, M. J. 1936. Report of work done in the mycological section for the year 1934-35. *Rep. Dept. Agric. Mysore*, pp. 19-22.

Narayanarao, A. And S. Suryanarayanan. 1974 a. Studies on the toxins of *Pyricularia*. I. Detection of Pyriculol in cultures of *Pyricularia* spp. *Proc. Indian Acad. Sci.* **80** : 31-40.

Narayanarao, A. and S. Suryanaryanan. 1974 b. Studies on the toxins of *Pyricularia* spp. II. Detection of Pyriculol in blast diseased leaves of Gramineae. *Curr. Sci.* **43** : 348-349.

Narayanarao, A. and S. Suryanarayanan. 1975. Studies on the toxins of *Pyricularia.* spp. - the phytotoxicity of culture filtrates. *Proc. Indian Acad.* Sci., **82** : 77-91.

Narayanarao, A., K. Manibhushanrao and S. Suryanarayanan. 1972 a. Effect of some phenolic compounds on sporulation and growth of *Pyricularia* spp. *Proc. Indian Acad. Sci.*, **45** : 140-144.

Narayanarao, A., K. Manibhushanrao and S. Suryanarayanan. 1972 b. Sporulation of *Pyricularia* spp. in culture. Effect of some aromatic compounds. *Indian Acad. Sci.* **45** : 105-110.

Narayanarao, A., K. Manibhushanrao and S. Suryanarayanan. 1972 c. Sporulation of *Pyricularia* spp. on detached leaves of Gramineae and in media supplemented with leaf extracts. *Proc. Indian Acad. Sci.*, **45** : 145-151.

Narita, S. and T. Iwata. 1959. Disease and pest forecasting in crops. *Rep. Hokkaido Pref. Agric. Expt. Stn.* (Mimeographed).

Narita, S., T. Iwata and S. Yamanuki. 1956. Studies on the host range of *Piricularia oryzae* Cav. *Rep. Hokkaido Pre. Agric. Expt. Stn.*, **7** : 1-33.

Neufeld, H.A., F. M. Laterell, L.F. Green and R. L. Weintraub. 1958. Oxidation of Meta-polyphenols by enzymes from *Piricularia oryzae* and *Polyporus versicolor*. *Arch. Biochem. Biophysics*, **76** : 317-327.

Nizeki, H. 1960. Resistant genes for rice blast disease in variety Aichiashi *Agric. Hort.*, **35** : 1321-1322.

Nilsson, H.E. 1911. Resistant gene Gelbroast beim Weizen, in Kreuzungsunter-suchungen an Hafer and Weizen. *Lond. Univ. Arsskr. Nf. Afd.*, **2** : 57-82.

Nishimura, M., K. Kogen, S. Okuda, K. Ohata and T. Kozaka. 1970. Isolation of the flourescent substance specially detected at the blast - infected rice leaves. *Ann. Phytopathol. Soc. Japan*, **36** : 360.

Nisikado, Y, 1917. Studies on the rice blast fungus. *Ohara Inst. Land. Forsch., Okayama Univ.*. **1** : 171-218.

Nisikado, Y. 1926. Studies on rice blast disease. *Bull. Bureaux Agric, Mini. Agric, For. Japan*, **15** : 1-211.

Nisikado, Y. 1927. Studies on rice blast disease. *Japan J. Bot.*, **3** : 239-244.

Nomura. K. and Ishii, 1985. Varietal difference in the response of rice leaf protoplasts to the mycelial components from *Pyricularia oryzae. Japan J. Breed.* **35** : 326-327.

Notteghem, J. L. 1985. Definition of a strategy for use of resistance through genetic analysis of the host-pathogen relationship : The case of the rice *Pyricularia oryzae* relation. *Agronomic Tropicale*, **40**: 129-147.

Nozu, M. and K. Yokogi, 1924. On a host plant of rice blast fungus. *Byochugai Zasshi*, **11** : 565-567.

Nukina, M. and T. Sato. 1992. Microbial production of a norvalin-analog of tenuazonic acid by *Pyricularia oryzae* Cavara. *Biosci. Biotech. & Biochem.* **56** : 1314-1315.

Nuqu, F. L., S.H. Ou, T.T. Ebron, J. Bandong and O. Garcia. 1971. Chemical control of rice blast. IRRI, Los Baños, Philippines, Saturday seminar. p. 15. (Mimeographed).

Ogasawara, N., K. Tamari and J. Kaji. 1957. Biochemical studies on the blast disease of rice plants. Piricularian-binding protein I. Presence of piricularin binding (PBP) produced by *Piricularia oryzae*. *J. Agric. Chem. Soc. Japan*, **31** : 460-463.

Ogasawara, N., K. Tamari and J. Kaji 1961. Biochemical studies on the blast disease of rice plants, Part 20. Mechanism of spore germination of *Piricularia oryzae* (2). Presence in spores of α–picolinic acid and the changes occurring during spore germination. *J. Agric. Chem. Soc. Japan*, **35** : 1412-1416.

Ogawa, M. et al. 1953. Studies on blast control by Ceresan lime. *Chugoku-Shikoku Agric Res.*, **3** : 1-5.

Ohata, K. and T. Kozaka. 1967. Interactions between two races of *Piricularia oryzae* in lesion formation in rice plants and accumulation of fluorescent compounds associated with infection. *Bull. Natl. Inst. Agric. Sci. Tokyo*, **21** : 111-132.

Ohata, K., S. Chitra and K. Shimada. 1982. Simulation of rice leaf blast using BASTCAST, a plant disease simulator. *Ann. Rep. Pl. Prot. North Japan*, **33** : 9-11.

Ohata, K., K. Goto and T. Kozaka, 1963. Observations on the reaction of rice cells to the infection of different races of *Piricularia oryzae*. *Ann. Phytopathol. Soc. Japan*, **28** : 24-30.

Ohata, K., K. Goto and T. Kozaka, 1966. Effect of low temperature on the susceptibility of rice plants to blast disease with special reference to some chemical components in the plants. *Bull. Natl. Inst. Agric. Sci., Tokyo*, **20** : 1-65.

Ohata, K.,K. Shimada and J. I. Chiba. 1987. A new improvement of the simulation model of rice leaf blast epidemics. *Bull. Aomori Agric. Exp. Stn., Japan*, **30** : 31-58.

Ohmori, K. and M. Nakajima, 1970. Effect of light on sporulation of *Pyricularia oryzae* Cavara. *Ann. Phytopathol. Soc. Japan*, **36** : 319-324.

Ohno, N., S. Inoue and H. Yoshioka. 1983. Recent fungicides for control of rice blast disease and new antifungal natural products, podoblastins. *J. Jap. Chem.*, **37** : 516-526.

Oka, H. I. 1968. V. Physiological races of *Pyricularia oryzae* found in Taiwan and gene conditioning the resistance to them. In : *Genetic evolution and breeding in rice. 12th Int. Cong. Genet. Tokyo, Japan*, **2** : 1-12.

Oka, H. I. and K. M. Lin. 1957. Genetic analysis of resistance to blast disease in rice by biometrical genetic method. *Japan J. Genet.*, **32** : 20-27.

Okada, M. and H. Maeda. 1956. Inheritance of resistance to leaf blast in crosses between foreign and Japanese varieties of rice. *Bull. Tohoku Agric. Expt. Stn.*, **10** : 59-68.

Okamoto, H. 1950. The influence of phosphate application on the susceptibility to blast disease of rice plants. *Ann. Phytopathol. Soc. Japan*, **14** : 6-8.

Okamoto, H. 1965. Chemical control of rice blast in Japan. In : *The rice blast disease*, John Hopkins Press, Baltimore, Maryland. pp. 399-407.

Okamoto, H. 1972. On the characteristics of Kasumin - antibiotic fungicide. *Japan Pest. Inform.*, **10** : 66-69.

Okamoto, H. and E. Hamaya, 1962. Studies on the volatilization of organomercuric compounds and their absorption by rice leaves with special reference to their phytotoxicity and effectiveness against rice blast. *Bull. Chugoku Agric. Expt. Stn.*, **7** : 273-298.

Okamoto, H., K. Matsumoto and T. Yamamoto. 1959. Studies on the influences of environmental conditions and application methods on the efficacy of fungicides applied against rice blast in the field. VII *Chugoku Agric. Res.*, **15** : 31-40.

Okamoto, H., K. Matsumoto, T. Yamamoto and Y. Sekiguchi. 1958. Studies on the influences of environmental conditions and application methods on the efficacy of fungicides applied against rice blast in the field. I-V. *Chugoku Agric. Res.*, **12** : 1-174.

Oku. H. 1967. Role of parasitic enzymes and toxins in development of characteristic symptoms in plant disease. In : *The dynamic role of molecular constituents in plant-parasite interaction* (C.J. Mirocha and I. Uritani, eds). *Proc. Ann. Phytopathol. Soc.*, pp. 237-255.

Ono, K., 1953, Morphological studies on blast diseases and sesame leaf spot of rice plants. *J. Hokuriku Agric.*, **2** : 75.

Ono, K. 1965. Principles, methods, and organization of blast disease forecasting. In : *The rice blast disease.* Johns Hopkins Press, Baltimore, Maryland. pp. 173-194.

Ono, K. 1966. Dry form of spores of several fungi. *Proc. Kanto-Tosan Pl. Prot. Soc.*, **13** : 22 (Abstr.).

Ono, K. and K. Nakazato. 1958. Morphology of the conidia of *Piricularia* from different host plants and produced under different conditions. *Ann. Phytopathol. Soc.* Japan, **23** : 1-2.

Ono, K. and H. Suzuki. 1959. Spore formation of blast fungus. *Proc. Ass. Pl. Prot. Hokuriku*, **7** : 6-19.

Ono, K. and H. Suzuki, 1959a. Spore formation of blast fungus on water surface. I. *Ann. Phytopathol. Soc. Japan*, **24** : 3-4.

Ono, K. and H. Suzuki. 1960. Studies on mechanism of infection and ecology of blast and stem rot of rice plant. *Spl. Res. Rep. Disease and Insect Forecasting, Min. Agri. & Forest. Japan*, **4** : 1-60.

Ono, K., et al. 1961. Studies of rice blast disease. *Hokuriku Agric. Expt. Stn. Res. Pl. Dis. Reps. for* 1959 and 1960. p. 54 and p. 64 (Mimeographed).

Ono, K., H. Suzuki, M. Onuma and E. Inoue. 1962, 1963. Data of investigations concerning spore flight of blast fungus. *Ann. Rep. Pl. Path.* 2nd *Lab. Natl. Hokuriku Agric. Expt. Stn.* (Mimeographed).

Onodera, I. 1917. Chemical studies on rice blast (*Dactylaria parasitans* Cavara). *J. Sci. Agric. Soc.*, **180** : 606-617.

Oota, K. 1982. The forecasting of rice blast disease. *Kongetu no nooyaku*. **26** : 12-19.

Orton, C.R. 1931. Seed-borne parasites. *Bull. W. Va. Agric. Exp. Stn.*, **245** : 47.

Otani, V. 1948. Studies on the relation between the principal components of the rice plant and its susceptibility to blast disease. I. Experiments with the rice seedings raised in different kinds of nursery beds. *Agric. Sci. North Temp. Reg.*, **2** : 269-280.

Otani, Y. 1950. Studies on the relation between the principal components of the rice plant and its susceptibility to the blast disease. II. *Ann. Phytopathol. Soc. Japan*, **14** : 37 (Abstr.).

Otani, Y. 1952a. Studies on the relation between the principal components of the rice plant and its susceptibility to the blast disease. III. *Ann. Phytopathol. Soc. Japan*, **16** : 97-102.

Otani, Y. 1952b. Growth factors and nitrogen sources of *Piricularia oryzae* Cavara. *Ann. Phytopathol. Soc. Japan*, **17** : 9-15.

Otani, Y. 1953. Carbon sources of *Piricularia oryzae* Cavara. *Ann. Phytopathol. Soc. Japan*, **17** : 119-120.

Otani, Y. 1955. On the proteolytic enzymes of *Piricularia oryzae* Cavara. *Jublee Pub. Comm.* 60th Birthday Prof. Y. Tochinai and Prof. T. Fukushi. Sapporo, Japan. pp-. 316-322.

Otoma, T. 1988. Damage caused by major plant diseases and plant pest forecasting program in Japan. In : *Int. sym. on Crop losses disease outbreaks in tropics & counter measures. Trop. Agric. Res. Centre Japan*, p. 14 (Abstr.).

Otsuka, H. 1959. Biochemical studies on rice blast disease. X. The biochemical classification of *Piricularia oryzae* Cav. (7) On the production of vitamin B group by *Piricularia oryzae* Cav. *J. Agric. Chem. Soc. Japan*, **33** : 1013-1018.

Otsuka, H. 1961. The biochemical classification of *Piricularia oryzae* Cav. *Memoirs Nigata Univ. Fac. Ed.* No. 3, p. 105.

Otsuka, H. K. Tamari and N. Ogasawara. 1957a, b,c, d. Biochemical studies on the rice blast disease. X. Biochemical classification of *Piricularia oryzae* Cav. (1-4). *J. Agric. Chem. Soc. Japan*, **31** : 791-794; 794-798; 886-890; 890-893.

Otsuka, H., K. Tamari and N. Ogaswara. 1965. Variability of *Piricularia oryzae* in culture. In : *The rice blast diseases.* Johns Hopkins Press, Baltimore, Maryland. pp. 69-109.

Otsuka, H., K. Tamari and N. Ogasawara. 1958a. Biochemical studies on rice blast disease. X. Biochemical classification of *Piricularia oryzae* Cav. (5). *J. Agric Chem. Soc. Japan*, **32** : 890-893.

Otsuka, H., K. Tamari, N. Ogasawara and R. Honda. 1958b. Biochemical studies on rice blast disease. X. Biochemical classification of *Piricularia oryzae* Cav. *J. Agric Chem. Soc. Japan*, **32** : 893-897.

Ou, S.H. 1961. Some aspects of rice blast and varietal resistance in Thailand. *Int. Rice Comm. Newsl.*, **10** : 11-17.

Ou, S.H. 1963. Rice blast disease and breeding for its resistance in Thailand, with notes on other rice diseases. *FAO Expanded Tech. Assist. Prog. Rep.* No. **1673**, p. 26.

Ou, S.H. 1965. A proposal for an international program of research on the rice blast disease. In : *The rice blast disease*. Johns Hopkins Press, Baltimore. Maryland. pp. 441-446.

Ou, S. H. 1980. Pathogen variability and host resistance in rice blast disease. *Ann. Rev. Phytopath.* **18** : 167-189.

Ou, S. H. 1985. *Rice diseases*, 2nd ed. Commonwealth Agric. Bureaux, Central Sales, Farnham Royal, Slough, UK. p. 380.

Ou, S. H. and M. R. Ayad. 1968. Pathogenic races of *Piricularia oryzae* originating from single lesions and monoconidial cultures. *Phytopathology*, **58** : 179-182.

Ou, S. H. and P. R. Jennings. 1969. Progress in the development of disease-resistant rice. *Ann. Rev. Phytopath.* **7** : 383-410.

Ou, S. H. and K. M. Lin. 1958. Preliminary report on rice breeding for blast disease resistance in Taiwan *J. Agric. Assoc. China*. N. S. **21** : 50-58.

Ou, S. H. and F. L. Nuque. 1963. The relation between leaf and neck resistance to the blast disease. *Int. Rice Comm. Newsl.*, **12** : 30-35.

Ou, S. H. V., A. Awoderu and T. T. Ebron, Jr. 1974. Daily periodicity of the conidia release of *Pyricularia oryzae* over a rice blast nursery. *Pl. Dis. Reptr.* **58** : 544-548.

Ou, S. H., F. L. Nuque and J. M. Bandong, 1975. Relation between qualitative and quantitative resistanc to rice blast. *Phytopathology*, **65** : 1315-1316.

Ou, S.H., F.L. Nuque, T.T. Ebron and V.A. Awoderu. 1971. A type of stable resistance to blast disease of rice. *Phytopathology*, **61** : 703-706.

Padhi, B. and N. K. Chakrabarti. 1972. Occurrence of new pathogenic races of *Pyricularia oryzae* in nature. *Proc. Ind. Sci. Cong.*, **59** : 32 (Abstr.).

Padhi, B. and N. K. Chakrabarti. 1982. Variation in the occurrence of blast disease incidence at monthly intervals in relation to meteorological factors. Z. Pflkrankh. *Pflschutz.*, **89** : 144-150.

Padmanabhan, S. Y. 1953. Effect of nitrogenous fertilization on the incidence of blast on rice varieties. *Curr. Sci.* **22** : 271-272.

Padmanabhan, S. Y. 1959. The present position and control of rice diseases in India. *Proc. Ind. Acad. Sci.*, **49** : 349-362.

Padmanabhan, S. Y. 1963. The role of therapeutic treatments in plant disease control with special reference to rice diseases. *Bull. Indian Phytopathol.* Soc., **1** : 79-84.

Padmanabhan, S. Y. 1965. Studies on forecasting outbreaks of blast disease of rice. I. Influence of meteorological factors on blast incidence at Cuttack. *Proc. Indian Acad. Sci.*, **62** : 117-129.

Padmanabhan, S. Y. 1965a. Estimating losses from rice blast in India. In : *The rice blast disease*. Johns Hopkins Press, Baltimore, Maryland. pp. 203-221.

Padmanabhan,. S. Y. 1965b. Breeding for blast resistance in India. In : *The rice blast disease*. Johns Hopkins Press. Baltimore, Maryland. pp. 343-359.

Padmanabhan, S. Y. 1965c. Physiologic specialization of *Piricularia oryzae* Cav., the causal organism of blast disease of rice. *Curr. Sci.*, **34** : 307-308.

Padmanabhan, S. Y. 1965d. Studies on the rice disease. D. Sc. Thesis. Utkal University, Bhubaneshwar, India, p. 524.

Padmanabhan, S. Y. 1966. Prospects of control of blast disease by forecasting and direct application of fungicides and antibiotics. *Bull. Indian Phytopathol.* Soc. **3** : 121-124.

Padmanabhan, S.Y. 1967. Blast disease of rice. *Pest articles news summ.*, **13** : 62-69.

Padmanabhan, S. Y. and G. H. Chandwani. 1960. Forecasting outbreaks of rice disease in the field. *Rice News Teller*, **8** : 8-9.

Padmanabhan, S. Y. and D. Ganguly. 1953. The effect of date of planting on the incidence of blast disease of rice. *Rice News Teller,* **1** : 107-110.

Padmanabhan, S. Y. and D. Ganguly. 1954. Relation between the age of rice plant and its susceptibility to helminthosporium and blast diseases. *Proc. Indian Acad.* Sci. **39** : 44-50.

Padmanabhan, S. Y. and D. Ganguly. 1959. Breeding rice varieties resistant to blast disease caused by *Pyricularia oryzae* Cav., I. Selection of resistant varieties from genetic stock. *Proc. Indian Acad. Sci.,* **50** : 289-304.

Padmanabhan, S. Y. and K. V. S. R. K. Row. 1966. Control of blast disease of rice through application of antibiotics. *Bull. Indian Phytopathol, Soc.,* **3** : 125-127.

Padmanabhan, S. Y., N. K. Chakrabarti and K. V. S. R. K. Row. 1971. Forecasting and control of rice disease. *Proc. Indian Acad. Sci.,* **37** : 423-429.

Padmanabhan, S. Y., S. C. Mathur and R. K. Misra. 1973. Inheritance of resistance of rice varieties to *Pyricularia oryzae. Oryza* **10** : 45-53.

Padmanabhan, S.Y., S. C. Mathur and R. K. Misra. 1974. Breeding for blast resistance in India : Genetics of resistance. In : *Breeding researches in Oceania. Indian Soc. Genet. & Pl. Breed. New Delhi.* pp. 424-429.

Padmanabhan, S. Y., N. K. Chakrabarti, S. C. Mathur and J. Veeraraghavan. 1967. Physiologic specialization in *Pyricularia oryzae* Cav., *Proc. Indian Acad. Sci.,* **64** : 63-73.

Padmanabhan, S. Y., N. K. Chakrabarti, S. C. Mathur and J. Veeraraghavan. 1970. Identification of pathogenic races of *Pyricularia oryzae* in India. *Phytopathology,* **60** : 1574-1577.

Padmanabhan, S. Y., D. Ganguly, G.H. Chandwani and J. Veeraraghavan. 1963. Control of blast disease of rice with spray fungicides. III. An economic spray schedule. *Indian Phytopathol.,* **16** : 55-61.

Padmanabhan, S. Y., D. Ganguly, G. H. Chandwani, J. Veeraraghavan and C. S. Krishnamurty. 1962. Control of blast disease of rice with fungicides-II. *Indian Phytopathol.,* **15** : 28-36.

Padwick, G. W. 1950. *Manual of rice diseases.* Commonwealth Mycol. Inst. Kew, Surrey, London, p. 198.

Panzer, J. D. and R. D. Beier. 1958. A simple field inoculation technique. *Pl. Dis. Reptr.,* **42** : 172-173.

Panzer, J. D. and E. C. Tullis and E. P. Van Arsdel. 1957. A simple 24 hour 1 slide spore collector. *Phytopathology,* **47** : 512-514.

Park, M. 1937. Report of the mycological division. *Ceylon Dir. Agr., Admin. Rep. 1936.* pp.28-35. (Rev. Appl. Mycol., **17** : 293-294).

Parsons, K. A., F. G. Chumley and B. Valent. 1986. Genetic transformation of the fungal pathogen responsible for rice blast disease. *Proc. Natl. Acad. Sci.* **84** : 4161-4165.

Patanayak, B. K., D. N. Rout and G. N. Mahapatra. 1978. Studies on oxazoles. I. Synthesis and chlorination of some new 2-amino-4- substituted oxazoles and their use as fungicides. *J. Indian Chem. Soc.* **55** : 264-267.

Pawer, M. S. and S. A. Kulkarani. 1954. *Pyricularia* species on *Setaria intermedia* Roem and Schult. *Curr. Sci.,* **23** : 229-230.

Peng. Y. L. and J. Shishivama. 1988. Temporal sequence of cytological events in rice leaves infected with *Pyricularia oryzae. Can. J. Bot.* **66** : 730-735.

Peng, Y. L. and J. Shishiyama. 1989. Timing of a cellular reaction in rice cultivars associated with differing degree of resistance to *Pyricularia oryzae. Can. J. Bot.* **67** : 2704-2710.

Podhradszky, J. 1961. Provocative methods for the assessment of resistance in rice varieties to *Pyricularia. Novenytermeles.* **10** : 67-76 (*Rev. Appl. Mycol.,* **40** : 606-607).

Podhradszky, J. and J. Sudi. 1957. Field Assay of resistance to 'brusone' in rice varieties. *Novenytermeles.* **6** : 239-248. (*Rev. Appl. Mycol.* **38** : 791).

Prabhu, A. S. and J. C. De Faria. 1982. Quantitative relationship between leaf blast and neck blast and their effect on grain filling and grain weight in upland rice. *Pergui Agropecn. Bras.,* **17** : 219-224.

Prabhu, A. S. and O. P. Moris, 1986. Blast management in upland rice in Brazil. In : Progress in upland rice research. IRRI, Box 933, Manila, Philippines. (Mimeographed).

Prasad, N. N. and V. Raghunathan. 1972. A study of physiological changes in resistant and susceptible rice varieties following nitrogen fertilization and infection by blast pathogen. *Proc. Annamalai Univ. Tamil Nadu*, p. 163.

Prasad, N. N., M. Kotireddy and D. Purshotam. 1972. A study of certain biochemical and physiological changes in resistant and susceptible rice varieties following infection by blast and bacterial blight pathogens. *Proc. Annamalai Univ. Tamil Nadu.* p. 267.

Purba, D., S. Kiyosawa, I. Ando and R. Furutani. 1994. Estimation of functional values of field resistance genes to blast disease in some foreign rice varieties. *Japan J, Breed.* (in press).

Quamaruzzaman, M. and S. H. Ou. 1970. Monthly changes of the pathogenic races of *Pyricularia oryzae* Cav. in blast nursery. *Phytopathology*, **60** : 1266-1269.

Radhaikar, P. W. and P. Neergard. 1969. Studies on Aureofungin as seed treatment in controlling seed-born fungal diseases. *Hindustan Antibiotic Bull.*, **11** : 163-165.

Rahnema, I. 1979. Simulation of the effect of different weather regime on germination and formation of appressoria of *Pyricularia oryzae. Z. Pflkrankh. Pflchutz.*, **86** : 315-329.

Ramakrishnan, K. 1948. Studies on the morphology, physiology and parasitism of the genus *Pyricularia* in Madras. *Proc. Indian Acad.* Sci., **38** : 174-193.

Ramakrishnan, L. 1966a. Studies on the host-parasite relations of blast disease of rice. I. *Phytopath. Z.*, **55** : 297-308.

Ramakrishnan, L. 1966b. Studies on the host-parasite relations of blast disease of rice. II. *Phytopath. Z.*, **56** : 279-287.

Ramakrishnan, L. 1966c. Studies on the host-parasite relations of blast disease of rice. III. The effect of night temperature on the infection phase of blast disease. *Phytopath. Z.*, **57** : 17-23.

Ramakrishnan, L. 1971. Diseases of rice. *Indian Coun. of Agric. Res.* (ICAR), New Delhi, p. 150.

Ramalingam, A. 1966. Spore dispersal in *Pyricularia oryzae* Cav. *Indian Phytopathol.*, **19** : 72-75.

Ramalingam, A 1970. Epidemiology and disease forecasting. In : *Plant disease problems. Proc. First Int. Symp. Pl. Path.* IARI., New Delhi, p. 895.

Ramalingam, P., K. Manibhushanrao and S. Suryanarayanan. 1977. Isolation of diauxotrophic mutants of *Pyricularia oryzae* Cav. with N-methyl-N'-nitro-N-notrosoguanidine (MNNG). *Microbiol. Genet. Bull.*, **42** : 1.

Ramaswamy, V. 1979. Studies on the role of phenolics and lignin in the rice blast disease. Ph. D. Thesis Univ. of Madras, Madras, India, p. 106.

Ramiah, K. and K. Ramaswamy. 1936. Breeding for resistance to *Pyricularia oryzae* in rice (*Oryzae sativa*). *Proc. Indian Acad. Sci.*, **3** : 450-458.

Ranganathaiah, K. G. and S. B. Mathur, 1976. Seed health testing of *Eleusine coracana* with special reference to *Drechslera nodulosa* and *Pyricularia grisea. Res. Rep. Danish Govt. Seed Path.* Copenhegen, Denmark (Mimeographed).

Ranganathaiah, K. G., S. M. Srinivasaiah and S. S. Gowda. 1979. Seed borne infection of rice by *Pyricularia oryzae* in Karnataka. *Curr. Sci.* **48** : 952.

Rangaswami, G., and T. V. Subramanian, 1958. The correlation between the leaf and neck infections of the blast disease of rice. *Indian Phytopathol.*, **11** : 179-185.

Rao, Y. M. and A. N. Rao. 1981. Sporulation of *Pyricularia* spp. stimulation by detached leaves and leaf extracts of *Commelina benghalensis Proc. Indian Acad. Sci.* **90** : 389-393.

Rapillay, F., B. Foucault and A. Bonnet. 1975. Realisation d'un appareil permettant l'etude an champ de la dispersion des spores de champignons par des gouttes d' eau; application a *Kobatiella zeae, Septoria nodorum* et *Fusarium roseum. Ann. Phytopathol.*, **7** : 45-50.

Rath, G. C. and S. Y. Padmanabhan, 1972. Studies on the inheritance of leaf blast resistance in rice. *Proc. Indian Acad. Sci.* **76** : 106-116.

Rath, G. C. and S. Y. Padmanabhan. 1973. Symbolization of *Pyricularia* resistance genes. *Int. Rice Commi. Newsl.*, **21** : 33-35.

Reddy, A. P. K. and A. O. Bastawisi. 1990 Resistance to blast (Bl) in Egyptian rice varieties. *Int. Rice Res. Newsl.*, **15** : 3.

Reddy, A. P. K. and J. M. Bonman. 1987. Recent epidemics of rice blast in India and Egypt. *Pl. Dis.*, **71** : 850.

Reddy, M. K. and N. N. Prasad. 1974. Effect of certain phenolic compunds on the growth of three isolates of *P. oryzae*. *Curr. Sci.* **43** : 612-614.

Reddy, M. K., R. Kothandaraman and A. Mahadevan. 1969. A comparison of certain biochemical characters of etiolated and normal rice seedlings in relation to susceptibility to blast disease. *Indian Phytopathol.*, **22** : 149-150.

Reddy, P. C., A.P. K. Reddy, and K. Krishnaiah, 1988. Rice blast disease management on farmer's field. In : *Current status of blast and sheath blight diseases of rice in India*. Directorate of Rice Research (DRR), Hyderabad, India. (Mimeographed).

Refatti, E. 1955. Observations on the morphology, germination and partial pigmentation of the spores of *Pyricularia oryzae* Br. et Cav. in culture. *Annali Sper. Agr.* **9** : 7-21. (Rev. App. Mycol., **35** : 924).

Revilla, V. A. 1953. El., 'quemado' o 'brusone' del arroz en el Peru. *Estac. Agric.* La molina Informa (86). p. 16. (Rev. Appl. Mycol., **33** : 378).

Rodas, C. J. E. and M. J. Rosero. 1975. Yield losses in rice due to neck rot infection caused by *Pyricularia oryzae* Cav. *Ravista ICA*, **10** : 115-126.

Rodriguez, M. and C. E. Galvez. 1975. Indications of partial resistance of rice to the fungus *Pyricularia oryzae* Cav. In : *Horizontal resistance to blast disease of rice*. CIAT, Cali, Colombia, pp. 137-154.

Roldan, E. F. 1938. New or noteworthy lower fungi of the Philippine Islands. II. *Philippine J. Sci.*, **66**: 7-13.

Rosenberg, N. J., B. L. Blad and S. B. Verma. 1983. *Microclimate: The Biological Environment*, Wiley, New York.

Rosero, M. J. 1967a. Inheritance of resistance in rice to four international races of *Pyricularia oryzae* Cav. Ph. D. Thesis. Baton Rouge, La. Diss (*Pl. Breed., Abstr.* **39** : 106).

Rosero, M. J. 1967b. Inheritance of resistance in rice to four international races of *Pyricularia oryzae* Cav. *Science*, **66** : 7-13.

Rossman, A. Y., R. J. Howard and B. Valent 1990. *Pyricularia grisea*, the correct name for the rice blast disease figures. *Mycologia*, **82:** 509-512.

Rotem. J. 1983. Quantitative assessment of inoculum production dispersal, deposition, survival and infectiousness in airborne diseases. In : *Experimental techniques in plant disease epidemiology*, (J. Kranz and J. Rotem, eds.) pp. 69-83. Springer-Verlag, New York.

Rotem, J. and J. Cohen. 1974. Epidemiological patterns of *Phytophthora infestans* under semi-arid conditions. *Phytopathology*. **64** : 711-714.

Row, K. V. S.R.K. and J. P. Crill. 1979. Cytological bases of mechanism of variability in the rice blast fungus. Int. Rice Res. Inst. (IRRI) Saturday seminar, Los Baños, Philippines.

Row, K. V. S. R. K., J. R. Aist and J. P. Crill. 1985. Mitosis in the rice blast fungus and its possible implications for pathogenic variability *Can. J. Bot.*, **63** : 1129-1134.

RDA, (Rural Development Authority), Korea. 1989. *Ann. Rep.* 1988.

Ryu, J. D., W. H. Yeh, S. S. Han, Y. H. Lee and E. J. Lee. 1987. Regional and annual fluctuation of races of *Pyricularia oryzae* during 1978-1985 in Korea. *Korean J. Pl. Pathol.*, **3** : 174-179.

Ryu, J. D., S. S. Han, Y. H. Lee and E. J. Lee. 1988. Development of perfect stage of *Pyricularia* spp. isolated from cereals and grasses. *Korean J. Pl. Pathol.*, **4** : 284-289.

Saccardo, P.A. 1880. Conspectus generum fungorum italiae inferiorum. *Michelia*, **2** : 1-135.

Sadasivan, T. S. 1968. Nitrogen metabolism and resistance to facultative parasitism. In : *Biochemical regulation in disease plants or injury. Phytopath. Soc. Japan*, pp. 239-252.

Sadasivan, T. S., S. Suryanarayanan and L. Ramakrishnan. 1965. Influence of temperature on rice blast disease. In: *The rice blast disease*, Johns Hopkins Press, Baltimore, Maryland. pp. 163-171.

Sakai, Y., T. Kohguchi and Y. Uehera. 1990. Some informations getting from the forecast model BLASTAM on the disease progress and their use for chemical control of rice leaf blast. *Bull. Hiroshima Agric. Expt. Stn.*, **53** : 21-32.

Sakamoto, M. 1941. Relation between blast resistance and the resistance to physical penetration by the rice leaves. *Ann. Phytopathol. Soc. Japan,* **10** : 359-360.

Sakamoto, M. 1948. On the relation between nitrogenous fertilizer and resistance to rice blast. *Ann. Phytopathol. Soc. Japan,* **13** : 53. (Abstr.).

Sakamoto, M. 1949. On inoculation of the leaf sheath of rice with the blast fungus. *Bull. Inst. Agric Res. Tohoku Univ.,* No. **1**. pp. 120-129.

Sakamoto, M. 1951. On the new method of inoculation on rice plants with the blast fungus, *Pyricularia oryzae* Cav. *Bull. Inst. Agric. Res. Tohoku Univ.,* No. **2**. pp. 15-23.

Sako, M. and T. Takamori. 1958. Studies on the relation between inorganic nutrition of rice plants and rice blast. I. Antagonistic action of potassium and magnesium. *Chuygoku Shikoku Agr. Res.,* **4** : 47-48.

Sakurai, Y. and K. Toriyama. 1967. Field resistance of the plant to *Pyricularia oryzae* and its testing method. In : *Rice disease control-growing resistant varieties and counter measures. Min. Agric. Fores., Japan,* pp. 123-135.

Samoto, S. and Owchi. 1968. Studies on augmenting resistance to rice blast *Rep. Hokkaido Agric. Expt. Stn.* **73** : 59-61.

Sasaki, R. 1922. Existence of strains in rice blast fungus. I. *J. Pl. Prot. Tokyo,* **9** : 631-644.

Sasaki, R. 1923. Existence of strains in rice blast fungus. II. *J. Pl. Prot. Tokyo,* **10** : 1-10.

Sasaki, T. and H. Kato. 1972. A statistical method of predicting outbreaks of rice panicle blast. *Phytopathology,* **62** : 1126-1132.

Sato, Z. and T. Kozaka. 1970. New toxic substances produced by *Pyricularia oryzae. Physiol. Pl. Path.,* **1** : 319-328.

Sawada, K. 1917. Blast of rice plants and its relation to the infective crops and weeds, with description of five species of *Dactylaria. Sp. Bull. Taiwan Agric. Expt. Stn.,* No. **16**, p. 78.

Sawada, K. 1919. Materials for a fungus flora of Formosa. I. *Formosa Agric Expt. Stn. Spl. Bull.,* No. **19**. pp. 615-616.

Sawada, K. 1940. Rice blast disease and manuring. *Jap. J. Bot.,* **10** : 62-63.

Sawada, K. 1959. Descriptive catalogue of Taiwan (Formosa) Fungi. XI. *Pub. Colle. Agric. Natl. Taiwan Univ.,* No. **8**, p. 268.

Schnelle, F., L. P. Smith and J. F. Wallin 1963. Report on instruments recording the leaf wetness period. The influence of weather conditions on the occurrence of apple scab. Append., 15-41. Tech. Note 55, *World Meteorol. Organ, Geneva,* p. 41.

Seem, R. C. 1988. The measurement and analysis of the effects of crop development on epidemics. In : *Experimental techniques in plant disease epidemiology* (J. Kranz and J. Rotem, eds.) Springer-Verlag, New York, pp. 51-83.

Sekido, H. and T. Akatsuka. 1987. Mode of action of oryzalexin D against *Pyricularia oryzae. Agric. Biol. Chem.,* **51** : 1968-1971.

Sekido, H., T. Endo, R. Suga, O. Kodama, T. Akatsuka, Y. Kono and S. Takeuchi. 1986. Oryzalexin D (3, 7-dihydroxy (+) sandaracopimaradiene), a new phytoalexin isolated from blast-infected rice leaves. *J. Pest. Sci.* **11** : 369-372.

Sekodo, H., R. Suga O. Kodama, T. Akatsuka, Y. Kono, Y. Esumi and S. Takeuchi. 1987. Qualitative and semi quantitative analysis of oryzalexins in blast and brown spot diseased rice leaves by mass chromatography. *J. Pest. Sci.,* **12** : 739-740.

Sekiguchi, Y. and T. Furuta. 1970. Influence of temperatures during incubation period for infection of rice plants by blast fungus on occurrence of leaf blast. *Ann. Phytopathol. Soc. Japan,* **36** : 350. (Abstr.)

Sekizawa, Y. and S. Mase. 1980. Recent progress in studies on non-fungicidal controlling agent, probenzole, with reference to the induced resistance mechanism of rice plant. *Rev. Pl. Protec. Res.* **13** : 114-121.

Shanghai Institute of Plant Physiology. 1975. Studies on Qingfengmycin. II. Isolation, purification and characterization. *Acta Microbiol. Sinica,* **15** : 101-109.

Shanmugam, N. and K. Ramakrishnan. 1965. Selectivity of toxins produced by *Pyricularia oryzae. Rice Newsl.*, **13** : 6-8.

Sharma, J. P., S. Kumar, and R. N. Verma. 1988. A linear model for predicting the yield loss in rice due to the neck blast. *Oryza.* **25** : 91-93.

Shigemura, C. 1955. Some aspects of rice breeding for blast resistance in Japan. 6th meeting of FAO-IRC working party, on rice breeding. (Mimeographed).

Shen, Z., W. Zhang, S. He, S. Sun, R. Tao and De Shi. 1986. Genetic analysis for blast resistance in some *indica* rice varieties. (*Oryza sativa* L.) *Chinese J. Rice Sci.*, **1** : 1-7.

Shimada, S. 1937. Infection weise der blatter der reispflanzen durch *Pyricularia oryzae. Agric. & Hort.*, **33** : 1106-1108.

Shimizu, S. 1980. Observation of conidium liberation by tyndall illumination and terminal velocity of conidia of *Pyricularia oryzae* Cav. *Ann. Phytopathol. Soc. Japan*, **46** : 322-328.

Shimura, M. M. Iwata, N. Tashiro, Y. Sekizawa, Y. Suzuki, S. Mase and T. Watanabe. 1981. Anti-conidial germination factors induced in the presence of probenazole in infected host leaves. I. Isolation and properties of 4 active substances. *Agric. Biol. Chem.*, **45** : 1431-1436.

Shindo, K. 1977. Effect of mixed cropping of varieties differred in major gene on blast incidence. *Ann. Rep. Pl. Prot. North Japan*, **28** : 55.

Shindo, K. and O. Horino. 1989. Control of rice blast disease by mixed plantings of isogenic lines as multiline cultivars. *Bull. Tohoku Natl. Agric. Exp. Stn.*, **79** : 1-13.

Shinoda, H., K. Toriyama, Y. Sekiguchi., and T. Yunoki. 1971. Linkage group of newly found blast resistance gene *Pi-s. Chugoku Nat. Agric. Res.*, **43** : 1-2.

Shinoda, H., K. Toriyama, T. Yunoki, A. Ezuka and Y. Sakurai. 1971. Studies on the varietal resistance of rice to blast. 6. Linkage relationship of blast resistance genes. *Bull. Chugoku Nat. Agric. Exp. Stn.*, **20** : 1-24.

Shirai, M. 1896. Notes on plants collected in Suruga, Totomi Yamato and Kii. *Bot. Magazine*, Tokyo, **10** : 111-114.

Shirai, M. 1905. Supplemental notes on fungus which causes the diseases so-called Imochibyo of *Oryza sativa* L. *Bot. Mag. Tokyo*, **19** : 19-28.

Shiyomi, M. 1970. Mathematical models representing the increase in the number of disease lesions. *Bull. Natl. Inst. Agric. Sci. Ser.*, No. **17**. pp. 103-116.

Shouji, H., T. Takeda, T. Kimura, H. Suzuki and Y. Fujita. 1982. *Bull. Yamagata Perf. Agric Expt. Stn.*, No. **17**. pp. 59-69.

Silue, D. and J.L. Notteghem. 1991. Compatibilite et fertilite de souches de *Magnaporthe grisea* agent de la pyriculariose du riz. *Cryptogamic Mycologie*, **12** : 87-95.

Silue, D. and J. L. Notteghem. 1992. Identification of a cross between two compatible isolates of *Magnaporthe grisea* (Herbet) Barr and genetic analysis of avirulence/virulence of rice. *J. Phytopathology*, **135** : 77-83.

Silue, D., J.L. Notteghem and D. Tharreau. 1992. Evidence of a gene-for-gene relationship in the pathosystem *Oryza sativa - Magnaporthe grisea. Phytopathology* , **82** : 577-580.

Simmonds, J. H., 1958. Queensland Dept. Agric. Science Branch, Plant Pathology Section. *Rep.* 1957-1958 : 58-59. (*Rev. Appl. Mycol.*, **38** : 296).

Singh, N. and I. S. Kunene. 1980. Cellulose decomposition by four isolates of *Pyricularia oryzae. Mycologia*, **72** : 182-190.

Singh, R. A. 1983. Studies on somatic nuclear division in *Pyricularia oryzae. Indian Phytopathol.*, **36**: 707-711.

Sinha, A. K. and T. K. Sengupta. 1986. Use of unconventional chemicals in the control of blast of rice. In : 2nd Intl. conf. in plant protection in the tropics. pp. 213-219.

Sitch, L. A., A. D. Amante. R. D. Dalmacio and H. Leung, 1989. *Oryza minuta* a source of blast and bacterial blight resistance for rice improvement In : Review of advances in plant biotechnology, Cimmyt, 1985-88. (A. Mujeebkazi and L. A. Sitch, eds.), Mexico, pp. 315-322.

Situe, D. and J. L. Notteghem. 1990. Production of perithecia of *Magnaporthe grisea* on rice plants. *Mycol. Res.* **94** : 1151-1152.

Smith, J. A., and J.P. Mcfauz. 1991. *Pseudomonas syringae* pv. *syringae* induces systemic resistance to *Pyricularia oryzae* in rice. *Physiol. & Mol. Pl. Path.* **39** : 451-461.

Sprague, R. 1950. Diseases of cereals and grasses in North America. Ronald Press, New York. p. 538.

Sridhar, R. 1967. Production of indole acetic acid by *Pyricularia oryzae*. *Indian Phytopathol.*, **20** : 379-380.

Sridhar, R. 1972a. Influence of nitrogen fertilization and *Pyricularia oryzae* development on some oxidases, their substrates and respiration of rice plants. *Acta. Phytopath. Acad. Sci. Hung.*, **7** : 57-70.

Sridhar, R. 1972b. Phenolic compounds and the rice blast disease as influenced by nitrogen fertilization *Riso*, **21** : 21-31.

Sridhar, R. 1978. Changes in peroxidase isoenzymes in blast diseased rice leaves. *Acta. Phytopath. Acad. Sci. Hung.*, **13** : 161-164.

Sridhar, R. and A. Mahadevan. 1979. Physiology and biochemistry of rice plants infected by *Pyricularia oryzae, Helminthosporium oryzae, Xanthomonas oryzae* and *Xanthomonas translucens*. f. sp. *oryzicola*. *Acta. Phytopath. Acad. Hung.*, **14** : 49-82.

Sridhar, R. and S. H. Ou. 1972. Extracellular enzymes produced by *Pyricularia oryzae* Cav. *Philippine Phytopath*ology, **8** : 52-56.

Sridhar, R. and S. H. Ou. 1974. Biochemical changes associated with development of resistant and susceptible types of rice blast lesions. *Phytopath. Z.*, **79** : 222-230.

Sridhar, R., I. W. Buddenhagen and S. H. Ou. 1973a. Phenol localization in rice leaf tissue. *Acta. Phytopath. Acad. Sci. Hung.*, **8** : 13-18.

Sridhar, R., I. W. Buddenhagen and S. H. Ou. 1973b. Presence of lipid bodies in rice leaves and their discolouration during pathogenesis. *Experentia*, **28** : 958-960.

Sridhar, R., S. Kumar and N. Mandakini. 1988. In : *Current trends in physiological plant pathology.* Akhtar Hussain 60th Birth day commoration. Lucknow, pp. 159-183.

Srinivasan, K. V. and U. Vijayalakshmi. 1957. *Pyricularia oryzae* Cav. on *Arundo donax* L. *Sci. & Cult.*, **23** : 490-491.

Srivastava. S. N. S. 1962. The systemic control of blast disease of rice. *Curr. Sci.* **31** : 243-244.

Stakman, E. C. and J. G. Harrar. 1957. *Principles of plant pathology*. The Ronald Press, New York.

Stehle, H. 1954. Quelgues notes sur la botanique et l'ecologie vegetate de l'Archipel des Caraibes. *J. d'Agric. Tropi. et de Bot. Appl.*, **1** : 71-110. (*Rev. Appl. Mycol.* **34** : 515).

Stevens, F. L. 1913. *The fungi which cause plant disease*. Macmillan, New York, p. 754.

Steyaert, R. L. 1948. Contribution a l' etuide des parasites des vegetaux du congo Belge. *Bull. Soc. Belg.*, **30** : 11-58.

Subbarao, N. S. and S. Suryanarayanan. 1957. Exosmosed substances on leaf blades of rice (*Oryza sativa* L.). *Curr. Sci.*, **26** : 186-187.

Subramanian, C. V. 1967. Nitrogen-utilization by some plant pathogenic fungi. *Bull. Nat. Inst. Sci. India.*, **35** : 11-22.

Subramanian, C. V., K. Manibhushanrao and C. Rajendran. 1978. Development of race ID-1 of *Pyricularia oryzae* Cav. *Phytopath. Z.*, **91** : 340-435.

Subramanian, S. 1967. Rice blast, nyctotemperature and metabolic changes. *Phytopath. Z.* **60** : 41-50.

Subramanian, S. and T. S. Sadasivan. 1964. Rice blast, nyctotemperature and nitrate reductase. *Experientia*, **20** : 508.

Subramanian, S., N. Sethunathan and B. Santhi. 1965a. Effect of growth regulators on the incidence of blast disease (*Pyricularia oryzae* Cav.) on rice (*Oryza sativa* L.) *Riso*. **14** : 199-203.

Subramanian, S., N. Sethunathan and B. Santhi. 1965b. Effects of foliar application of uracil and thiouracil on the incidence of blast disease *Pyricularia oryzae* Cav. on rice (*Oryza sativa* L.). *Riso*, **14** : 281-282.

Sudo, I., H. Nagayama and K. Tamari, 1973. Occurrence and some properties of cellulase in the filtrate of conidiopores and mycelia of *Pyricularia oryzae* Cav. *Agric. Biol. Chem.* **37** : 1651-1659.

Sueda, H. 1928. Studies on the rice blast disease. *Rep. Dept. Agric. Res. Inst. Formosa.*, **36** : 1-130.

Suematsu, N. 1916. On the artificial culture of rice blast fungus (*Dactylaria parasitans* Cav.) *Bot. Mag. Tokyo*, **30** : 97-99.

Sumina, I. 1967. Effect of β-acylphenylhydrazine derivatives on the respiratory chain of *Pyricularia oryzae*. *Ann. Phytopathol. Soc. Japan*, **33** : 207-215.

Sun, G. C., D. Shi, G. Z. Zhuge and S. Sun. 1990. Some components of partial resistance to blast (Bl) in *indica* rices. *Int. Rice Res. Newsl.*, **15** : 11-12.

Sun, S., R. Tao and S. De. 1987. Studies on rice breeding for blast resistance in China. In : The developments in integrated control of rice diseases and insect pests in China. (ed. GSPP, Min. Agric. Animal Husb. and Fisheries PRC.) *Zhejiang Sci. Tech. Pub.* pp. 33-45.

Sun, S., M. Jin, S. Zhang, X. Tao, R. Tao and D. Fang. 1981. Rice blast disease and its control. *Shangai Sci. & Tech. Pub.* p. 182.

Sundararaman, S. 1921. Board of Agriculture, Proceedings of 3rd meeting of mycological workers in India, 1921.

Sundararaman, S. 1925. Longivity of paddy *Pyricularia*. *Madras Dept. Agric. Year Book*, 1924. p. 9.

Sundararaman, S. 1927. Administration report of the Government mycologist, Coimbatore, for 1926-27. pp. 326-344.

Sundararaman, S. 1928. Administration report of the Government mycologist, Coimbatore, for 1927-28. pp. 355-372.

Supad, M. A. 1982. A. Rep. 1981. Race division. MARDI. Malaysia.

Surin, A. P. Arunyanart, R. Dhitikiakttipong. W. Rojanahusdin, K. Soontrajarn, S. Munkong and S. Disthaporn. 1989. Yield loss due to rice blast (Bl) disease at different crop stages. *Int. Rice Res. Newsl.*, **14** : 34-35.

Surin, A., P. Arunyanart, W. Rojanahusdin, S. Mukong, R. Dhitikiattipong and S. Disthaporn 1990. Using empirical blast models to establish disease management recommendations in Thailand. In : *Rice blast modelling and forecasting*. Intl. Rice Res. Inst., Box. 933, Manila, Philippines pp. 69-744.

Suryanarayanan, S. 1958a. Mechanism of resistance of paddy (*Oryza sativa* L.) to *Pyricularia oryzae* Cav. I. General consideration. *Proc. Natl. Inst. Sci. India*, **24** : 285-292.

Suryanarayanan, S. 1958b. Thiamine, its intermediates and growth of *Pyricularia* spp. and *Selerotium oryzae*. *Phytopath. Z.*, **33** : 341-348.

Suryanarayanan, S. 1958c. Growth factor requirements of *Pyricularia* spp. and *Sclerotium oryzae*. *Proc. Indian Acad. Sci.*, **48** : 154-158.

Suryanarayanan, S. 1958d. Role of nitrogen in host susceptibility to *Pyricularia oryzae* Cav. *Curr. Sci.* **27** : 1447-1448.

Suryanarayanan, S. 1967. Environment and the blast disease. *Bull. Indian Phytopathol. Soc.* **3** : 110-114.

Suryanarayanan, S. and S. H. Ou. 1978. Role of epicuticular waxes in blast resistance. *Int. Rice Res. Newsl.* **3** : 7.

Sutton, T. B. and A. L. Jones. 1976. Evaluation of four spore traps for monitoring discharge of ascospores on *Venturia inaequalis*. *Phytopathology*, **66** : 453-456.

Sutton, J. C., T. J. Gillespie and P. D. Hildebrand. 1984. Monitoring weather factors related to plant disease. *Pl. Dis.*, **68** : 78-84.

Sutton, J. C., T. J. Gillelspie and T. D. W. James. 1988. Electronic monitoring and use of microprocessors in the field. In : *Experimental techniques in plant disease epidemiology* (J. Kranz and J. Rotem, eds.) Springer-Verlag, New York, pp. 99-113.

Suzuki, H. 1930. Experimental studies on the possibility of primary infection of *Pyricularia oryzae* and *Ophiobolus miyabeanus* internal of rice seeds. *Ann. Phytopathol. Soc. Japan.* **2** : 245-275.

Suzuki, H. 1933a, b, c. On the relation of soil moisture to development of blast disease of rice with special reference to the inoculation experiments on :
(a) the resistance and susceptible varieties of the paddy rice and the upland rice.
(b) seedlings and pedicels of spikes of plants grown on soils differing in the time and deviation of drying and irrigation.
(c) plants grown on soils differing in moisture and amount of nitrogenous manure.
Frosch. Geb. Pfkrankh., Kyoto, **2** : 78-97; 172-185; 279-291.

Suzuki, H. 1934. Studies on the influence of some environmental factors on the susceptibility of the rice plant to blast and *Helminthosporium* diseases and on the anatomical characters of the plant. I. Influence of differences in soil moisture. *J. Coll. Agric. Tokyo Imp. Univ.*, **13** : 45-108.

Suzuki, H. 1935a, b. Studies on the influence of some environmental factors on the susceptibility of the rice plant to blast and *Helminthosporium* disease on the anatomical characters of the plant.
(a) Influences of differences in soil moisture and in the amount of nitrogen fertilizer given.
(b) Influence of differences in soil moisture and in the amount of nitrogenous fertilizer given.
J. Coll. Agric. Tokyo, Imp. Univ. **13** : 235-237; 277-331.

Suzuki, H. 1937. Studies on the relation between the anatomical characters of the rice plant and its susceptibility to blast disease. *J. Coll. Agric. Tokyo Imp. Univ.*, **14** : 181- 264.

Suzuki, H. 1951. Studies on the classification of strains in *Pyricularia oryzae. Rep.*, 1951 (Mimeographed).

Suzuki, H. 1953. Studies on the classification of strains in *Pyricularia oryzae*. Rep., 1953 (Mimeographed).

Suzuki, H. 1954. Studies on antblastin (I-IV). *Ann. Phytopathol. Soc. Japan*, **18** : 138 (Abstr.)

Suzuki, H. 1965. The falling of *Pyricularia oryzae* - spore by rain. *Proc. Assoc. Pl. Prot. North Japan*, **16** : 135-137.

Suzuki, H. 1965a. Heterokaryosis in *Pyricularia oryzae* and its cytological evidence with interference phasecontrast microsocpe. *Trans. Mycol. Soc. Japan*, **5** : 65-74.

Suzuki, H. 1965b. Origin of variation in *Pyricularia oryzae*. In : *The rice blast disease*, Johns Hopkins Press, Baltimore, Maryland. pp. 111-149.

Suzuki, H. 1967a. Studies on biologic specialization in *Pyricularia oryzae* Cav. *Tokyo Univ. Agric. & Tech. Tokyo, Japan*, p. 235.

Suzuki, H. 1967b. Studies on media for sporulation and stock culture of *Piricularia oryzae* Cavara. *Ann. Phytopathol. Soc. Japan*, **33** : 101-114.

Suzuki, H. 1968. Rain drops, dew and guttation drops related to the spore germination appressorium formation of rice blast fungus. *Proc. Assoc. Pl. Prot. Hokuriku*, **16** : 13-16.

Suzuki, H. 1969a. Studies on the behaviour of rice blast fungus spore and application to outbreak forecast of rice blast disease. *Bull. Hokuriku Agric. Exp. Stn.*, **10** : 114-118.

Suzuki, H. 1969b. Effect of temperature on germination of conidia and appressorium formation of *Pyricularia oryzae. Ann. Rep. Pl. Prot. North. Japan.* **17** : 6-9.

Suzuki, H. 1970. Interrelationship between the occurrence of rice blast disease and meteorological conditions. *Rev. Pl. Prot. Res.*, **3** : 1-11.

Suzuki, H. 1973. Germination ability of the spore on the lesion and the dispersion spore of rice leaf blast. *Proc. Assoc. Pl. Prot. Hokuriku*, **21** : 4-5.

Suzuki, H. 1974. Behaviour of rice blast fungus spore and application to outbreak forecast of rice blast disease. *J. Agric. Res. Q.*, pp. 78-83.

Suzuki, H. 1975. Meteorological factors in the epidemiology of rice blast. *Ann. Rev. Phytopathol.*, **13** : 239-256.

Suzuki, H. and Y. Hashimoto. 1953. Pathogenicity of the rice blast fungus to plants other than rice. I; II. *Ann. Phytopathol. Soc. Japan*, **17** : 94-168.

Suzuki, N. 1965c. Nature of resistance to blast. In : *The rice blast disease.* Johns Hopkins Press, Baltimore, Maryland, pp. 277-301.

Suzuki, N., Y. Doi and S. Toyoda. 1953. Histochemical studies on the lesions of rice blast caused by *Piricularia oryzae* Cav. II. On the substance in the cell membrane of rice reacting red in colour with diazo reagent. *Ann. Phytopathol. Soc. Japan,* **17** : 97-101.

Suzuki, T. and Y. Fujita. 1977. The effect of number of *Pyricularia oryzae* spore on the action for penetration and the lesion development. *Trans. Mycol. Soc. Japan,* **18** : 385-390.

Suzuki, Y. and S. Yoshimura. 1963. Effect of light on sporulation of the rice blast fungus. *Ann. Phytopathol. Soc. Japan,* **28** : 62-63. (Abstr.).

Sweigard, J.A., F.G. Chumley, and B. Valent 1992a. Cloning and analysis of CUTI, a cutinase gene from *Magnaporthe grisea. Mol. & Genet.* **232** : 174-182.

Sweigard, J.A., F.G. Chumley, and B. Valent 1992b. Disruption of *Magnaporthe grisea* cutinase gene. *Mol. & Gen. Genet.* **232** : 183-190.

Sydow, H., P. Sydow and E. J. Butler. 1911. Fungi idae orientalis. III. *Ann. Mycol.,* **9** : 372-421.

Taga, M., H. Nakagawa, M. Tsuda and A. Ueyama. 1978. Ascospore analysis of kasugamycin resistance in the perfect state of *Piricularia oryzae. Phytopathology,* **68** : 815-817.

Taga, M., H. Nakagawa, M. Tsuda and A. Ueyama. 1979. Identification of three different loci controlling kasugamycin resistance in *Pyricularia oryzae. Phytopathology,* **69** : 463-466.

Takahashi, Y. 1951. Phytopathological and plant breeding investigations on determining the degree of blast resistance in rice plants. *Rep. Hokkaido Pref. Agric. Exp. Stn.,* **3** : 1-61.

Takahashi, Y. 1956. Studies on the mechanism of the resistance of rice plants to *Piricularia oryzae.* II. Pathological changes microscopically observed in host cells in which fungus hyphae do not grow well. *Bull. Yamagata Univ. Agric. Sci.,* **2** : 83-100.

Takahashi, Y. 1957a. Fundamental data for the forecasting of blast disease. *Min. Agric. Fores. Japan, Pl. Protect. Sec.,* p. 20.

Takahashi, Y. 1957b. The mechanism of resistance of rice plants to blast disease. *Yamagata Agric. Fores. Soc.,* **13** : 17-28.

Takahashi, Y. 1958a. A method of forecasting blast disease using leaf sheath inoculation technique. *Pl. Prot. Tokyo,* **12** : 339-349.

Takahashi, Y. 1958b. Studies on the mechanism of resistance to blast disease (*Piricularia oryzae*) paddy. *Bull. Yamata Univ. Fac. Agric.* p. 13.

Takahashi, Y. 1959. Studies on the mechanism of resistance of rice plant to *Piricularia oryzae. J. Yamagata Agric. For. Soc.,* **13** : 17-28.

Takahashi, Y. 1965. Genetics of resistance to rice blast disease. In : *The rice blast disease.* Johns Hopkins Press, Baltimore, Maryland, pp. 303-329.

Takahashi, Y. 1967. Sheath inoculation method to assess the grade of resistance or susceptibility of rice plants to *Piricularia oryzae. Ann. Phytopathol. Soc. Japan,* **33** : 89-100.

Takahashi, Y., M. Uchiyama and K. Izawa. 1969. Production of Phytoalexin - like substance as the phenomenon of blast resistance. *Ann. Phytopathol. Soc. Japan,* **35** : 367.

Takai, A., K. Kanai, M. Kobayashi, K. Murata and R. Komori. 1985. Studies on the forecasting of rice blast by system dynamic model. Forecast for number of lesions and infection area of rice leaf blast by application of computer simulation. *Bull. Ibarkai Agric. Exp. Stn.* **25** : 97-131.

Takasahi, T. 1982. Forecasting method of rice blast disease using computer in Fukuoka prefecture. *Kongetu no Noyaku,* **26** : 33-37.

Tamari, K. and J. Kaji. 1954. Biochemical studies of the blast fungus (*Piricularia oryzae Cav.*), the causative fungus of the blast disease of rice plants. I. Studies on the toxins produced by blast fungus. *J. Agric. Chem. Soc. Japan,* **28** : 254-258.

Tamari, K. and J. Kaji. 1955. Biochemical studies of the blast fungus (*Piricularia oryzae Cavara*). II. Studies on the physiological action of piricularin, toxin produced by the blast fungus on rice plants. *J. Agric. Chem. Soc. Japan,* **29** : 185-190.

Tamari, K. and J. Kaji. 1957. Biochemical studies of the blast disease of rice plant. IV. The functions of oxygen and nitrogen in the piricularin molecule. *J. Agric. Chem. Soc. Japan,* **31** : 387-390.

Tamari, K. and J. Kaji. 1959. Biochemical studies of the blast disease of rice plant. XIII. The accumulation of coumarin in stunted rice plants caused by the ill-effects of piricularin. *J. Agric. Chem. Soc. Japan.* **33** : 181-183.

Tamari, K., N. Ogasawara and J. Kaji. 1965. Biochemical products of the metabolism of *Piricularia oryzae*. In : *The rice blast disease*, Johns Hopkins Press, Baltimore, Maryland, pp. 35-68.

Tanaka, S. 1961. Analytical studies on the environmental conditions related to epidemics of rice blast disease in Hokkaido. *Hokkaido Natl. Expt. Stn.*, **55** : 1-58.

Tanaka, S. 1965. Nutrition of *Pyricularia oryzae in vitro*. In : *The rice blast disease*. Johns Hopkins Press, baltimore, Maryland. pp. 23-33.

Tanaka, S. and H. Katsuki. 1951 a. Growth factors of the fungus causing rice blast disease. I. Biotin as the principal factor. *J. Chem. Soc. Japan.* **72** : 231-233.

Tanaka, S. and H. Katsuki 1951b. Growth factors of the fungus causing rice blast disease. II. Thimine and pteroyl-glutamic acid as supplementing growth factors. *J. Chem. Soc. Japan*, **72** : 285-287.

Tanaka, S. and J. Tsuji. 1952. Studies on pentose metabolism of *Piricularia oryzae*. *Ann. Phytopathol. Soc. Japan*, **16** : 31.

Tanaka, S., H. Katsuki and F. Katsuki,. 1952. Biochemical studies of the blast disease of rice plants. IV. Effect of different nitrogen sources in the culture media upon the nutritional absorption of the blast disease fungus. *Ann. Phytopathol. Soc. Japan*, **16** : 103-106.

Tanaka, S., H. Katsuki and F. Katsuki. 1953. Biochemical studies of the blast disease. V. Effect of the different nitrogen sources in the culture media upon the mycelial constitutents and amylase activity of the blast fungus. *Ann. Phytopathol. Soc. Japan*, **17** : 54-56.

Tanaka, S., F. Katsuki and K. Kimura. 1956. Biochemical studies on susceptibility of rice plants to the blast disease. IV. Presence of organic acids of the TCA cycle in the leaves of rice plants. *J. Chem. Soc. Japan*, **77** : 1066-1069.

Tanaka, T. 1987. Rice cultivation in Egypt. *Agric. & Hort.* **62** : 125-134.

Tanaka, Y. and Y. Yamasaki. 1973. Studies on the variation of the rice blast fungus, *Piricularia oryzae* Cav. I. Variation of DNA content per nucleus in conidial cell and mycelial cell of *Piricularia oryzae* Cav. *Japan J. Breed.* **23** : 164-165.

Tanaka, Y. and Y. Yamasaki. 1974. Studies on variation of rice blast fungus, *Piricularia oryzae* Cav. II. Variation of chromosome number. *Japan J. Breed.*, **24** : 118-119.

Tanaka, Y. and Y. Yamasaki. 1975. Studies on the variation of the rice blast fungus, *Piricularia oryzae* Cav. III. Chromosome number and DNA content. *Japan J. Breed.*, **25** : 19-20.

Tanaka, Y., N. Murata and H. Kato. 1979. Behaviour of nuclei and chromosomes during ascus development in the mating between either rice-strain or weeping love-grass strain and ragi-strain of *Pyricularia*. *Ann. Phytopathol. Soc. Japan*, **45** : 182-191.

Tastra, I. K., R. Irmansyah and Z. H. Yunizar. 1987. Epidemiology of leaf blast (*Pyricularia oryzae*) under various abiotic and biotic conditions. Rep. of the SARIF - team case study.

Tasugi, H. and K.Yoshida. 1958. On the relation between silica and resistance of rice plants to rice blast. *Monbusho Shiken Kenkyu Hokoku*. **48** : 31-36.

Tasugi, H. and L. Yoshida. 1959, 1960. Relation between rice blast resistance and temperature environmental. *Ann. Phytopathol. Soc. Japan*, **24** : 1; **25** : 1.

Templeton, G. E., T. H. Johnson and S. E. Henry. 1961. Rice blast. *Arkansas Farm Res.* **10** :12.

Teng. S. C. 1932. Fungi of Nanking. II. Contributions from the Biological Laboratory of the Science Society of China. *Bot. Ser.*, **8** : 5-48.

Teng, P. S. 1990a. Integrated pest management in rice: An analysis of the status quo with recommendations for action. Rep. submitted to the USAID/ACIAR/IDRC/FAO/ODNRI/IPM Task force p. 86.

Teng, P. S. 1990b. Assessing the impact of planned release of bioengineered organisms using modelling and non-modelling approaches. *Phytopathology*, **81** : 353-356.

Teng, P. S., H. W. Gebbinck and H. Pinnschmidt. 1991. An analysis of the blast pathosystem to guide modelling and forecasting. In : *Rice blast modelling and forecasting*. Int. Rice Res. Inst. (IRRI) Box. 933, Manila, Philippines, pp. 1-30.

Teng. P. S., T. W. Mew, J. M. Bonman and H. Leung. 1988. Methods for estimating yield losses in tropical rice disease. In : Crop losses disease outbreaks and counter measures. *Min. Agric. Fores. Fisher., Japan*, p. 21.

Tepora, J. N. and P. R. Jennings. 1966. Inheritance of resistance to *Pyricularia oryzae* (Philippine race 15 and 20). *Philippine Agric.* **50** : 42-55.

Terui, M. 1940. Internal formation of conidia of the rice blast fungus. *Ann. Phytopathol. Soc. Japan.* **10** : 265-268.

Tewari, S. N. and A. P. Dath. 1984. Effect of leaf extract media of some plants on the growth of three fungal pathogens of rice. *Indian Phytopathol.*, **37** : 458-461.

Thirumalachar. M. J. 1968. Antibiotics in the control of plant pathogens. *Adv. Appl. Microbiol.*, **10** : 313.

Thirumalachar, M. J. 1974. Aureofungin a systemic antifungal antibiotic for plant disease control. *Ann. Micro.*, **24** : 265-279.

Thirumalachar, M. J. and J. N. Misra. 1953. Some disease of economic plants in Bihar, India. *FAO Pl. Prot. Bull.*, **1** : 145-146.

Thirumalachar, M. J., N. B. Kulkarni and M. K. Patel. 1956. Two new records of *Piricularia* species from India. *Indian Phytopathol.* **9** : 48-51.

Thomas, K. M. 1930. Some aspects of the control of blast disease of paddy. *Madras Agric. J.*, **18** : 596-604.

Thomas, K. M. 1938. Detailed administration report of the government mycologist, Madras for the year 1937-38. p. 21.

Thomas, K.M. 1939. Detailed administration report of the government mycologist, Madras for the year 1938-39. p. 30.

Thomas, K. M. 1940. Detailed administration report of the government mycologist, Madras for the year 1939-40. p. 18.

Thomas, K. M. 1941. Detailed administration report of the government mycologist, Madras for the year 1940-41. pp. 53-74.

Thyagarajan, S. 1971. Studies on *Pyricularia* with reference to host - parasite relationship. Ph. D., Thesis, Univ. of Madras, Madras, p. 95.

Tien, C. M. and P. M. Huang. 1975. Report of the crop losses caused by rice blast disease (*Pyricularia oryzae*) in 1974 in Taiwan. *Taiwan Agric.* Q., **11** : 139-144.

Tochinai, Y. and S. Komuya. 1939. Studies on the infection of *Piricularia oryzae* Br. and Cav. on maltreated rice plants. *Rev. App. Mycol.*, **18** : 546. (Abstr.).

Tochinai, Y. and I. Nakano. 1940. Studies on the nutritional physiology of *Piricularia oryzae* Cavara. *J. Fac. Agric. Hokkaido Imperial Univ.*, **44** : 183-229.

Tochinai, Y. and M. Shimamura. 1932. Studies on the physiologic specialization in *Piricularia oryzae* Br. et Cav. *Ann. Phytopathol. Soc. Japan*, **2** : 414-441.

Togashi, K. 1961. Chlamydospore and appressorium formation in *Piricularia oryzae*. *Ann. Phytopathol. Soc. Japan*, **26** : 60. (Abstr).

Togashi, K., et al., 1946. A survey of the disease of cultivated plants in Iwate prefecture in 1943-44. *Iwate Agric. Expt. Sta. Bull.*, **10** : p. 48.

Togashi, K., N. Ogaswara and K. Tamari. 1960. Biochemical studies on the blast disease of rice plants. XVII. Influence of piricularin, a toxic substance of the blast fungus, on the resistance of rice plants to the blast disease. *Ann. Phytopathol. Soc. Japan*, **25** : 142-145.

Tokousbalides, M. C. and H. D. Sisler. 1978. Effect of tricyclazole on growth and secondary metabolism in *Pyricularia oryzae*. *Pestic. Biochem.* and *Physiol.*, **8** : 26-32.

Tokunaga, Y. 1959. Studies on the relationship between metabolism of the rice plant and its resistance to blast disease. I. Correlation of nitrogen and sugar contents of rice plant to blast disease. *Bull. Tohoku Nat. Agric. Expt. Stn.*, No. **16**, pp. 1-5.

Tokunaga, Y., T. Furuta and T. Shimoyama. 1953. Process and occurrence of stunting in rice plants caused by blast disease. *Ann. Phytopathol. Soc. Japan*, **18** : 76-77.

Tokunaga, Y., T. Katsube and Y. Koshimizu. 1966. Studies on the relationship between metabolism of rice plant and its resistance to blast disease. III. Correlation of nitrogen metabolism and blast disease in rice plant. *Bull. Tohoku Nat. Agric. Expt. Stn.*, **34** : 37-79.

Tometa, H. and S. Yamanaka. 1983. Studies on resistance reaction in the rice blast disease caused by *Pyricularia oryzae* Cav. I. The pathological changes in the early infection stage of the inner epidermal cells of leaf sheath. *Ann. Phytopathol. Soc. Japan.*, **49** : 514-521.

Tomizawa, C. 1953. Studies on the nutritional physiology of the rice blast fungus (*Piricularia oryzae*). *Ann. Phytopathol. Soc. Japan*, **17** : 113-118.

Tomizawa, C. 1956. Influence of purine antagonists on the growth of the rice blast fungus *Piricularia oryzae*. *Bull. Natn Inst. Agric. Sci. Tokyo*, **6** : 111-122.

Toriyama, K. 1972. Breeding for resistance to major rice disease in Japan. *Rice Breeding*. Int. Rice Res. Inst. (IRRI), Box 933, Manila, Philippines. pp. 253-280.

Toriyama, K. 1975. Recent progress of studies on horizontal resistance in rice breeding for blast resistance in Japan. In : *Horizontal resistance and blast disease of rice*. CIAT, Cali, Colombia, pp. 65-100.

Toriyama, K., T. Yunoki and H. Shinoda. 1968. Breeding rice varieties for resistance to blast II. Inheritance of high field resistance of Chugoku 31. *Japan J. Breed.*, **18** : 145-146.

Torres, C.Q. 1986. Effect of plant age on the expression of resistance to *Pyricularia oryzae* Cav. in upland rice varieties. Ph.D. dissertation Univ. of the Philippines at Los Baños, Laguna, Philippines p. 82.

Torres, C.Q. and P. S. Teng. 1989. Analysis of blast effects on rice growth and yield. *Phytopathology*., **79** : 1526.

Toyoda, S. and N. Suzuki. 1952. Histochemical studies on the lesions of rice blast caused by *Piricularia oryzae* Cav. I. Some observations on the sporulation of lesions of different types occurring on leaves of the same variety. *Ann. Phytopathol. Soc. Japan*, **17** : 1-4.

Toyoda, S. and N. Suzuki. 1958. Lesion-like browning of rice leaves induced by several metabolic inhibitors. *Pl. Prot. Tokyo*, **12** : 346-348.

Toyoda, S. and N. Suzuki. 1960. Histochemical studies of rice blast lesions caused by *Piricularia oryzae* Cav. IV. Changes in the activities of oxidases in infected tissue. *Ann. Phytopathol. Soc. Japan*, **25** : 172-177.

Trevtathan, L. E. 1982. Pathogenicity of ryegrass and cultural variability of Mississippi isolates of *Pyricularia grisea*. *Pl. Disease*, **66** : 592-594.

Tsai, W. H. 1986. Prediction of rice leaf blast. III. Meteorological variables and percentage of leaf area infected by *Pyricularia oryzae*. *Pl. Prot. Bull.*, **28** : 111-118.

Tsai, W. H., H. J. Su. 1984. Prediction of rice leaf blast. I. Meteorological variables and the development of the number of lesions. *Pl. Prot. Bull. Taiwan*, **26** : 171-180.

Tsai, W. H. and H. J. Su. 1985. Prediction of rice leaf blast. II. The relationships of meteorological variables, conidial numbers and the development of the number of lesions. *J. Agric. Res. China*, **34**: 71-78.

Tseng, T. C. and L. C. Wu. 1967. The carbon nutrition of *Pyricularia oryzae*. *Bot. Bull. Acad. Sci Taipai*, **8** : 73-79.

Tsing, F., Y. S. Ma, K. M. Tsao, H. H. Hau and C. T. Li. 1955. Experiments on the control of rice blast by organic mercurial dusts. *Acta Phytopath. Sinica*, **1** : 79-89.

Tsuda, M., H. Nakagawa, M. Taga and A. Ueyama. 1977. Cultural conditions and formation of the perfect state of *Pyricularia oryzae* Cavara. *Trans. Mycol. Soc. Japan*, **18** : 161-169.

Tsumara, T. 1976. Fuji-one, new systemic fungicide. *Japan Pest. Inform.*, **27** : 20-23.

Tsuzuki, H., H. Koide, M. Nishioka and I. Nakahishi. 1975. Tricyclazole (EL-291) : a new systemic fungicide for control of rice blast and its characteristics. *Res. Bull. Aichi-Ken Agric Res. Cent. Ser.*, **47** : 55-63.

Tullis, E. C. 1951. Disease of rice. *US Dept. Agric. Farmer's Bull.* No. 1854. p. 16.

Turner, W. B. 1971. *Fungal metabolites*. Academic Press, Lond., & N. Y.

Uchijima, Z. 1962. Studies on the micro-climate within plant communities II. The scale of turbulent and the momentum transfer within plant layer. *Nogyo Kisho*, **18** : 14-21.

Uchiyama, T., N. Ogasawara., Y. Nanba and H. Ito. 1979. Conidial germination and appressorial formation of the plant pathogenic fungi on the coverglass or cellophane coated with various lipid components of plant leaf waxes. *Agric Biol. Chem.* **43** : 383-384.

Uehara,K. 1958. On the production of phytoalexin by the host plant as a result of interaction between the rice plant and the blast fungus *Piricularia oryzae* Cav. *Ann. Phytopathol. Soc. Japan*, **23** : 127-130.

Uehara, K. 1962. Studies on the production and the mechanism of action of phytoalexin. *Hiroshima Agri. Coll. Lab. Pl. Path. Sci. Rep.*, 1-87.

Uehara, Y. 1985. Regional classification of rice blast occurrence in Hiroshima prefecture by multivariate analysis. *Bull. Hiroshima Pref. Agric. Exp. Stn.*, **49** : 19-30.

Uesugi, Y., and K. Fukunaga. 1967. Fungitoxin action of higher alkyl compounds I. Mechanism of fungitoxic action of alkylphenols. *Ann. Phytopathol. Soc. Japan*, **33** : 168-175.

Uesugi, Y. and C. Tomizawa. 1972. Metabolism of S-benzyle O-ethyl phenylphosphonothioate (Inezin) in mycelial cells of *Pyricularia oryzae. Agric. Biol. Chem.*, **36** : 313-317

Uesugi, Y., M. Katagiri and K. Fukunaga. 1969. Resistance in *Pyricularia oryzae* to activities and organophosphorus fungicide. *Bull. Natl. Inst. Agric. Sci.*, **23** : 93-112.

Ueyama, A. and M. Tsuda. 1975. Formation of the perfect state in culture of *Pyricularia* sp. from some graminaceous plants (preliminary report). *Trans. Mycol. Soc. Japan*, **16** : 420-422.

Ueyama, A., M. Tsuda, M. Taga and H. Nakagawa. 1977. Mating types for perfect state of *Pyricularia oryzae. Trans. Mycol. Soc. Japan*,**18** : 312-317.

Ujihara, M. 1951. On the breeding process of blast resistant varieties in the Inshashi Experiment Farm. *Bull. Aichi Pref. Agric. Expt. Stn.*, No. **5**. pp. 11-24.

Ujihara, M. 1960. On the breeding for blast disease resistance varieties in rice. *Japan J. Breed*, **10** : 113-114.

Umeda, Y. 1972. Anti-blast fungicide "Hinosan" and its biological properties and metabolism in rice blast. *Japan Pest. Inform.*, **10** : 85-88.

Umetsu, N., J. Kaki, and K. Tamari. 1972. Investigation on the toxin production by several blast fungus strains and isolation of tenuazonic acid as a novel toxin. *Agric. Biol. Chem.*, **36** : 859-866.

Umetsu, N., J. Kaji and K. Tamari. 1973. Isolation of tenuazonic acid from blast-diseased rice plants. *Agric. Biol. Chem.*, **37** : 451-452.

Umestu, N., J. Kaji, K. Aoyama and K. Tamari. 1974. Toxins in blast diseased rice plants. *Agric. Biol. Chem.*, **38** : 1867-1874.

Valent, B. 1990. Rice blast as a model system for plant pathology. *Phytopathology*, **80** : 33-36.

Valent, B. and F. G. Chumley. 1986. Genetic analysis of host species specificity in *Magnaporthe grisea*. In : *Molecular strategies for crop protection* (C. J. Arntzen and C. Ryan. eds.) Alan R. Liss. Inc. New York. pp. 83-93.

Valent, B., and F. G. Chumley. 1989. Genes for cultivar specificity in the rice blast fingus, *Magnaporthe grisea* In : *Signal molecules in plants and plant microbe interactions*. (Ed. B. J. J. Lutenberg), Springer-Verlag, Birlin NATO ASI Series, **36** : 415-422.

Valent, B., L. Farall, and F. G. Chumley 1991. *Magnaporthe grisea* genes for pathogenicity and virulence identified through a series of backcrosses. *Genetics* , **12** : 88-101.

Valent, B., M. S. Crawford, C. G. Weaver and F. G. Chumley. 1986. Genetic studies of fertility and pathogenicity in *Magnaporthe grisea (Pyricularia oryzae). Iowa state J. Res.*, **60** : 569-594.

Van der Plank, J. E., 1968. *Disease resistance in plants*. Academic Press, New York, p. 206.

Van der Plank, J. E. 1975. Horizontal resistance : six suggested projects in relation to blast disease of rice. In : *Horizontal resistance and blast disease of rice*. CIAT, Cali, Colombia, pp. 21-26.

Varga, M. 1970. Correlation between the phenol content, polyphenolase and peroxidase activity and the growth of rice seedlings with particular respect to flooding conditions. *Riso*, **19** :353-360.

Veeraraghavan, J. 1975a. Stability in *Pyricularia oryzae* Cav. *Proc. Indian Acad. Sci.*, **82** : 127-131.

Veeraraghavan, J. 1975b. A new method of classification and nomenclature of physiologic races of *Pyricularia oryzae* Cav., *Int. Rice Comm. Newsl.*, **24** : 128-138.

Veeraraghavan, J. 1985. Aggressive races of *Pyricularia oryzae* Cav. and a method of their identification. *Acta. Phytopath. Acad. Sci. Hung.*, **20** : 119-123.

Veeraraghavan. J. 1986. Specialization in pathogenicity of *Pyricularia oryzae* Cav. *Proc. Indian Natl. Acad. Sci.*, **52** : 300-307.

Veeraraghavan, J. 1988. Nature of interaction between immune varieties of rice (*Oryza sativa*) and *Pyricularia oryzae*. *Acta Phytopath. Entamol. Hung.*, **23** : 33-37.

Veeraraghavan, J. and A. P. Dath. 1975. Current position of Physiologic races of *Pyricularia oryzae* Cav. in India. *Curr. Sci.*, **44** : 19-20.

Veeraraghavan, J. and A. P. Dath. 1976. Host pathogen equilibrium between rice (*Oryza sativa*) and its pathogen *Pyricularia oryzae* Cav. *Curr. Sci.*, **45** : 333-334.

Veeraraghavan, J. and A. P. Dath, 1981. Physiologic specialization in *Pyricularia oryzae* in India 1976-78. *Proc. Indian Natl. Sci. Acad.*, **47** : 226-228.

Veeraraghavan, J. and S.Y. Padmanabhan. 1965. Studies on the host range of *Piricularia oryzae* Cav. causing blast disease of rice. *Proc. Indian Acad. Sci.*, **61** : 109-120.

Veeraraghavan, J. and S. Y. Padmanabhan. 1976. Effect of application of growth regulators on resistance to rice *Pyricularia oryzae* Cav. *Riso.*, **25** : 75-78.

Veeraraju, V. and N. N. Prasad. 1971. Studies on the chemical nature of inhibitor occurring in rice leaves. *Labdev. J. Sci. Tech.*, **9** : 224-225.

Veeraraju, V. and N. N. Prasad. 1972. Effect of Nitrogen fertilization on inhibitor level and blast disease resistance in rice. AUARA. **4/5** : 82-87.

Venkatachalam, S., C. P. Natarajan and A. Marikulandai. 1957. Study of the free amino acids in paddy plants in relation to the incidence of blast disease. *Madras Agric. J.*, **34** : 157-158.

Venkatakrishnaiah, N. S. and M. H. Delvi. 1960a. Blast on rice in Mysore : trials with fungicides for the control of the disease. *Rice News Teller*, **8** : 16-18.

Venkatakrishnaiah, N. S. and M. H. Delvi. 1960b. Blast on rice in Mysore : fungicidal treatment on rice seeds. *Mysore Agric. J.* **35** : 74-79.

Venkatarao, G. and A. Apparao. 1972. Cross inoculation studies on *Pyricularia* spp. from different hosts. *Andhra Agric J.* **19** : 134-138.

Venkatarao, G. and A. Apparao. 1974. Pathogenic races in *Pyricularia oryzae* in Andhra Pradesh. *Oryza*, **11** : 53-60.

Ventakarao, G. and K. Muralidharan. 1982a. Effect of meteorological conditions on the incidence and progress of blast disease on rice. *Z. Pflkrankh. Pflschutz.*, **89** : 219-223.

Venkatarao. G. and K. Muralidharan. 1982b. Plant population density and severity of rice blast disease. *Indian Phytopathol.*, **35** : 318-320.

Venkatarao, G. , C. P. P. Rajan and K. Muralidharan. 1988. Current status of blast and sheath blight disease in Andhra Pradesh. In : *Current status of blast and sheath blight diseases of rice in India.* Directorate of Rice Research (DRR), Hyderabad, India. (Mimeographed).

Villareal, R. L., D. R. MacKenzie., R. R. Nelson and W. R. Coffman. 1980. Apparent infection rates of *Pyricularia oryzae* on different rice cultivars. *Phytopathology*, **70** : 1224-1226.

Volk, R., R. P. Kahn and R. Weintraub. 1958. Silicon content of the rice plant as a factor influencing its resistance to infection by the blast fungus *Pyricularia oryzae*. *Phytopathology*, **48** : 179-184.

Wakimoto, S. and H. Yoshii. 1958. Relation between polyphenols contained in plants and phytopathogenic fung I. Polyphenols contained in rice plants. *Ann. Phytopathol. Soc. Japan*, **23** : 79-84.

Walde, E. L. and R. Roberts. 1947. Japanese food crop disease control. Rep. No. 94 and Rep. 2268. *Tech. Inf. Document Unit Lond.*, UK.

Wallace, G. B. 1950. Annual report of the plant pathologist, 1948. *Tanganykia Dept. Agric. Rep.* 1948. 145-147.

Wan, W., H. Luo, H. Shenyi and G. Wang. 1986. Relationship between resistance of rice varieties to blast disease and the isoenzymes of peroxidase. *J. Hunnan Agric College*. **3** : 19-25.

Wang, S. K. and Z. Chen. 1987. Studies on slow-blasting resistance of rice varieties and its mechanism. *J. Hunan Agric. College*, **3** : 48-55.

Wang. Z., D. J. Mackill and J. M. Bonman. 1989. Inheritance of partial resistance to blast in indica rices. *Crop Sci.*, **29** : 848-853.

Wardlaw, C. W. 1940. Banana diseasees. XIII. Further observations on the condition of banana plantations in the Republic of Haiti. *Tropi. Agric. Trinidad*, **17** : 124-127. (*Rev. Appl. Mycol.*, **19** : 661-663).

Watanabe, K. et al., 1957. Blastymycin, a new antibiotic from *Streptomyces* sp. *J. Antibiotics*, **10** : 39-45.

Watanabe, M., S.Sakaniwa, M. Uchiyama and H. Abe. 1979. Antimicrobial activities of momilactones against *Xanthomonas oryzae* and *Pyricularia oryzae*. *Ann. Phytopathol. Soc. Japan*, **45** : 509-511.

Watanabe, T. 1957. Studies on the vaccine therapy of the blast disease of rice plants. X- The detection of the plant hormone. *Ann. Phytopathol. Soc. Japan*, **22** : 143-147.

Watanabe, T. 1977. Effect of probenazole (Oryzaemate) on each stage of rice blast fungus (*Pyricularia oryzae* Cavara) in its life cycle. *J. Pesticide Sci.* **2** : 395-398.

Watanabe, T., Y. Sekizawa. M. Schimura, Y. Suzuki, K. Matsumoto, H. Iwata and S. Mase. 1979. Effects of probenazole (Oryzaemate) on rice plants with reference to controlling rice blast. *J. Pesticide Sci.*, **4** : 53-59.

Webster, J. 1965. The perfect stage of *Pyricularia aquatica*. *Trans. Br. Mycol. Soc.* **48** : 449-452.

Webster, J. 1980. *Introduction to Fungi*. 2nd Ed., Cambridge Univ. Press. Cambridge, UK. p. 669.

Wells, H. D., G. W. Burton and J. B. Powell. 1969. *Piricularia* leaf spot of pearl millet. *Phytopathol.* **59** : 1057. (Abstr.).

Wheeler, M. H. and G. A., Greenblatt. 1988. The inhibition of melanin biosynthetic reactions in *Pyricularia oryzae* by compounds that prevent rice blast disease. *Expt. Mycol.* **12** : 151-160.

Wili, G. M. 1985. Comparison of the designs of two volumetric spore traps. *Phytopathology*, **75** : 380.

Willis, G. M., R.D. Allowitz and E. S. Menvielte. 1968. Differential susceptibility of rice leaves and panicles of *Piricularia oryzae*. *Phytopathology*, **58** : 1072. (abstr).

Wislon, K. I. 1962. New record of *Piricularia* on *Rhyncheletum repens* (Wild) Hubbard. *Indian Phytopathol.*, **15** : 76-77.

Woloshuk, C. P., H. D. Sisler and E. L. Vigil. 1983. Action of the antipenetrant, tricyclazole on appressoria of *Pyricularia oryzae*. *Physiol. Pl. Path.* **22** : 245-259.

Woloshuk, C. P., H. D. Sisler, M. C. Toukousbalides and S. R. Dutky. 1980. Melanin biosynthesis in *Pyricularia oryzae*. Site of tricyclazole inhibition and pathogenicity of melanin-deficient mutants. *Pest Biochem. & Physiol.* **14** : 256-264.

Woloshuk, C. P., P. M. Wolkow and H. D. Sisler. 1981. The effect of three fungicides specific for the control of rice blast disease, on the growth and melanin biosynthesis by *Pyricularia oryzae* Cav. *Pesti. Sci.* **12** : 86-90.

Woo, S. C. 1965. Some experimental studies on the inheritance susceptibility of leaf blast disease *Pyricularia oryzae* Cav. *Bot. Bull. Acad. Scinica*, **6** : 208-217.

Wood, R.K.S. 1967. *Physiological plant pathology*. Blackwell Scientific Publications, Oxford, UK, p. 570.

Wu, L.C. and C. L. Shih. 1967. Studies on the fungal proteins in *Piricularia oryzae* Cav. *Bot. Bull. Acad. Sinica*, **8** : 54-65.

Wu, H. K. and M. M. Tsai. 1974. Studies on variation in pathogenicity of *Pyricularia oryzae* Cav. I. The mechanism of the genetical variation. *Memoir. Coll. Agric. Natl. Taiwan Univ.*, **15** : 7-21.

Wu, H. K. and T. H. Tsao. 1967. The ultrastructure of *Piricularia oryzae* Cav. *Bot. Bull. Acad. Sinica*, **8** : 353-363.

Xia, J.Q., F.N. Lee, H.A. Scott and L.R. Raymond. 1992. Development of monoclonal antibodies specific for *Pyricularia grisea*, the rice blast pathogen. *Mycol. Res.* **96** : 867-873.

Yaegashi, H. 1977. On the sexuality of blast fungi *Pyricularia* spp. *Ann. Phytopathol. Soc. Japan,* **43** : 432-439.

Yaegashi, H., and K. Asaga. 1981. Further studies on the inheritance of pathogenicity in crosses of *Pyricularia oryzae* with *Pyricularia* sp. from finger millet. *Ann. Phytopathol. Soc. Japan,* **47** : 677-679.

Yaegashi, H. and T. T. Hebert. 1976. Perithecial development and nuclear behaviour in *Pyricularia. Phytopathology,* **66** : 122-126.

Yaegashi, H. and T. Kobayashi. 1976. Factors affecting the racial distribution of *Pyricularia oryzae.* II. Comparison of multiplication of races C-1 and C-3 on rice varieties with Kanto 51 type resistance. *Ann. Phytopathol. Soc. Japan,* **42** : 272-278.

Yaegashi, H. and N. Nishihara. 1976. Production of the perfect stage in *Pyricularia* from cereals and grasses. *Ann. Phytopathol. Soc. Japan,* **42** : 511-515.

Yaegashi, H. and S. Udagawa. 1978. The taxonomical identity of the perfect state of *Pyricularia grisea* and its allies. *Can. J. Bot.,* **56** : 180-183.

Yaegashi, H. and M. Yamada. 1984. Possible genetic factor conditioning the size of blast lesions induced rice leaves with *Pyricularia oryzae. Ann. Phytopathol. Soc. Japan,* **50** : 664-667.

Yaegashi, H. and M. Yamada, 1986. Pathogenic races and mating types of *Pyricularıa oryzae* from Soviet Union, China, Nepal, Thailand, Indonesia and Colombia. *Ann. Phytopathol. Soc. Japan,* **42** : 556-562.

Yaegashi, H. and K. Yanagita. 1972. Factors affecting the racial distribution of rice blast fungus *Pyricularia oryzae* Cav. I. A comparison of hyphal growth temperature among the isolates. *A. Rep. Pl. Prot. North Japan,* **23** : 39-44.

Yaegashi, H., T. Higashi and H. Koga. 1990. Varietal resistance to blast : A Japanese prespective. In : *Int. Rice Res. Conf.* Seoul, S. Korea. p. 13.

Yamada, M. and M. Iwano. 1975. Variation in the prevalence of rice blast races in Japan. In : *Proc. Int. Cong. Int. Assoc. Microb. Sci. Coun. Tokyo, Japan,* pp. 425-435.

Yamada, M., S. Matsumoto and T. Kozaka. 1969. Groupings of Japanese rice varieties on the basis of the reaction to pathogenic races of *Piricularia oryzae* Cav. *Bull. Natl. Inst. Agric. Sci.* No. **23**, pp. 37-62.

Yamada, M. S. Kiyosawa, T. Yamaguchi, T. Hirano, T. Kobayashi, K. Kushibuchi and S. Watanabe. 1976. Proposal of new method for differentiating races of *Pyricularia oryzae* Cavara in Japan. *Ann. Phytopathol. Soc. Japan,* **42** : 216-219.

Yamaguchi, T. 1970. Forecasting techniques of rice blast. *J. Agric. Res. Q.,* **5** : 26-30.

Yamaguchi, T. 1977. Life cycle of blast fungus. *Pyricularia oryzae. Heridity. Japan,* **31** : 59-64.

Yamamoto, R. 1958. Method of studies on new agricultural chemicals. Nankado, Japan, pp. 673-677.

Yamanaka, S. 1975. Studies on competition among the isolates of rice blast fungus, *Pyricularia oryzae* Cavara. In : *Int. Assoc. Microb. Sci. Coun. Tokyo, Japan,* pp. 462-466.

Yamanaka, S. and T. Kobayashi. 1962. Temperature in relation to the size of conidia in *Piricularia oryzae. Ann. Phytopathol. Soc. Japan,* **27** : 57-58.

Yamanaka, S., T. Namai and S. Misawa. 1976. Studies on sporulation of rice blast fungus *Pyricularia oryzae* Cavara I. Influence of light on sporulation. *Trans. Mycol. Soc. Japan,* **17** : 483-491.

Yamasaki, Y. and S. Kiyosawa. 1966. Studies on inheritance of resistance of rice varieties of blast I. Inheritance of resistance of Japanese varieties to several strains of the fungus. *Bull. Natl. Inst. Agric. Sci. Japan,* No. **14** : 39-69.

Yamasaki, Y. and H. Niizeki. 1963. Studies on resistance of cultivated and wild rice to the blast disease. I. Resistance of seedlings to 4 races of fungus. *Japan J. Breed.,* **14** : 1-10.

Yamasaki, Y. and H. Nizeki. 1965. Studies on variation of the rice blast fungus *Piricularia oryzae* Cav. I. Karyological and genetical studies on variation. *Bull. Natl. Inst. Agric. Sci,* **D 3**. pp. 231-274.

Yamasaki, Y., N. Murata and T. Suwa. 1964a. Effect of mutations for nutritional requirements on the pathogenicity of two pathogens of rice. *Proc. Japan Acad.*, **40** : 266-331.

Yamasaki, Y., S. Tsuchiya, H. Nizeki and T. Suwa. 1964b. Studies on drug-resistance of the rice blast fungus, *Piricularia oryzae* Cav. I. Acquisition of the resistance in successive passages through media containing drugs and the nature of such acquired resistance. II. Cross-resistance of drug-resistant strains. III. Search for $CuSO_4$-resistant cells in the population of parental fungus P-2. IV. Reversing effects of various chemicals on permanent resistance to $CuSO_4$-resistant strains. V. Physiological studies on the resistance to $CuSO_4$. VI. Physiological resistance to H_3BO_4. VII. Pathogenicity of drug-resistant strains. *Bull. Natl. Inst. Agric. Sci.*, No. **11**, pp. 1-110.

Yeh, W. H. and J. M. Bonman. 1986. Assessment of partial resistance to *Pyricularia oryzae* in six rice cultivars. *Pl. Path.*, **35** : 319-323.

Yin, T. P. and G. Varghese. 1970. Physiologic races of *Pyricularia oryzae* Cav. in West Malaysia. *Malaysian Agric.*, **9** : 1-6.

Yokoo, M. 1974. Multiline cultivars for disease control. *Japan J. Breed.*, **24** : 104-111.

Yorinori, J. T. and H. D. Thurston. 1975. Factors which may express general resistance in rice to *Pyricularia oryzae* Cav. In : *Horizontal resistance and blast disease of rice*. CIAT, Cali. Colombia. pp. 117-135.

Yoshida, S., Y. Ohnishi and K. Kitagishi. 1962a,b,c,d. Histochemistry of silica in rice plants. (a) The presence of cuticle-silica double layer in the epidermal tissue. (b) A new method for determining the localization of silicon within plant tissues. (c) Localization of silicon within rice tissues. (d) Chemical forms, mobility and deposition of silicon in rice plants. *Soil Sci. Pl. Nutrition.* **8** : 1-5; 30-35; 36-41; 107-113.

Yoshii, H. 1933. How does *Piricularia oryzae* penetrate into the host? *J. Pl. Prot.*, **20** : 841-844.

Yoshii, H. 1936. Pathological studies of rice blast caused by *Pyricularia oryzae*. I. Some studies on the physiology of the pathogen. II. Mode of infection of the pathogen. *Ann. Phytopathol. Soc. Japan*, **6** : 199-218.

Yoshii, H. 1937. Pathological studies on rice blast caused by *Pyricularia oryzae* III. Pathohistological observations of diseased plants. *Ann. Phytopathol. Soc. Japan*, **6** : 289-304.

Yoshii, H., Y. Asada, A. Kiso and N. Tanabe. 1958. Studies on the drug resistance in plant pathogenic fungi against agricultural chemicals. I. On the drug resistance of blast fungus, *Piricularia oryzae*, against phenyl mercuric acetate. *Ann. Phytopathol. Soc. Japan*, **23** : 215-218.

Yoshii, K. 1949. Studies on *Cephalothecium* as a means of artificial immunization of agricultural crops. *Ann. Phytopathol. Soc. Japan*, **13** : 37-40.

Yoshii, K. 1950. Studies on *Cephalothecium* as a means of the artificial immunization of agricultural crops. II. On the effect of treatment by the dried mycelium power of *Cephalothecium* on the development of leaf blast in rice seedlings. *Ann. Phytopathol. Soc. Japan*, **14** : 9-10.

Yoshino, R. 1971. A method for detecting hyphae (*Pyricularia oryzae*) of rice blast in leaf blade in epidermis. *Proc. Assoc. Pl. Prot. Hokurkiku*, **19** : 14-17.

Yoshino, R. 1974. Ecological studies on the infection in rice blast epidemics. Dew, temperature and sequent change of infection rates. *Ann. Phytopathol. Soc. Japan*, **39** : 186. (Abstr.).

Yoshino, R. 1979. Ecological studies on penetration of rice blast fungus, *Pyricularia oryzae*, into leaf epidermal cells. *Bull. Hokuriku Agric. Exp. Stn.* **22** : 163-221.

Yoshino, R. and T. Yamaguchi. 1970. Relation between air temperatures after colonization and diseases development. *Ann. Phytopathol. Soc. Japan*, **36** : 156. (Abstr.).

Yoshino, R., C. K. Kim and K. M. Chin. 1990. Use of chemicals in the control of blast. In : *Int. Rice Res. Conf.* Seoul, S. Korea. p. 18.

Yu, Z., D. J. Mackill, J. M. Bonman and S. D. Tanksley. 1990. RFLP tagging of blast resistance genes in rice. In : *Rice Genetics*, Int. Rice Res. Inst. (IRRI) Box, 933, Manila, Philippines.

Yunoki, T., A. Ezuka, H. Sakurai, H. Shinoda and K. Toriyama. 1970a. Studies on the varietal resistance to rice blast. II. Testing methods for field resistance on young seedlings growth in greenhouse. *Bull. Chugoku Natl. Agric. Expt. Stn.* **6** : 1-19.

Yunoki, T., A. Ezuka, T. Morinaka, Y. Sakurai. H. Shinoda and K. Toriyama. 1970b. Studies on the varietal resistance to rice blast. IV. Variation of field resistance due to fungus strains. *Bull. Chugoku Natl. Agric. Expt. Stn.* **6** : 21-41.

Zhang, W. and Z. Shen. 1986. Inheritance of resistance to blast and bacterial blight in rice (*Oryza sativa* L.). *Acta. Phytopath. Sinica.* **13** : 103-108.

Zhuge, G. Z., G. C. Sun, S. Sun and M. Iwano. 1990. Progress in blast resistance research in China. In : *Intl. Rice Res. Conf.* Seoul, S. Korea. p. 20.

Zhuge, G. Z. and J. M. Bonman. 1989. Rice blast in pure and mixed stands of rice varieties. *Chinese J. Rice Sci.*, **3** : 11-16.

Zsoldos, E. 1959. Quantitative changes in aminobutryic acid induced by low temperature in rice plants. *Nature,* **184** : 280.

Zuber, M. and K. Manibhushanrao. 1979. *Pyricularia* sp. on ragi seeds. In : *Physiology of host-pathogen interaction.* Today & Tomorrow Publishers, New Delhi, pp. 473-478.

Index

[Note: Rice cultivars and B1 pathogen *P. oryzae* are not listed]